C 7/76

TRADE POLICY AND
ECONOMIC WELFARE

J6

TRADE POLICY
AND
ECONOMIC
WELFARE

BY

W. M. CORDEN

CLARENDON PRESS · OXFORD

1974

Oxford University Press, Ely House, London W. I

GLASGOW NEW YORK TORONTO MELBOURNE WELLINGTON
CAPE TOWN IBADAN NAIROBI DAR ES SALAAM LUSAKA ADDIS ABABA
DELHI BOMBAY CALCUTTA MADRAS KARACHI LAHORE DACCA
KUALA LUMPAR SINGAPORE HONG KONG TOKYO

CASEBOUND ISBN 0 19 828199 4
PAPERBACK ISBN 0 19 828401 2

© OXFORD UNIVERSITY PRESS 1974

PRINTED IN NORTHERN IRELAND AT THE UNIVERSITIES PRESS, BELFAST.

IN MEMORY OF MY FATHER

PREFACE

THIS book is a companion to *The Theory of Protection*, published in 1971. While a reader need not be familiar with the earlier book to follow the present one, the two books together result from my work over many years on the theory and practice of trade policies and the principles of government intervention. The present book has turned out larger than originally intended, and perhaps says more on the subject than every reasonably interested reader would want to read. Thus I should like to stress that Chapters 2 and 3 present much of the basic method of analysis, while the other chapters can be regarded as being more or less self-contained: the reader can select from them according to his interests.

My debts are many. My late father, to whom I have dedicated this book, first kindled my interests in the general field of trade restrictions and economic policy. I began work on tariffs in 1955, publishing an article on "The Calculation of the Cost of Protection" in the *Economic Record* in 1957. From 1958 to 1967 I was at Melbourne University and the Australian National University, and developed many of my ideas while working on Australian tariff policy. Thus I owe considerable debts to my colleagues and friends of those days, and to the two Universities which allowed me time and facilities. I continued working on the subject when I came to Nuffield College in 1967, and benefited from an environment in which trade policies of less-developed countries were much under discussion.

It will be obvious to any reader where my principal intellectual debts are owed: to James Meade's *Trade and Welfare* (1955), and to Harry Johnson's article "Optimal Trade Intervention in the Presence of Domestic Distortions" (1965). Much of the book is a variation upon the themes of these two works, though I hope I have carried their analyses a little further. My debt to Harry Johnson is much greater than through the paper just cited. During the period that I have worked in this field he has completely dominated it, as is evident from numerous citations in this book. I have built on his work, sometimes criticised or gone beyond it, but always benefited from it. Not least, he has helped me with constant encouragement. In addition, I have been influenced by, and have built on, the work of many other contributors to the literature of trade theory and policy during recent years, particularly Jagdish Bhagwati, and also Bela Balassa, Murray Kemp, Ronald Jones and the late V. K. Ramaswami.

I owe a great deal to many good friends who, at some stage, read various parts of the manuscript and led me to make many changes, to clarify points, eliminate errors, and rearrange chapters. The whole manuscript was read by John Black, Rattan Bhatia, John Martin, and Jean van der Mensbrugghe. John Flemming particularly helped me with some problems in Chapters 4 and 11 (especially the discussion of risk avoidance), and I am also indebted for help or criticisms to Heinz Arndt (Chapter 12), Stanley Engerman (Chapter 5), Ronald Jones (Chapter 5), Vijay Joshi (Chapters 10, 14), P. R. Narvekar, Maurice Scott (Chapters 6, 10, 14), Richard Snape (Chapter 8), K. P. Teh and David Wall. Graduate students of Oxford University, Princeton University and the University of Minnesota also innocently influenced this book with their often very perceptive, if disconcerting, questions.

The book was mainly written in Nuffield College, but also in the Woodrow Wilson School of Princeton University; in addition, I wrote little bits of it in the University of Minnesota, Monash University and La Trobe University, and to all these institutions I am much indebted. The manuscript was typed competently and cheerfully by Penny Sylvester at Nuffield College and by Mary Leksa at the Woodrow Wilson School. I am deeply grateful to them for their patience and efficiency. Finally, my wife and daughter again put up with a husband and father still dreaming about theories, still drawing diagrams, still hammering away at a typewriter.

In conclusion I might add that this book was completed and handed to the printer in July 1973. Six months later the world seems to have changed a little, and no doubt it will change further. A book which is essentially theoretical ought not to need revision just because oil prices have gone up and many exchange rates are floating. But perhaps, if I were writing it now, I would say a few things relevant to oil in Chapter 7 and would not dismiss exporting cartels, or retaliation possibilities, so readily on p. 185. Further, I would trouble even less about protectionist arguments which have assumed fixed exchange rates (notably pp. 322–4).

Nuffield College, Oxford
January 1974

ACKNOWLEDGMENTS

Chapter 8, section V is a revised version of "The Efficiency Effects of Trade and Protection" which appeared in I. A. McDougall and R. H. Snape (eds.), *Studies in International Economics: Monash Conference Papers*, North-Holland Publishing Company, Amsterdam 1970. An earlier version of Chapter 10 (excluding sections II(3) and III) appeared in Luis Di Marco (ed.), *International Economics and Development*, Academic Press, Inc., New York 1972. Much of Chapter 12 is based on "Protection and Foreign Investment", *The Economic Record*, 43, June 1967. This chapter also includes some passages from "The Multinational Corporation and International Trade Theory", in John Dunning (ed.), *The Multinational Corporation and Economic Analysis*, Allen and Unwin, London 1974.

I am indebted to the editors and publishers of all these publications for permission to make use of this work.

CONTENTS

1

INTRODUCTION

THE AIM of this book is to expound, review, and develop the
normative theory of trade policy. The concern is mainly with
tariffs, but all forms of government policies affecting trade,
notably import quotas, export taxes, and taxes and subsidies
on the production of particular products or industries, or on
the use of factors, come within the scope of the study. More
generally, its subject-matter is really the principles of government intervention in the economy at the micro-level, especially
government subsidization of private industry. While the book
rests heavily on international trade theory it goes beyond
some of the theory's common assumptions—two goods,
perfect competition, fixed factor supplies, and availability
of non-distorting lump-sum transfers. It puts particular
emphasis on income distribution effects of trade policies and on
the fiscal implications of various kinds of government
intervention.

It is hardly necessary to stress the significance of the
subject-matter. The issue of free trade versus protection
dominated much economic discussion in Britain, France,
Germany, and the United States in the nineteenth century.
As this is being written, the United States appears to be going
through a new protectionist phase, and the subject is a matter
of daily newspaper discussion. Tariffs are now less important
in developed countries than they used to be, and in Western
Europe agricultural industries are now the main beneficiaries
of tariff protection. But there are also many forms of non-
tariff intervention, for example, through preferences to local
producers in government procurement, and through subsidies
to so-called 'advanced technology' industries.[1]

[1] Robert E. Baldwin, *Nontariff Distortions of International Trade*, The
Brookings Institution, Washington, 1970. On tariff barriers, see Bela Balassa,
Trade Liberalization Among Industrial Countries, McGraw-Hill, New York,
1967.

It is for less-developed countries that the issue of protection, whether by tariffs or by import quotas, is really important. In some less-developed countries tariffs and import quotas are the principal means by which government policy influences the pattern of economic development, and rates of protection for manufacturing industries are very high.[2] For this reason much of the book is directed to illuminating economic policies in less-developed countries and issues that have been debated by economists advising these countries. Of particular interest for these countries is the modern issue of the optimum structure of a protectionist system.

I
THREE STAGES OF THOUGHT

The recent transformation of the normative theory of trade policy dominates the whole of this book, and its essentials are expounded and somewhat further developed in the next chapter, while various difficulties and refinements are discussed in Chapter 3.

One could, rather crudely, classify the evolution of modern post-mercantilist economic thought about tariffs and other trade-restricting devices into three stages.

In the first stage, the gains from trade and, more specifically, the benefits from completely free trade, came to be appreciated. The great tool of analysis to emerge was the law of comparative costs. This stage is, above all, associated with the names of Adam Smith and David Ricardo, but it dominated economic thought in Britain for the whole of the nineteenth century, and found clear expression in that 'Samuelson's *Economics*' of the latter half of the nineteenth century, John Stuart Mill's *Principles of Political Economy*. The relevant point is that the case for free trade was developed simultaneously with the case for laissez-faire. Indeed, the case for free trade was really a special case of the argument for laissez-faire.

Gradually, more and more reasons have come to be seen why laissez-faire may not lead to an optimum for a country. Perfect competition does not necessarily rule, there may not

---

[2] Bela Balassa et al., *The Structure of Protection in Developing Countries*, Johns Hopkins University Press, Baltimore, 1971; Ian Little, Tibor Scitovsky, and Maurice Scott, *Industry and Trade in Some Developing Countries*, Oxford University Press, London, 1970.

be full employment, the income distribution that is yielded by the laissez-faire solution may not be a desirable or just one, necessary structural changes may not take place, and so on.

Parallel with the qualifications to laissez-faire more and more qualifications to the argument for complete free trade were developed. This has been the second stage of thought. One of the arguments for tariffs, the infant industry argument, was even sanctified by the approval of John Stuart Mill. Another, the argument that a country can improve its terms of trade at the expense of other countries and so obtain net gains by some degree of trade restriction, was first systematically expounded by Bickerdike early this century, and was much refined and emphasized in more recent years, though indeed the basic idea can also be found in Mill's *Principles*.

The result was that, until a few years ago, the whole subject of free trade and protection was typically expounded in the textbooks as follows. First, the case for free trade, resting on the law of comparative costs, would be developed. Then, in subsequent chapters, various 'arguments for protection'—that is, for possible qualifications to the basic case—would be presented. Some of the more popular arguments would be found fallacious (such as the 'pauper labour' argument) but others would be found valid provided various specified and possibly not unrealistic assumptions held. The arguments for protection emerged pari passu with the qualifications to the case for laissez-faire, and since there are numerous qualifications to the latter, there are also numerous arguments for protection.

As the free trade case has been more and more overlaid by numerous sophisticated protectionist arguments, faith in the gains from relatively free trade has gradually been lost by many students of economics, and in many countries, especially the less-developed ones. The exception to this generalization can be found mainly in some countries of Western Europe, where a strong belief in the market economy has revived as a result of post-war successes. But during the nineteen fifties the case for free or freer trade was at a discount in most of the less-developed countries. Laissez-faire had failed to develop them, so it was argued; hence free trade had failed.

In the third stage of thought *the link between the case for free trade and the case for laissez-faire has been broken*. One can believe that there are many reasons for the government to intervene in the economy—to maintain full employment, to bring about a desirable distribution of income, to adjust resource allocation and consumption patterns in the light of external economies and diseconomies, and so on—and yet one can also believe that, broadly, 'free trade is best'.

The point is simply that intervention in trade may not be the best way of dealing with the various problems mentioned since they are all essentially caused by 'domestic distortions' rather than 'trade distortions'. The best way may be to deal with them in some direct 'first-best' way, and at the same time to allow trade to flow freely. Of course, if, for some reason, the first-best ways of correcting the distortions are not used, or cannot be used, then trade intervention may still be better than nothing.

This approach has been called the *theory of domestic distortions*, though the term 'distortions' is somewhat loaded and hence will be used in a more limited sense in this book.

The new approach has radically transformed the theory and has three implications. Firstly, it restores the argument for free or freer trade, though always subject to some exceptions, as we shall see. Secondly, it focuses attention on the choices between different policies—such as the choice between tariff and subsidy—as distinct from the choice between trade intervention and *no* intervention. Thirdly, it provides a greatly improved method of analysing the effects of trade policies.

Essentially, the new approach, and the techniques of analysis which it uses, stem from J. E. Meade's *Trade and Welfare*. The most important single proposition was high-lighted and expounded systematically in an article by Harry G. Johnson, 'Optimal Trade Intervention in the Presence of Domestic Distortions'.[3] Appendix 2 of the next chapter gives

---

[3] James E. Meade, *Trade and Welfare* (Vol. II of *The Theory of International Economic Policy*), Oxford University Press, London, 1955; Harry G. Johnson, 'Optimal Trade Intervention in the Presence of Domestic Distortions', in Baldwin et al., *Trade, Growth and the Balance of Payments*, North-Holland, Amsterdam, 1965 (reprinted in J. Bhagwati (ed.), *International Trade*, Penguin Modern Economics, 1969).

some fuller notes on the history of these ideas and more references. In this book an attempt is made to expound and develop the new approach, especially with respect to its practical applications, to show its wide applicability and at the same time to point out its limitations.

II
THE SCOPE OF THE BOOK

This book is addressed to professional economists, especially the more practical-minded amongst them, and to students who have some familiarity with international trade theory. Familiarity with elementary welfare economics would also be helpful. It does not begin at the beginning. The beginning, one would think, is a simple exposition of the positive theory of trade policy, and especially of what a tariff and—of more interest—a structure of tariffs, export taxes, and so on, may do to production and consumption patterns and to factor prices. This was expounded in the author's earlier book, *The Theory of Protection*, though the degree of detail developed there is not a prerequisite for the present book. Some elementary familiarity with the positive theory as it can be derived from any modern basic text on international economics will suffice.

Upon positive theory rests *normative* theory—the theory of policy. The foundation of the normative theory is the familiar argument that countries can mutually gain from trade. For each country the opportunity to trade extends its choice— its frontier of consumption (and investment) possibilities.[4] This simple idea rests essentially on the law of comparative costs or is a modern development of it. The proposition that there are potentially some gains from trade to a country is perhaps helpful in disposing of some elementary fallacies, but it is not in itself sensational. Some might even say that it is obvious.

The more important proposition is that, given certain assumptions, free trade is optimal. The assumptions upon

[4] P. A. Samuelson, 'The Gains from Trade Once Again', *Economic Journal*, 72, December 1962, pp. 820-29 (reprinted in *International Trade*, Penguin Modern Economics, 1969).

which this proposition rests are notably that there is perfect competition, that all goods and factor services pass through the market, that there are no distorting taxes and other interventions, and that a country cannot affect its terms of trade. In that case free trade will, for any pair of goods, automatically bring about equality between the marginal rate of transformation in domestic production, the marginal rate of substitution in domestic consumption, and the marginal rate of transformation through trade (or, to be more precise, will ensure the absence of certain inequalities). The fulfilment of all the marginal conditions will suffice to ensure 'Pareto-optimality'. A concern with the latter implies that domestic income distribution effects are ignored.

These elementary or basic propositions, which can be found in many general economic textbooks, and all textbooks of international trade theory, are not expounded here.[5] The aim, rather, is to clarify or elaborate what texts sometimes leave obscure or inadequately interpreted, and to carry on where the texts leave off. Further, some issues that have not yet systematically entered the textbooks are pursued.

All aspects of trade policy will not be covered. The book does not deal at all with *country* (as distinct from *product*) discrimination in tariffs—with preferences, customs unions and so on. This topic is a large one of its own, and there are good books and surveys available.[6] Also, it does not deal with the theory of the measurement of protection—notably the concept of the average tariff, the measurement of the cost of

[5] The literature on the existence of competitive equilibrium and the requirements for its 'Pareto-optimality' is relevant. See James Quirk and Rubin Saposnik, *Introduction to General Equilibrium Theory and Welfare Economics*, McGraw-Hill, New York, 1968, Ch. 4. If a competitive equilibrium exists, which requires (among other things) 'convexity', and if the closed-economy Pareto-optimality requirements are fulfilled, then free trade will (broadly) be Pareto-optimal from a world point of view—and with given terms of trade, from a national point of view. See also pp. 104–5 below for further discussion of 'Pareto-optimality'.

[6] See J. E. Meade, *The Theory of Customs Unions*, North-Holland, Amsterdam, 1955; R. G. Lipsey, 'The Theory of Customs Unions: A General Survey', *Economic Journal*, 70, Sept. 1960, pp. 498-513; Jaroslav Vanek, *General Equilibrium of International Discrimination*, Harvard University Press, Cambridge, 1965; M. B. Krauss, 'Recent Developments in Customs Union Theory: An Interpretive Survey', *Journal of Economic Literature*, 10, June 1972, pp. 413-436.

protection and of effective protection.[7] Rather, a number of topics are selected—mainly on the basis of their policy relevance—and are analysed thoroughly.

In addition, this is not a book of rigorous general equilibrium theory. Readers who are interested in rigorous formal theory where everything ties into a single general equilibrium system which is fully expounded and all assumptions are always carefully specified should stop here. Every economist must always have a general equilibrium system in his mind and fit a particular partial analysis into it. The partial analysis is unlikely to tell the whole story; it will be a piece of a jigsaw, and the approach here is sometimes to look at one or two pieces on their own with some care. But it should never be forgotten that it is only a piece. Commonsense and an appreciation of the limitations of partial analysis must be used in interpreting some of the piecemeal analysis.

Finally, and most importantly, this book is only theory. It is meant to be relevant theory, guided in its selection of issues and variables by the author's knowledge of the world of economic affairs, but it is not a direct guide to economic policy in particular countries. At particular stages, in particular chapters or sections, arguments about optimal policies are developed subject to specified assumptions, and perhaps subject to some unspecified ones. But these only say: if the following assumptions hold—and ignoring many matters discussed in other chapters—the following is 'optimal' policy.

The author has views about economic policy in some countries at some times, but these views are influenced not only by an appreciation of the relevant theory but also by a knowledge of the relevant facts and likely reactions to policies, and by value judgements about such matters as income distribution. Theory is vital, but it is not enough. Theory

[7] On the cost of protection, see W. M. Corden, 'The Calculation of the Cost of Protection,' *Economic Record*, 33, April 1957, pp. 29-51; H. G. Johnson, 'The Cost of Protection and the Scientific Tariff,' *Journal of Political Economy*, 68, August 1960, pp. 327-45; S. P. Magee, 'The Welfare Effects of Restrictions on U.S. Trade', *Brookings Papers on Economic Activity*, 1972(3), pp. 645-701. On the average tariff, see W. M. Corden, 'The Effective Protective Rate, the Uniform Tariff Equivalent and the Average Tariff', *Economic Record*, 42, June 1966, pp. 200-216. On effective protection, see W. M. Corden, *The Theory of Protection*, Clarendon Press, Oxford, 1971; and Balassa, *The Structure of Protection in Developing Countries*, 1971.

does not 'say'—as is often asserted by the ill-informed or the badly taught—that 'free trade is best'. It says that, *given certain assumptions*, it is 'best'. Appreciation of the assumptions under which free trade or alternatively any particular system of protection or subsidization is best, or second-best, third-best, and so on, is perhaps the main thing that should come out of this book.

III
READERS' GUIDE

The plan of the book is simple. The most important chapter, setting the theme for the whole book, is the next chapter. Almost as central is Chapter 3, which qualifies the arguments of Chapter 2. Chapter 15 summarizes the main 'first-best' results of earlier chapters, listing the circumstances when the use of tariffs might be first-best for a country.

Every other chapter can essentially stand on its own and so be read on its own. There are, of course, inevitable cross-references. A reader interested in the main lines of the argument and not fascinated by theoretical technicalities and geometric exposition of fine points could pass over (or just skim through) the Appendices to various chapters and the lengthy footnotes.[8]

Chapters 4 and 5, dealing respectively with the revenue role of trade taxes and with the income distribution effects of protection, might be regarded as central to the book, along with Chapters 2 and 3. A number of later chapters, such as Chapter 8 on monopoly, Chapter 11 on dynamic aspects, and Chapter 14 on the relation between protection and cost-benefit analysis, open up some areas that are essentially new for international trade theory. Chapter 9 (infant industry argument), Chapter 11 (dynamic aspects), and Chapter 12 (foreign investment) are concerned with growth, or at least with models in which the factor supply is not given. Chapter 6 (employment and industrialization) and Chapter 9 (capital accumulation) are particularly concerned with less-developed countries. Chapter 14 (on cost-benefit analysis) might be regarded as somewhat peripheral to the book.

[8] He might also do this with the following passages: Chapter 5, Sec. II (the Stolper-Samuelson model); Chapter 7, Sec. I(3) and (4) and Sec. VIII (all concerned with the terms of trade and optimum tariff theory); and Chapter 12, Sec. V (foreign investment and the terms of trade).

2

THE THEORY OF DOMESTIC DIVERGENCES

I

A SIMPLE MODEL: MARGINAL DIVERGENCE AND THE OPTIMUM SUBSIDY

A VERY SIMPLE diagram can be used to illustrate the argument that in certain circumstances a subsidy is to be preferred to a tariff as a method of protection.[1]

In Figure 2.1 the quantity of a particular product, an *importable* product, is shown along the horizontal axis, and its

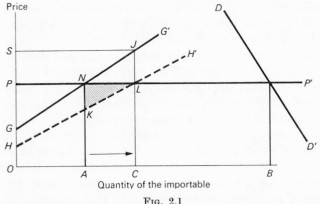

FIG. 2.1

price along the vertical axis. The domestic demand curve for the product is DD', the foreign supply curve for imports of the product is PP' and the supply curve of the domestic import-competing producers is GG'. In the absence of any tariffs, subsidies or other intervention, domestic production would be OA, demand would be OB, and the excess of demand over production, AB, would be imports.

[1] See Appendix 2 for some notes on the history of these ideas and full references.

The next, and crucial, step is to introduce a *marginal divergence between private and social cost*. This will be the justification for some kind of government intervention in the economy. GG' can be regarded as indicating the marginal private cost of production for various levels of output. But the *social* cost is assumed to be less. We can imagine that external economies of some kind attach to production of this product: there are social benefits that are not taken into account in the private cost calculus. The value of these benefits should be subtracted from the costs from a social point of view. Hence we obtain a curve showing the marginal social cost of production at various outputs, namely HH'. Its general characteristic is that it is below GG'. It is of course possible that external *dis*economies attach to this particular type of output; in that case HH' would be above GG'.

Two more assumptions must be made at this stage. (i) The demand curve, DD', indicates not only the private but also the social value at the margin of various quantities consumed. Thus there is *no* marginal divergence on the demand side. (ii) The price of imports, OP, facing private producers and consumers correctly indicates not only the private but also the social cost of imports. The price is assumed to be unaffected by the quantity of imports the country wishes to purchase, so that average and marginal social costs of imports coincide. This is the *small country assumption*. The assumption that OP correctly indicates the marginal social cost of imports also means, among other things, that the exchange rate is in equilibrium. We can assume that any reductions in imports of this particular product have only a minor effect on the balance of payments, so that exchange rate alteration would not be required. We shall come to the exchange rate complication later. Essentially the approach now is partial equilibrium but this assumption does not limit the applicability of the argument to follow, since it can and will be extended to general equilibrium.

The marginal value of extra consumption is equal to the marginal cost of imports when consumption is OB. Hence OB is the socially desirable level of consumption; it is obtained without intervention. The marginal social cost of production is equal to the marginal cost of imports at output OC. This is greater than actual output in the absence of intervention. Hence intervention, designed to increase output or *protect* the industry, is required. But this intervention should not, ideally, alter the level of consumption.

The aim would be achieved by a subsidy on output at the rate PS per unit, or alternatively, the *ad valorem* rate PS/OP. It would raise the price received by producers and lead them to raise output to OC. The marginal private cost of production would become CJ. The total cost of the subsidy to the Treasury would be $PSJL$. Consumers would continue to pay a price of OP for the product.

In what sense is this subsidy 'optimum'? And what are the nature and size of the gains that it brings about? Much of the next chapter will be devoted to examining the assumptions implicit in this type of analysis, and then modifying the analysis appropriately. In addition to the two assumptions already listed, four important assumptions are involved. (a) The act of financing the subsidy through taxation does not upset any marginal conditions, for example through making leisure more attractive at the margin than the rewards from effort, and hence reducing incentives. (b) There are no collection costs of taxation. (c) There are no disbursement costs of the subsidy. (d) The redistribution of income from the relevant tax-payers towards the factors of production that produce the protected product and that will gain in income from the subsidy does not represent a net social gain or loss and so can be neglected. From a social point of view, pure income distribution effects either cancel out or are costlessly offset in some way.

We now have the following principal result. The social cost of the protected output is the area under the social marginal cost curve, $AKLC$. The cost of the imports that are replaced is $ANLC$. Hence the import replacement cost is *less* than the value of the imports by the shaded area KNL. This is the social gain brought about by the subsidy. A higher subsidy would reduce this total gain since at the margin there would be a social loss indicated by the excess of the social cost curve over the import price OP. A lower subsidy would not fully exploit the potentialities of gain.

There are other effects of a redistributive nature, but these are peripheral to the main argument. The relevant taxpayers lose *PSJL*, which is the cost of the subsidy to them (but not to society). Producers of the product gain *PSJN*. The beneficiaries of the external economies created by the extra output, whoever they might be, gain *KNJL*, which is the extent to which social cost of the protected output falls below private cost.

II
BY-PRODUCT DISTORTION EFFECT OF TARIFF

The same degree of protection could be brought about by a tariff at the rate *SP/OP* (Figure 2.2). It would raise the

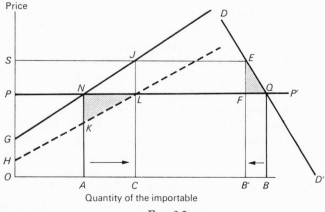

Fɪɢ. 2.2

domestic price of the product to *OS*. Production would increase to *OC*, consumption would fall to *OB'* and imports would fall to *CB'*. The *production* or *protection effect* would be exactly the same as in the case of a subsidy, yielding a social gain represented, as before, by the shaded area *KNL*. The difference from the subsidy is that this time there would also be a *consumption effect*, inflicting a loss. The effort to correct the effects of a *marginal divergence* on the production side would have as a by-product the creation of a new divergence on the consumption side.

Let us look at this by-product effect more closely. A tariff that raised the domestic price to OS would reduce consumption to OB'. On the last unit of consumption forgone, (represented by point B'), the private and social value, as indicated by the demand curve, is EB'. But the social cost, as indicated by the cost of imports, is only FB'. Hence the conditions for equality of social costs and benefits at the margin required for social optimization would no longer be fulfilled. The consumption of the product concerned which is forgone as a result of a tariff is the amount $B'B$. If one accepts the simple 'consumers' surplus' approach the value to consumers of this amount is $B'EQB$ while its cost is only $B'FQB$, so that consumers' surplus of FEQ (shaded in Figure 2.2) is obtained. This would be lost as a result of the tariff. One need not regard the 'consumers' surplus' triangle as measuring the loss precisely, but the basic idea remains that the tariff would impose a new divergence between private and social cost in the process of offsetting an existing divergence.

When such a loss is imposed as part of a corrective policy we shall call it a *by-product distortion*. This does not mean that it is unimportant. In Figure 2.2, the loss FEQ could well be greater than the gain KNL. It is obvious that a subsidy is preferable to a tariff because the subsidy does not—at least with our assumptions so far—have an adverse by-product effect.

It might also be noted that in the discussion of the tariff we have implicitly made various assumptions similar to those made in the case of the subsidy. Income distribution effects have been neglected. Various effects connected with the use of the customs revenue are ignored. There are assumed to be no collection or administration costs of the tariff.

Before going on, let us look at the matter of terminology for a moment. We are using the term *marginal divergence* to refer to any divergence between marginal private and marginal social costs, or marginal private and marginal social benefits, however caused. If a divergence is caused by government policy of some kind, such as a tariff or other form of tax, then it is a *distortion*. Thus a *distortion* is a particular kind of *divergence*. If a divergence is caused by 'market failure' which is not obviously the result of government policy—for example, a monopoly situation or an externality—then it is not a distortion. We could, if we liked, call it an 'endogenous' divergence, though in general it will just be described as a divergence.

To return to *distortions*, a special type of distortion is the *by-product distortion*. This is the by-product of a government policy designed to correct, or partially correct, a divergence of some kind, defining the latter in the broadest sense. Presumably most distortions would be of this kind, but one can also envisage distortions that are not the results of corrective policies but are historical relics, are irrational, "political" in motive, and so on.

In the literature on this subject the term *distortion* has tended to be used in a very broad way to include everything we call here a *divergence*. We depart here from this broad use of the term *distortion* in order to narrow it meaning, because it gives the false impression that monopolistic restrictions and externalities are in some sense not endogenous or natural to the economic system and could easily be removed.[2]

III
ELABORATIONS OF THE SIMPLE ANALYSIS

The partial equilibrium model used is extremely simple, indeed naive. It certainly rests on very many assumptions, some of which will be reviewed in the next chapter, and others later in the book. But nevertheless it is a powerful, and most significant, analysis which can later be generalized. Before doing so it is well worthwhile manipulating it to squeeze out of it several other simple results.

(1) *Consumption Divergence*

The marginal divergence may be on the consumption side. Consumption or usage of the product concerned may have external effects, so that the private and social valuation curves diverge. The product might be gasoline, the use of which

---

[2] Bhagwati, in 'The Generalized Theory of Distortions and Welfare' (in J. Bhagwati et al. (eds.), *Trade, Balance of Payments and Growth*, 1971) calls an *endogenous distortion* what we call an endogenous divergence, a *policy-imposed instrumental distortion* what we call a by-product distortion and a *policy-imposed autonomous distortion* what we just call a distortion that is not a by-product distortion. Essentially the system of classification is the same as Bhagwati's. The term *divergence* is used in Meade's *Trade and Welfare*.

pollutes the atmosphere, and hence creates an external diseconomy. The social valuation curve would then be to the left of the private demand curve DD', like RR' in Figure 2.3. Socially optimum consumption would be at OB'. If there is no marginal divergence on the production side, production should stay at the non-intervention level OA. First-best policy would be a tax on consumption of the product at the

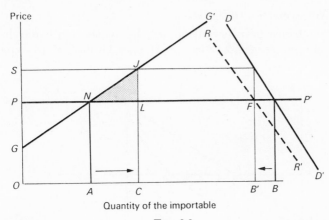

Quantity of the importable

Fig. 2.3

rate PS/OP. This tax should apply to imports and to domestic output at the same rate, so that no protection would be provided.

If a tariff alone were used, the desired decline in consumption could indeed be attained by setting a tariff rate of PS/OP. But a *by-product distortion* would be created by the protection effect. Domestic output would expand by AC. This extra output would cost $ANJC$ to produce (assuming that private and social cost coincide), but would replace imports that would cost only $ANLC$. Hence there would be excess cost of domestic production of NJL, this triangle (shaded in Figure 2.3) measuring the cost of the *by-product distortion*.

It is perfectly conceivable that an external economy on the production side coincides with an external diseconomy on the consumption side. It would then be desirable to eliminate both divergences. First-best policy would consist of a production subsidy equal to the externality on the production side combined with a consumption tax equal to the externality on the consumption side. A tariff is the equivalent, from the point of view of the present analysis, of a consumption tax combined with a production subsidy at the *same* rate. But it would be pure chance if the rate of production subsidy required were the same as the rate of consumption tax required, so it is unlikely that a tariff alone would be a first-best policy.

(2) *Non-economic Objectives*

The purpose of intervention may be to attain or come closer to some *non-economic objective*. It may be desired to expand

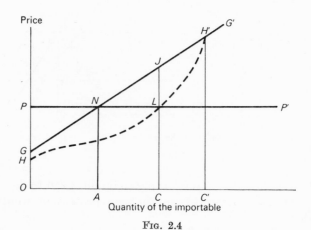

Fɪɢ. 2.4

domestic import-competing production of the product for military reasons, or because some emotional or 'political' value is believed to attach to its production. We need not concern ourselves with the nature of the non-economic objective, and certainly not with its rationality, but can define it as any objective that comes, essentially, from outside the model.

The attainment of such an objective will have an economic cost. There is likely to be, in principle at least, a trade-off between the cost and the non-economic benefits derived from coming closer to the objective. This can be represented as follows. In Figure 2.4 the private and social marginal cost of production is shown by GG'. The marginal value that society attaches to varying outputs of the product for non-economic reasons is shown by the vertical distance between GG' and HH'.

As drawn, the marginal social valuation of increments of output from the point of view of the non-economic objective first increases and then decreases. At output OC' the two curves coincide and the non-economic objective has been completely attained. At output OC society regards an extra unit of output from a non-economic point of view as worth forgoing JL in real resource cost. Attaining the whole of the non-economic objective would be worth to society the whole of the area HGH' between the two curves. HH' is thus, in fact, a special kind of net social marginal cost curve which takes non-economic benefits into account.

Output is optimum at OC, where the marginal excess *economic* cost JL is just equal to the marginal *non-economic* value of the output. The net economic cost of expanding output from OA to OC is NJL—the *production-distortion cost of protection*. This is the excess cost of protected output over the value of the imports it replaces. This assumes that a subsidy is used to bring about the necessary expansion of output.

Non-economic objectives can be assimilated in our simple analysis. If the non-economic objective has to do with domestic production, as in our example, a production subsidy or tax will be first-best. A tariff would create a by-product distortion. If the non-economic objective is associated with domestic consumption, a consumption tax or subsidy will be first-best.

(3) *Fixed Targets*

Sometimes it is convenient to express an objective of policy, whether economic or non-economic, as a *fixed target*. There is no question then of trading-off the target against the cost of achieving it. In practice no target is likely to be absolutely fixed, quite irrespective of costs. But here, let us assume such a case, since it is an assumption that is often made.

The target might be to attain domestic output OC (Figure 2.5). If a production subsidy is used to attain it, an inevitable production-distortion cost of NJL is imposed. The marginal cost of the target is JL. The target could also be attained with a tariff, but then, in addition to the inevitable production-distortion cost, a by-product consumption-distortion cost would be imposed. Similarly, the minimum cost way of attaining a consumption target would be by a consumption tax or subsidy; again, a tariff would be second-best because it would create a by-product production-distortion cost. The whole of our analysis applies in the case of a fixed target.

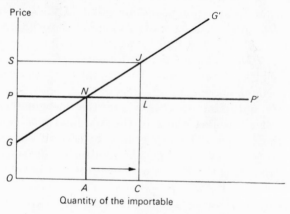

Quantity of the importable

Fig. 2.5

IV
THE SECOND-BEST OPTIMUM TARIFF

If there is a marginal divergence in domestic production, social cost being below private cost, a production subsidy is first-best. A tariff inflicts a by-product distortion. Hence, disregarding any other possible devices, a tariff is second-best. Let us now examine the properties of the second-best solution more closely. We shall suppose that there is a constraint that prevents the use of a production subsidy. Only a tariff can be used. We could think of the constraint as being non-economic, say political, in nature. Alternatively, we could imagine the administrative costs of a subsidy being so high as to outweigh any possible gain from correcting the production divergence.

A tariff could certainly have an adverse effect. The benefit it yields by correcting the marginal divergence in production could be more than offset by the cost of the by-product distortion in consumption. In Figure 2.6 a tariff that completely corrects the production divergence would yield a gain on the production side of *KNL* and a cost on the consumption side of *FEQ*. The latter cost could clearly exceed the gain.

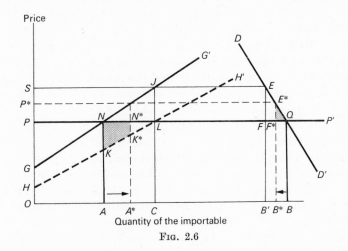

Fig. 2.6

While it is true that a tariff *could* have an adverse effect, there will be a particular second-best optimum tariff rate that maximizes the gain in the presence of a marginal divergence in production. Provided one makes the simple assumptions implicit in our diagrams, namely that the demand curve and the supply curve, as well as the social marginal cost curve *HH'*, all slope consistently negatively (demand curve) or positively (supply and social cost curves), something fairly precise and interesting can be said about the properties of the second-best optimum solution. Three interrelated properties of this optimum solution should be noted.

Firstly, the second-best optimum tariff will be positive but less than the rate of tariff that would correct completely the production divergence. In fact, the correction of the production divergence has to be 'traded-off' against the by-product consumption distortion that the process of making the correction inevitably inflicts.[3]

[3] *Trade and Welfare*, pp. 228-9.

In Figure 2.6 the optimum tariff rate is represented by the rate of tariff $PP*/OP$. How is it arrived at? We can imagine the rate of tariff to be gradually raised from zero. On the first increment in output there is a gain on the production side equal to the whole of the initial marginal divergence, namely NK; as output increases, the marginal gain on the production side gradually declines until, at output $OA*$, it reaches $K*N*$. If the tariff had been increased beyond that, output would eventually have reached OC, at which point the marginal gain would have been zero. At the same time the rise in the tariff inflicts a consumption cost. At first the marginal consumption distortion cost is zero; when consumption has fallen as far as $OB*$ the marginal consumption cost is $F*E*$.

The net gain consists of the gain on the production side, resulting from the partial correction of the marginal divergence, less the consumption distortion cost. The net gain will be maximized when marginal production gain is equal to marginal consumption cost. In Figure 2.6 this is at a tariff rate of $PP*/OP$ since $K*N*$ (marginal production gain) is equal to $F*E*$ (marginal consumption cost). Hence the second-best optimum tariff must be (a) positive and (b) less than PS/OP, which would correct the divergence completely.

Secondly, it is better to impose the second-best optimum tariff than to do nothing at all. There is some rate of tariff that must (given our assumptions) lead to a welfare gain. The size of the gain in Figure 2.6 is the production gain up to output $OA*$, namely $KNN*K*$ (shaded), minus the consumption distortion cost up to $OB*$, namely $F*E*Q$ (also shaded). This net gain must be positive, and the second-best optimum tariff maximizes it since at that rate the *marginal* net gain is zero.

Thirdly, the gain from the second-best solution (a tariff) must be less than the gain from the first-best solution (a subsidy). In Figure 2.6, the first-best solution of a fully corrective production subsidy yields a gain, all on the production side, of KNL. The second-best solution falls short of it by the sum of two areas: the production gain forgone by deliberately failing to make a full correction, namely $K*N*L$, and the consumption distortion cost inflicted, $F*E*Q$.

V
MARGINAL DIVERGENCE IN FACTOR COST

It might be thought that much of the intricate theorizing so far is rather irrelevant. It hinges completely on the importance of the by-product consumption distortion that a tariff inflicts. It might be argued that such consumption distortions are simply not important. This depends, of course, on elasticities of substitution in demand. But, in any case, the real interest in the argument is in the scope it offers for generalization. The simple case where the marginal divergence is associated with production of an import-competing product, and where the only adverse effect of a tariff is the consumption distortion, is heuristically convenient. Now let us go beyond this case.

The marginal divergence between private and social cost might be associated with the use of a factor in a particular industry. For example, the wage that an industry has to pay for its labour may exceed the social opportunity cost of this labour. Other industries are assumed to face no such divergence. This case has provided a popular argument for protection in less-developed countries, and will be more fully discussed in Chapter 6. Alternatively, external diseconomies may result from the use of a particular factor or component by an industry, for example, the use of a certain type of polluting fuel. First-best correcting policy then requires a subsidy or tax directly related to the relevant factor input—a wage subsidy to the industry in the first case, and a tax on the use of the fuel by the industry in the second.

Let us take the first case as our example. The wage subsidy will have two favourable effects. Firstly, labour will be substituted for other factors in the industry to an appropriate extent. This assumes that there are not technically fixed proportions between them. Secondly, the costs of the industry will fall, and hence its output will expand relative to other industries.

A production subsidy to the industry would be second-best. It represents, in effect, a subsidy to all factors used by the industry, both labour and non-labour, and hence would lead to the by-product distortion of over-use of non-labour factors in the industry. It would fail to make the industry's production more labour-intensive. Even the first-best solution may require some inflow of non-labour factors into the industry, though proportionately less than the inflow of labour (unless there are fixed proportions). But the production subsidy would lead to a greater inflow of non-labour factors for any given expansion of the industry's use of labour.

The second-best optimum production subsidy will result from trading-off the gains from offsetting the marginal divergence in labour cost against the by-product distortion, along the lines of our earlier analysis of the second-best optimum tariff. It will normally involve a lesser stimulus to the industry's use of labour than the first-best policy would have done; it will lead to some welfare gain, but this gain will be less than the gain from the first-best wage-subsidy.

A tariff is third-best, since it imposes not only the same by-product factor distortion that the production subsidy imposes, but in addition inflicts a by-product consumption distortion. There will be a third-best optimum tariff. It will normally involve a lesser stimulus to the industry's use of labour and to its output than the second-best optimum production subsidy would have done, and it will lead to some welfare improvement, but less than the second-best optimum policy.

It might be noted that one could interpret the concept of a marginal divergence so widely as to incorporate all possible considerations which call for some kind of social intervention, not just the ones we have mentioned so far. There might be monopolistic pricing. There might be a distorting tax which, for one reason or another, must be taken as given. The divergence may be, not between private and social interest, but between *privately perceived* private interest and socially perceived private interest, the latter based perhaps on more mature judgement or better information (as in the case of smoking or drugs).

VI
GENERAL EQUILIBRIUM AND MANY COMMODITIES

The method of exposition so far has been partial equilibrium. But this has only been a matter of exposition. The analysis is fully applicable in general equilibrium. Indeed, if one stays within the constraints of a two-commodity, two-factor general equilibrium model of the type generally used in international trade theory, the translation to general equilibrium is easily made. There is a large literature which expounds, with much elaboration and sophistication, many of the arguments advanced in this chapter in terms of two-good two-factor geometry. A two-sector general equilibrium exposition of the main arguments of Sections I, II and IV is given in Appendix 1.

Even our partial equilibrium diagrams can be translated into general equilibrium. The product can be defined as the economy's sole importable, and price and marginal cost can be redefined respectively as the relative price ratio in a two-good model and the marginal cost of the importable in terms of the exportable. Similarly, the demand curve can be interpreted as showing the marginal value to consumers of the importable in terms of the exportable. One would have to be careful about using the consumers' surplus method, (for which purpose one would need to use a constant utility demand curve, hence excluding income effects), but the main ideas would stand.

In the real world of many commodities and many factors everything becomes, of course, much more complicated. One could describe numerous possible complex cases, with a variety of interacting divergences, to show what, in each case, the appropriate first-best package of policies would be, and how some other package would compare with it. The optimal correction required for one particular divergence may well depend on whether another divergence has been corrected.

If a second-best policy is used to correct one divergence, hence creating a by-product distortion, it may create the need for a supplementary policy designed to correct, at least partially, the newly created distortion. In general, one can say that the the first-best package of corrective policies will consist of policies that get as close as possible to the sources of the various marginal divergences, and hence that minimize the by-product distortions that are created.

The case where it is desired to foster domestic production of an import-competing intermediate good is rather interesting, and illustrates some of the issues.

External economies may be believed to attach to domestic production of an import-competing intermediate product A which is an input into another import-competing product, B. Let us assume that this externality is initially the only marginal divergence and that *the only policy instruments available are tariffs*. Hence a tariff is imposed on A. This raises the costs of producers of B, imposing negative effective protection on them, so leading to the by-product distortion of underproduction of B. A second corrective policy, consisting of a tariff for B, needs then to be imposed. The second-best optimum tariff rate will not be so high as to restore the effective rate of B to zero since now the tariff will impose a by-product consumption distortion. But finally there will be an optimum set of a tariff on A and a partially compensating tariff on B, the whole exercise being designed to correct the marginal divergence in A, taking account of the various by-product effects.

VII
HOME-MARKET BIAS

A country may foster the development of its manufacturing industry with tariffs or import restrictions and create a large and sophisticated industrial sector, one which might be expected to export a significant part of its output. But if the protection is only provided for the home market, with little special encouragement for exporting, lopsided development, with very few exports of manufactures may result. This indeed has been the case in a number of Latin American countries, notably Brazil and Argentina, and also in India and Australia. It has come to be gradually realised that one of the main weaknesses of the so-called import substitution strategy is *home market bias*: its failure to encourage exports of manufactures. Hence opinion has swung in favour of measures which in some form or other subsidize or encourage these exports.[4]

---

[4] See Little et al., *Industry and Trade in Some Developing Countries*, especially pp. 128-134. Policies in India and Brazil have become more export-orientated in recent years.

It is of course assumed that there is some general case for fostering manufacturing, presumably at the expense of agriculture and mining.

(1) *A Three-product Model*

The nature of *home-market bias* can be analysed rigorously and related to our general analysis. It is an important by-product distortion created by a tariff system, one not brought out in our simple partial equilibrium analysis earlier.

Let us assume that three products are produced and consumed in the country. Two products are exported as well as consumed domestically, namely X_a, the agricultural product and X_m, a manufactured product. Another manu-factured product, M_m, is imported and also produced domesti-cally.

Now we make two assumptions. Firstly, external economies attach to production of both types of manufacturing, so that their marginal social costs of production fall below their private costs by a given percentage, the same for both of them. 'Cost' is defined in terms of agricultural production forgone. In the absence of any intervention production of X_a will be too high from a social point of view and of X_m and M_m too low. Secondly, relative price changes over the relevant range do not induce any consumption substitution between the three goods. This assumption is to isolate *production* effects; it will be removed later.

(2) *Subsidy and Tariff Compared*

The first assumption means that optimum output could be attained by a subsidy to production of manufactures equal to the rate of external economies, financed by a non-distorting tax. Alternatively agricultural production could be taxed, the revenue being redistributed in a non-distorting way. Essentially, both methods are equivalent to a tax on agricul-ture which finances a subsidy on manufacturing. This subsidy to manufacturing would bring about a resource movement out of agriculture into the two manufacturing industries as indicated by the arrows below.

Exports of X_a would fall, exports of X_m would increase and imports of M_m would fall.

If, instead of the production subsidy, a tariff were imposed, the domestic price of M_m would rise relative to the domestic prices of the two export products, and resources would be drawn into M_m out of both export products. The resource movements induced by the tariff are shown by the arrows below.

When we compare the tariff with the production subsidy we notice two differences: firstly, the production subsidy leads to a movement of resources out of agricultural exporting into manufactured exports, while the tariff fails to bring about such a shift in the pattern of exports; secondly, the tariff leads to a switch of resources within the manufacturing sector, out of manufactures for exports into import-competing manufactures.

If the rate of tariff is just sufficient to increase total manufacturing production to the same extent as the optimum rate of production subsidy would have done, the movement of resources out of X_a will be the same in both cases. The difference between the two cases will be in the pattern of manufacturing production. The tariff will lead to an expansion of import-competing manufacturing production M_m and a contraction of X_m, though production of the two combined (valued at world prices) will have expanded; the production subsidy will lead to a lesser expansion of M_m and in addition to an expansion of X_m.

(3) *Protection Versus Promotion*

Home-market bias would be removed if the tariff on M_m were supplemented by an export subsidy for the manufactured export. (To avoid re-import of the subsidized export, X_m, a tariff may also need to be put on imports of X_m.)

If the rate of export subsidy were the same as the rate of tariff for the import-competing manufacture, M_m, the domestic prices of both manufactured goods would rise to the same extent, and the only resource movements would be out of X_a into both manufactures. When a tariff alone is imposed, manufacturing is only *protected*; when the tariff is supplemented by an export subsidy for manufacturing, then manufacturing is *promoted*.[5] *Protection* creates a home-market bias; *promotion* avoids it.

The method of combining a tariff and an export subsidy at the same rate would have exactly the same effect as a production subsidy for manufacturing (given that we are assuming no consumption effects). Hence there are two methods of *promotion*. Both raise the prices received by manufacturers relative to the prices received by agricultural producers and so bring about the desired resource movement.

It is possible that in the absence of any intervention the country does not export any significant quantities of manufactured products. Hence, in terms of our formal model, initially X_m is an imported product. But when the optimum subsidy is imposed, its production would expand so much that it would begin to be exported. Thus optimal policy would require *trade reversal* with respect to this product. A tariff alone could not bring about this trade reversal. At the most, it could bring about complete replacement of imports of the product by domestic production.[6] But any increase in the tariff beyond the point where imports have ceased would have no effect. To bring about the optimum output of X_m and hence the necessary trade reversal it would be necessary to supplement the tariff with an export subsidy. Otherwise there will, as before, be under-expansion of X_m.

(4) *Consumption Effects*

We could now introduce the familiar consumption effects. Both tariffs and export subsidies will raise prices to the domestic consumers of the protected products and so bring

[5] The distinction between 'protection' and 'promotion' comes from *Industry and Trade in Some Developing Countries*.

[6] This is definitely true only in the formal model used here, with an upward-sloping supply curve and perfect competition. If the industry were monopolized and subject to economies of scale it might become possible that in the absence of a tariff a product would be wholly imported while the tariff caused domestic producers to supply the whole domestic market and also to export. See Chapter 8, pp. 242–3.

about an undesired switch in the consumption pattern. Our main argument that promotion is preferable to protection is not affected. But it does mean that the two forms of promotion—tariff-cum-export subsidy on the one hand, and production subsidy to manufacturing on the other—no longer have identical effects. The production subsidy does not create a consumption distortion and is thus clearly preferable. It could also be shown that, if one must use a tariff-cum-export subsidy as promotion, rather than using a production subsidy, the rate of export subsidy should not necessarily be identical with the rate of tariff. The aim of non-uniformity would be to minimize the inevitable consumption cost.

VIII
THE HIERARCHY OF POLICIES

For any given marginal divergence, or set of divergences, there is a first-best optimal policy or set of policies. Essentially this policy involves making the appropriate correction as close as possible to the point of the divergence. But many policies may be conceivable, and one should be able to order them in a hierarchy of policies, from first-best to second-best, and so on. This is an interesting exercise, for it brings out the logic of the general approach, and especially the effects of imposing constraints on the choice of policies which compel a movement down the hierarchy.

At each step down the hierarchy an additional by-product distortion is imposed, and the welfare level attainable with the appropriate optimal policy declines. Furthermore, given the appropriate optimal policy, the extent of the correction to the basic divergence will normally decline.

We shall now look at a particular case of a marginal divergence to illustrate this approach. But it must be stressed that the methodology is perfectly general. It can be applied to the analysis of economic policies in all spheres, not just trade policies. For any divergence between private and social costs or benefits there is a hierarchy of policies. It is always true that as one goes down the hierarchy an additional by-product distortion is imposed, the welfare level attainable declines, and the extent of the optimal correction to the original divergence will normally decline. This approach will be used continually throughout this book, and perhaps it should be used much more widely when economists describe or analyse alternative economic policies. For example it could be used in comparing various policies to deal with external diseconomies, such as pollution or traffic congestion. The approach will be qualified or elaborated somewhat in the next chapter.

Suppose that the marginal private cost of labour to the manufacturing sector in a less-developed country exceeds the social opportunity cost of this labour. Assume that this is the only significant marginal divergence. Chart 2.1 represents the relevant hierarchy of policies.

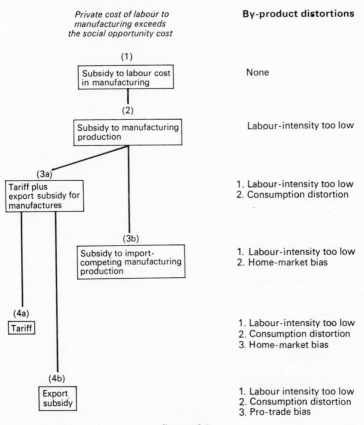

THE HIERARCHY OF POLICIES

Private cost of labour to manufacturing exceeds the social opportunity cost   **By-product distortions**

(1)

Subsidy to labour cost in manufacturing   None

(2)

Subsidy to manufacturing production   Labour-intensity too low

(3a)

Tariff plus export subsidy for manufactures

1. Labour-intensity too low
2. Consumption distortion

(3b)

Subsidy to import-competing manufacturing production

1. Labour-intensity too low
2. Home-market bias

(4a)

Tariff

1. Labour-intensity too low
2. Consumption distortion
3. Home-market bias

(4b)

Export subsidy

1. Labour intensity too low
2. Consumption distortion
3. Pro-trade bias

CHART 2.1

First-best policy is to subsidize the use of labour in manu-facturing. (Alternatively, if it were possible, labour in agriculture could be taxed.) Second-best policy is to subsidize manufacturing production; this, as already pointed out, will impose the by-product distortion of over-encouraging the use of non-labour factors (notably capital) in manufacturing. Third-best policy is to impose a tariff on imports of manu-factures combined with an export subsidy for manufactures at the same rate. This will add a by-product consumption distortion. Fourth-best policy is to impose a tariff on manu-factured imports on its own. This will add a third by-product distortion, namely home-market bias.

The tariff thus creates three by-product distortions. The fourth-best optimal tariff may need to be set at a level that will yield a stimulus to the employment of labour in manufacturing that is much less than the stimulus the first-best wage-subsidy policy would have provided. The gain from correcting the basic divergence has to be traded-off against no less than *three* by-product distortions.

There is an element of arbitrariness in the choice of policies to include in a hierarchy of policies. One could include all technically feasible policies. Alternatively, one could limit oneself to those policies which have been under consideration, or perhaps which have actually been used in the past, or which are politically conceivable. The example just given has not been comprehensive; the aim has been to display the special features of the tariff as a policy instrument. This is brought out in Chart 2.1, which shows two other policies that might be included in the hierarchy.

A subsidy on production of import-competing manufactures imposes a home-market bias but no consumption distortion. In the hierarchy it is thus graded as third-best, alongside the tariff-cum-export subsidy. Similarly, an export subsidy on manufactures on its own parallels a tariff on its own, both being fourth-best: the subsidy creates *pro-trade* bias and the tariff *home-market bias*.

When two policies are third-best in the hierarchy we are simply saying that they both add one more by-product distortion to the second-best policy. One of the 'third-best' policies may lead to a higher welfare level than the other, so

that one would have to renumber the policies if one wished to
order them strictly on the basis of welfare levels attainable,
and not just on the basis of the numbers of by-product
distortions imposed.

IX
TRADE DIVERGENCES

So far we have assumed that all divergences are *domestic*
divergences (and we shall maintain this assumption in the
next four chapters). Thus in our various diagrams we have
assumed that the import price OP indicates both marginal
private and marginal social cost of imports. But it is possible
that there is a *trade divergence*. In that case a trade intervention will indeed be first-best. The most important case is
where the country can affect its terms of trade. We shall
discuss this at length in Chapter 7, and here we continue to
make the *small country assumption* that the country faces
given import and export prices at any point in time. But there
may be other reasons why private and social costs of importing
(or returns from exporting) diverge.

Some kind of external diseconomy may attach to trade as
such. If this is so, in our simple model, a tariff will be a first-
best policy. It will be optimum to restrict trade by *both*
increasing import-competing production and reducing consumption of imports. In Figures 2.1 to 2.3, if the marginal
social cost of imports is OS, then a tariff of PS/OP will bring
about optimum domestic production and consumption. It is
consistent with our general approach that when a marginal
divergence is genuinely associated with trade as such, the
first-best policy is a trade intervention.

There might be a non-economic objective connected with
trade, namely to make the country more self-sufficient,
perhaps for nationalistic reasons. It is only by positing such an
objective that one can make sense of some seemingly irrational
policies. In terms of our model, the aim is then to reduce
imports. This will inevitably be attained at a cost—the sum of
the *production* and *consumption distortion costs of protection*. A
tariff will then be first-best. Since the country is indifferent
to the way in which imports are reduced—whether it is by

increasing domestic production or reducing domestic consumption—the reduction should be brought about so as to equate marginal production-distortion cost to marginal consumption-distortion cost: this will minimise the total cost of achieving any given degree of import restriction, and will indeed be brought about by a tariff.

One could suppose that some marginal value is attached to the nationalistic objective, perhaps declining as the volume of trade is reduced. This marginal value must then be equated to the marginal cost of attaining the objective, and this can be brought about by fixing the tariff equal to the marginal value of the objective.

A frequent misunderstanding should be avoided. The foreign suppliers of the product may be subsidized or taxed in some way so that the costs of their exports—our country's imports—are not equal to marginal social costs in the supplying country. Alternatively, there may be some externality in the foreign country. But this does not mean that there is a marginal divergence from *our* country's point of view. If the price is given to our country, then, unless some particular advantage or disadvantage attaches to foreign trade in the product as such, there is no marginal divergence on the trade side. Our country can ignore the marginal divergence abroad. This approach assumes that it wishes to apply a national, and not a cosmopolitan, welfare criterion. Production of the product in the foreign supplying country may pollute the latter's environment and so inflict external diseconomies. This may well present an argument for social intervention in the foreign country. But from the purely selfish point of view of the importing country, there is no argument for intervention.

It is particularly important to note—because it is often not understood—that a foreign country may be subsidizing its exports, and this may well create a marginal distortion abroad, but it does not mean that there is a marginal divergence for *our* country which is importing the product. If the import price is given, then it is indeed given, and it is not really relevant whether it has been brought down by the improved efficiency of foreign producers, a decline in foreign demand for the product, or foreign subsidies. It follows that a free trade policy may be optimal for our country even when its trading partners are intervening in trade. The central issue is whether the import price facing us is indeed given. It is possible that our country's tariff may cause the foreign country's intervention policy to change; but this is another matter to which we turn in Chapter 7.

Finally, there is the question of the exchange rate. Let us again make the small country assumption and assume that there are no domestic divergences of any kind, and no trade divergences except one. This is that the country is running a balance of payments deficit which has to be ended or reduced. We can think of this most conveniently as a fixed target. Hence foreign exchange is wrongly priced, so that there is a trade divergence. First-best policy is to raise the price of foreign exchange appropriately (devalue the domestic currency). This will cause the pattern of domestic production to shift from non-traded to traded goods and the pattern of consumption in the other direction. The use of consumption or production-switching measures alone (such as subsidies on import-competing production) would be second-best or worse. It is not too difficult to prove this rigorously, although we shall not do so here.

Since the divergence is in the price of foreign exchange, the use of a tariff or a set of tariffs will not be first-best. A tariff will only act on the import side, but will fail to stimulate exports. Hence trade intervention that uniformly raises the domestic prices of imports and exports is required. This can be attained by altering the exchange rate itself or by combining a uniform *ad valorem* tariff with a uniform *ad valorem* export subsidy at the same rate, tariff and subsidy applying to all trade, visible and invisible.

APPENDIX 1

THE THEORY OF DOMESTIC DIVERGENCES IN TWO-PRODUCT GEOMETRY

It is possible to represent some of the main ideas of the theory of domestic divergences by using the two-product geometric method familiar to all students of international trade theory. The method is elegant. It is also attractive that the approach is explicitly and clearly general equilibrium. Hence it is comforting to find that the results obtained from the simple demand-and-supply diagrams used

earlier are confirmed here. First we represent the central ideas of
Sections I, II and V of this chapter. At the end a special case, where
trade reversal would be the first-best optimum, is analysed.

In Figure 2.7 the importable, product M, is shown along the
vertical axis, and the exportable, product X, along the horizontal
axis. The quadrant contains a map of community indifference curves

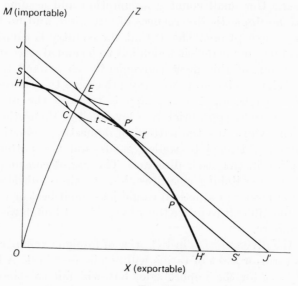

Fig. 2.7

which represent ordinal welfare levels as well as community preferences on the assumption that a given income distribution is maintained by an independent policy—that is, that a social welfare
function is being consistently and costlessly applied.

The *social* production transformation curve is HH'. It shows full
employment production possibilities. Its slope at any point indicates
the social marginal rate of transformation at that point. The
privately perceived transformation rate differs from this. Thus at
point P the latter is assumed to be given by the slope of SS' and at P'
by the slope of tt'. These private transformation rates are flatter than
the social ones; hence we are assuming that the marginal private cost
of transforming X into M exceeds the social cost. This is the case that
was analysed in section I; an external economy is associated with
production of the importable. It should be noted that in this case the
economy will be producing on its production-efficiency frontier. In

terms of the Edgeworth-Bowley box diagram, it is producing on the contract curve. The case where it is *not* producing on this frontier will be considered later.

(1) *Free Trade Equilibrium*

The slope of SS' is assumed to indicate the given world price ratio. In free trade, output is at P, and the country can trade along SPS', choosing to consume at C, on the income-consumption line OZ appropriate to the world price ratio. At C an indifference curve is tangential to SS'. The rate at which X can be transformed into M through foreign trade is the *foreign marginal rate of transformation* or *FRT*. The slope of the indifference curve is the *domestic marginal rate of substitution in consumption* or *DRS*. The slope of the domestic production transformation curve HH' at the relevant point gives the *domestic marginal social rate of transformation in production, or DRT*. In the free trade situation the foreign marginal rate of transformation is equal to the domestic marginal rate of substitution in consumption, but both differ from the domestic marginal social rate of transformation in production (i.e. $FRT = DRS \neq DRT$).

(2) *The Optimum*

In the first-best optimum situation the three marginal rates would all be equal. This is attained by a production subsidy on M (or a production tax on X) which shifts output to P', the rate of subsidy being given by the excess of the private transformation rate at P' (given by the slope of tt') over the social rate (slope of HH'). It is, of course, assumed that the subsidy is financed by costless non-distorting means. At point P' the country can trade along $JP'J'$ (parallel to SPS') and will choose consumption point E. At this point an indifference curve is tangential to $JP'J'$. The point E is on the same income-consumption line OZ as the free trade consumption point C. It is the best point attainable with the given production transformation curve HH' and the given world price ratio.

(3) *A Tariff*

Now consider a tariff (or export tax). We refer to Figure 2.8. Imagine the tariff to be raised from zero. The output point will move along HH' from P towards P' and perhaps beyond. This is the production effect of the tariff. For each production point the trading possibilities are given by a line through that point having the slope of the world price ratio. Thus, when production is at P' the trading possibilities are along $JP'J'$. At the same time, as the tariff is raised,

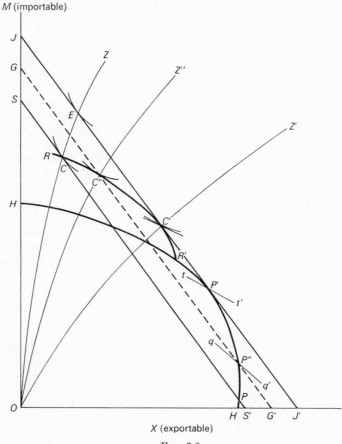

FIG. 2.8

the relative price of importables facing consumers increases above the free trade price ratio, so that the relevant income-consumption line swings to the right. We are assuming that tariff revenue is returned to consumers, and that all of it is spent in accordance with the demand pattern indicated by the community indifference curve map.

If a tariff were imposed that was sufficient to eliminate the production divergence and hence to bring production to P'—the first-best optimum output—the consumption pattern would move to the income-consumption line OZ', and the consumption point would be C'. It can be seen that the indifference curve through C' could be higher or lower than the curve through C (which was the free trade consumption point); in other words, if a tariff is used to eliminate the production divergence completely, there could be a gain *or a loss* relative to free trade.

(4) *The Meade Curve and the Second-Best Optimum*

There are many possible tariff levels and hence consumption points. Suppose a tariff shifts output to P''. The country can then trade along $GP''G'$. With the price facing consumers above the free trade price, they will choose to consume at C'', on the relevant income-consumption line OZ''. The slope of the indifference curve at C'' is flatter than at C, and is the same as the privately-perceived marginal rate of transformation in production at P'' (indicated by the slope of the line qq' through P'').

One can thus trace out a curve $CC''C'$—to be called the *Meade curve*—which shows the locus of consumption points as the tariff is raised from zero towards the rate at which the production divergence is eliminated.[7] The tariff could be raised further, so that this curve could be continued to R', the self-sufficiency point. (The curve could also be carried to the left of C, the range RC resulting from an import subsidy.)

Somewhere on this *Meade Curve CR'* an indifference curve must be tangential to it; this will indicate the best consumption that can be obtained with a tariff. It would result from the second-best optimum tariff. In the diagram this second-best optimum consumption point is C'', with the associated production point P''.

The optimum point on the *Meade curve* must be between C (the point of free trade consumption) and C' (the consumption point if the production divergence were completely eliminated). The production effect of a marginally positive tariff will bring the consumption point to a trading possibility line above SS', in fact extending consumption possibilities outside the free trade consumption possibility line SS'. Furthermore, the curve must be tangential to JJ' at C', and hence at C' an indifference curve must cross it (since the slope of the indifference curve at C' is equal to the slope of the privately-perceived transformation rate at P'). It follows that the second-best optimum tariff is positive and should be less than needed to bring production to P' where the whole of the production divergence would be eliminated.

In this second-best optimum situation $DRS > FRT > DRT$, and hence the equality $DRS = FRT$ which existed in free trade has

---

[7] It is termed the *Meade curve* here because it is used to represent ideas that originated in *Trade and Welfare* (even though the curve itself cannot actually be found in any of Meade's books).

been destroyed. It is evident from the diagram that the indifference curve through C'' must be lower than that through E. In other words, the second-best optimum policy attains a lower welfare level than the first-best (production subsidy) policy.

(5) *Trade Reversal*

The first-best optimum solution could call for a production subsidy on M which is sufficient to reverse the pattern of trade and turn M into the exported good. This case is represented in Figure 2.9. Free

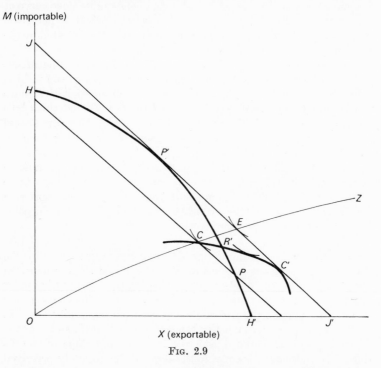

FIG. 2.9

trade production is at P and consumption at C. First-best optimum production is at P', as before, but the optimal consumption point, with given production at P', is at E to the right of P'. Hence in the optimal situation production of M would exceed consumption, and it would be exported.

What is the second-best optimum in this case? Imagine the tariff to be raised from zero until trade is eliminated. The resultant consumption points would trace out the *Meade curve* CR', derived as in Figure 2.8. A tariff on its own cannot carry consumption to the right of R', since once production and consumption have reached that point any increase in the tariff would be redundant. The optimum will now be at the point where an indifference curve is tangential to CR', or if none is, at the self-sufficiency point R' itself. A tariff on its own cannot, of course, bring about trade reversal. The most it can do is to eliminate trade. But it may not be optimal to eliminate trade even though the first-best optimum requires trade reversal.

Suppose we allow for both a tariff and an export subsidy as instruments of policy. The *Meade curve* can then be continued beyond R' to C', and even beyond that. Thus a tariff combined with an export subsidy (both at the same rate) could bring production to P', and hence consumption to C'. Any point to the right of R' represents trade reversal. It is interesting to note that in this case it may not be second-best optimal to bring about trade reversal even though trade reversal *could* be achieved and the first-best optimum requires trade reversal. This would be so if an indifference curve were tangential to the extended *Meade curve* $CR'C'$ to the left of R'. Some imports of M would then continue.

(6) *Factor Market Divergence*

The domestic divergence might be in the factor-market rather than associated with production of a product as such. An extensive literature has analysed the case of a fixed wage differential in the two-sector, two-factor model, one sector paying a higher real wage than the other.[8]

In the absence of intervention the economy will then produce *within* the social production transformation curve. The marginal rate of substitution of one factor for another along an isoquant will not be the same in both sectors; in the Edgeworth-Bowley box diagram, the dimensions of which show the country's fixed factor stocks, the country will not be producing on the contract curve which traces out points of tangency of the isoquants. Hence a Pareto-efficiency condition will not be fulfilled. There will be a distorted or 'shrunk-in' production-possibility curve within HH' (Figure 2.7) which traces out possible production points *given* the wage-differential, the actual production point being determined by the product price ratio. (The curve is not drawn in Figure 2.7.)

[8] See E. Hagen, 'An Economic Justification of Protectionism', *Quarterly Journal of Economics*, 72, November 1958, pp. 496–514; Bhagwati and Ramaswami, 'Domestic Distortions, Tariffs and the Theory of Optimal Subsidy', *Journal of Political Economy*, 77, December 1969, pp. 1001–13; and the extensive subsequent literature surveyed in S. P. Magee, 'Factor Market Distortions, Production and Trade: A Survey', *Oxford Economic Papers*, 25, March 1973, pp. 1–43.

A first-best policy of appropriate wage-subsidy will bring the economy to the first-best solution represented in Figure 2.7 (production at P'; consumption at E). But this time a production subsidy to the industry paying the excessive wage will be only second-best. The second-best optimum production subsidy will bring the economy to the point where the given world price line is tangential to the 'shrunk-in' curve.

APPENDIX 2
THEORY OF DOMESTIC DIVERGENCES: NOTES ON THE LITERATURE

The theory originated in Chapter XIV of J. E. Meade's *Trade and Welfare*, Oxford University Press, London, 1955, though this has often been forgotten. The chapter is entitled 'The Second-Best Argument for Trade Control: (3) Domestic Divergences'. The important proposition that a subsidy is better than a tariff to deal with a marginal divergence is made clearly on pages 230–231 of this book. Meade also expounded the concept of the second-best optimum tariff (pp. 228–230). Furthermore, he pointed out that a factor-subsidy is the best device to use when there is a factor-market divergence (pp. 232–234). Nevertheless, the modern rigorous analysis of factor-market distortions can be said to have begun with E. Hagen, 'An Economic Justification of Protectionism', *Quarterly Journal of Economics*, 72, November 1958, pp. 496–514, who noted that a subsidy per unit of labour, and not a tariff, is first-best to use with a wage-rate divergence (owing to a wage-differential).

A frequently-quoted article is by Jagdish Bhagwati and V. K. Ramaswami, 'Domestic Distortions, Tariffs and the Theory of Optimum Subsidy', *Journal of Political Economy*, 71, February 1963, pp. 44–50, which stressed the main point about the optimality of dealing directly with a 'distortion', and the crucial relevance of the relationships between domestic rate of substitution (DRS), domestic rate of transformation (DRT), and foreign rate of transformation (FRT). This was followed by the important article that really focused on the central issue, applied it to various cases, and set one of the main themes of the present book, H. G. Johnson's 'Optimal Trade Intervention in the Presence of Domestic Distortions' in R. E. Caves et al., *Trade, Growth and the Balance of Payments*, North-Holland, Amsterdam, 1965.

Many other articles, mainly by Bhagwati, Ramaswami, and Srinivasan, in various combinations, have been published on the subject. These have been consolidated in Bhagwati, 'The Generalized

Theory of Distortions and Welfare', in Bhagwati et al. (eds.), *Trade, Balance of Payments and Growth: Papers in Honor of Charles P. Kindleberger*, North-Holland, Amsterdam, 1971. At the end of this paper there is a comprehensive bibliography. To this list should be added V. K. Ramaswami, 'Optimal Policies to Promote Industrialisation in Less Developed Countries' in P. Streeten (ed.), *Unfashionable Economics*, Weidenfeld and Nicolson, London, 1971, which is probably one of the best summary statements of the general approach.[9]

The new approach has also clearly influenced Ian Little, Maurice Scott, and Tibor Scitovsky in *Industry and Trade in Some Developing Countries*, Oxford University Press, London, 1970.

[9] For the early history one might also note two other articles. (a) G. Haberler, in 'Some Problems in the Pure Theory of International Trade', *Economic Journal*, 60, June 1950, pp. 223–40, analysed systematically for the first time the gains from trade (no-trade compared with free trade) in the presence of various domestic divergences; this article influenced the later papers by Bhagwati and Ramaswami and by Johnson in their choice of problems, but was not concerned with the question of the optimal policy instrument. (b) The present author, in 'Tariffs, Subsidies and the Terms of Trade', *Economica*, 24, August 1957, pp. 235-242, showed geometrically that in the small-country case a subsidy is preferable to a tariff if a given amount of protected production is desired; this article, possibly for the first time, highlighted the superiority of subsidies over tariffs in this type of case, but was of course predated by *Trade and Welfare*.

3

THE FOUR ASSUMPTIONS OF THE
THEORY OF DOMESTIC DIVERGENCES

THE WHOLE argument of the previous chapter has involved four assumptions. (1) Subsidies can be financed by 'non-distorting' taxes. (2) Taxation involves no collection costs. (3) There are no costs of disbursement of subsidies. (4) The income distribution effects of various policies (such as the redistribution from taxpayers to subsidy recipients) can be neglected. These four assumptions are generally made in the numerous articles that have been published advancing the 'theory of domestic distortions'.[1]

The question is whether the central arguments of this approach still stand when the assumptions are removed, especially the argument that taxes and subsidies on *trade* are never first-best policies when the divergences between social and private cost or benefit are *domestic* in nature.

The issues involved are obviously important. If the basic approach really depended on these assumptions, then it could

---

[1] The two standard references are surprisingly strongly worded on this subject. 'The contention that the payment of subsidies would involve the collection of taxes which in practice cannot be levied in a non-distortionary fashion is fallacious. A tax-cum-subsidy scheme could always be devised that would both eliminate the estimated divergence and collect taxes sufficient to pay the subsidies'. (Jagdish Bhagwati and V. K Ramaswami, 'Domestic Distortions, Tariffs and the Theory of Optimum Subsidy', *Journal of Political Economy*, 71, February 1963, p. 50.)

'... it is assumed in this paper, in accordance with the conventions of theoretical analysis of these problems, that government intervention is a costless operation: in other words, there is no cost attached to the choice between a tax and a subsidy. This assumption ignores the empirical consideration, frequently introduced into arguments about protection, that poor countries have considerably greater difficulty in levying taxes to finance subsidies than they have in levying tariffs on imports. This consideration is of practical rather then theoretical consequence, and to constitute a case for tariffs requires supplementation by empirical measurement of both the relative administrative costs and the economic effects of promoting favored industries ...' (H. G. Johnson, 'Optimal Trade Intervention in the Presence of Domestic Distortions', in Bhagwati, *International Trade*, Penguin Modern Economics, pp. 188–9.)

be rather easily dismissed as academic—as indeed it has been
dismissed by many a practical-minded reader. It is rather
surprising that, in spite of the surfeit of articles in the economic
journals on the 'theory of domestic distortions', no one other
than Meade has found it necessary to remove or review these
assumptions, or, in the case of assumption (3), even to state it.
Meade, in *Trade and Welfare*, removed assumption (1) and
brought out some of the considerations discussed in Sections I
and II below.

Each of the assumptions will be considered here in turn.
It will be shown that the analysis certainly needs to be modified, but the central argument is unshaken by the removal of
the important assumptions (1) and (4). Removal of assumption
(2) slightly dents it, while removal of assumption (3) does
affect it. We begin with removing assumption (1). At the end
some doubts about the whole 'subsidy-biased' approach will
be introduced.

I
BY-PRODUCT DISTORTION COSTS
OF REVENUE-RAISING:
SUBSIDIES REMAIN FIRST-BEST

There is no practical way of collecting taxes that does not
impose costs. Income taxes distort the choice between work
and leisure. Taxes on commodities at various rates distort the
choices between these goods as well as between the taxed
goods and leisure. These used to be called the 'deadweight'
costs of taxation and result from production and consumption
distortions created by taxation. In addition there are collection
costs, which we shall consider more fully shortly. Economists
sometimes talk about 'non-distorting poll taxes'. This
concept is a useful theoretical fiction but no more than that.
Raising revenue imposes inevitable *by-product costs*—the
'deadweight' distortion costs and the collection costs.

It was the theme of the previous chapter that when a
domestic divergence is corrected by a subsidy at the point of
the divergence, no by-product distortion is imposed. By
contrast, a tariff imposed a by-product distortion, or possibly

even several kinds of them. On this basis it was argued that the subsidy was first-best. Now we find that the process of raising revenue to finance correcting subsidies also imposes inevitable by-product costs. Has the whole argument then been destroyed? Can one argue that, since each method imposes costs, there is no *a priori* way of deciding which method will make a given correction at minimum cost? Surprisingly, such an argument would be fallacious.

Let us assume for the moment that there are no collection costs of taxation, including tariffs, and no disbursement costs of subsidies. The correct approach appears then to be the following. The first-best way of correcting the consequences of an excess of private over social cost to a given extent is to provide a subsidy and then to finance this subsidy by a minimum-cost package of taxes. Alternatively, first-best policy may require a tax at the point of the divergence, with the revenue raised being remitted in a minimum-cost way. To return to the case where a subsidy is required and has to be financed, the minimum-cost tax package is one that raises a given amount of revenue in a way that minimizes the by-product distortion costs of doing so. By-product costs or distortions cannot be avoided, but they can be minimized. The considerations influencing the composition of the minimum-cost tax package will be discussed in more detail in the next chapter.

This first-best approach of protecting an industry with a subsidy and then financing the subsidy in a minimum-cost way can be compared with the second-best method of providing protection for a particular product wholly by a tariff on that product. It is then important to appreciate the characteristics of a tariff. It is *the equivalent of a tax on consumers of the product concerned, the revenue from which finances a subsidy to the domestic producers of this product, the rates of consumption tax and production subsidy being the same.* In addition, if some imports remain, the tariff would yield customs revenue; it would then be equivalent to a tax on consumers that is greater than is needed to finance the subsidy.[2]

---

[2] This could be illustrated with Figure 2,2 (p. 12). The tax on consumers per unit is *PS*. This consumer tax raises revenue of *PSEF*. This revenue finances the production subsidy of *PSJL* and raises customs revenue of *LJEF*.

Hence the use of a tariff on its own in order to protect the product means that the subsidy is financed in a very particular way—solely by a tax on consumers of that particular product, the rate of tax being the same as the desired rate of subsidy. By contrast, the first-best method of protection involves no such constraint: the subsidy is simply financed by a minimum cost tax package. It is most unlikely that the minimum cost way of raising given revenue would be solely by a consumption tax or tariff on *this* particular product at the rate of the desired subsidy. There may be far less distorting ways of raising the same revenue.[3]

Our main point can be made by considering an extreme case. Suppose that the country cannot or will not use any taxes other than tariffs. Thus the subsidy would have to be financed from customs revenue. Suppose, further, that there are only two importable products, X and Y. There is, as before, a marginal divergence in production of X, requiring it to be protected. Even then, a tariff on X to achieve the whole of the desired protection would probably not be first-best. First-best policy would be a subsidy on production of X financed by that combination of tariffs on X and Y that inflicts minimum distortion costs. Because there will be some tariff on X, the rate of subsidy required will indeed be less than in the absence of such a tariff, but the main point is that a tariff on the other product is likely to be part of the minimum-cost package.

II
COLLECTION COSTS

Let us now remove assumption (2) and allow for collection costs of taxes. These are not only the costs of tax administration to the Ministry of Finance but also the costs imposed on taxpayers in their efforts to comply with the tax law while minimizing their tax burdens, or to evade it. The by-product costs of taxation thus consist of distortion costs and of collection costs. The interesting question is whether collection costs alter the argument that domestic divergences should be

---

[3] *Trade and Welfare*, p. 231.

corrected by subsidies and taxes close to the points of the divergences, rather than by trade policies.

Suppose initially that collection costs of tariffs are low or zero while there are significant collection costs for other types of taxes. It will then remain optimal to subsidize, and to finance the subsidy with a minimum-cost tax package. This package must take collection costs into account and for this reason may well consist mainly of trade taxes. It will be optimal to allow some increase in distortion costs in order to keep down collection costs. For the limiting case where the tax package would consist wholly of trade taxes, the argument in favour of subsidizing and then using several tariffs to finance the subsidy rather than simply imposing a protective tariff, has been made already. This argument underlines that while high collection costs for non-trade taxes may make the use of trade taxes first-best for raising revenue, these costs do not justify the direct use of trade taxes to correct domestic divergences.

Once we allow for collection costs on trade taxes, the argument may have to be qualified, though this qualification is probably not of great importance. Suppose, as before, that product X is to be protected and that any subsidy would have to be financed by revenue tariffs on X and Y. If there are high collection costs for the tariff on Y, it may be more economical to achieve the whole protection with a tariff on X, where collection costs may be low because imports will be low. Since distortion-costs will exceed those that would result if X were subsidized and financed by tariffs on X and Y, the extra distortion costs must be set against the lower collection costs. The general argument is clearest when the optimum tariff on X would be a prohibitive one. Collection (or administration) costs would probably be very low, much lower than when non-prohibitive tariffs on X and Y have to be imposed to finance a subsidy for domestic producers of X. This argument also applies if the subsidy were financed by non-trade taxes. The collection costs incurred may be quite high compared to the collection or administration costs of a protective tariff, and so might make the latter first-best.

We have thus a qualification to the general argument that subsidies remain first-best in the presence of domestic divergences.

III
BY-PRODUCT COSTS OF REVENUE-RAISING: THE NEED FOR 'TRADING-OFF'

The first-best way of dealing with domestic divergences as set out in the previous chapter will be affected by the by-product costs of raising revenue. This is true even though (subject to qualifications noted above) the principal message of the previous chapter is unaffected—that domestic divergences should be corrected by subsidies (and taxes) as close as possible to the point of the relevant divergence.

It will be necessary to trade-off the gains from correcting the original divergence against the new by-product costs (distortion costs plus collection costs) that are imposed by the need to finance the subsidy. The subsidy should be just high enough for the marginal gain from partially correcting the divergence to be equal to the marginal by-product costs. Hence, if the marginal by-product costs are positive it will be optimal now *not* to make the subsidy so high that it corrects the divergence completely.[4] The principle is the same as in the case of the second-best optimum tariff situation. But this time we are talking about a first-best situation since we are assuming that for every given degree of correction of the divergence by a subsidy the by-product costs from raising the revenue to finance the subsidy are minimized.

One would normally expect the *marginal* by-product costs of raising revenue to be positive and so some trading-off to be necessary. This is true even though there are likely to be indivisibilities in tax collection costs, so that the marginal collection costs could indeed by zero over a range once the tax system has been set up, since costs may not increase significantly when tax rates are increased. But marginal distortion costs—resulting from the disincentive effects and other distortions created by taxes—are likely to be positive, and indeed to increase with the amount of revenue raised.[5]

---

[4] See *Trade and Welfare*, pp. 230–231.

[5] One could envisage a situation where it would be optimal not to trade-off but to make a full correction of the divergence even though the marginal by-product distortion cost is positive. This would require the marginal gain curve to have a kink and drop vertically to zero. We shall generally ignore the possibility of such a 'corner-solution' here.

How is the hierarchy of policies described in the previous chapter affected by the by-product costs of raising revenue? At every stage in the hierarchy the gains to be derived from a particular policy will be less than in the absence of these costs. When a subsidy is used there will always be some collection and some distortion costs to be added to the by-product distortion costs already taken into account when setting up the hierarchy. When a tariff is used there will be tariff collection costs that were not taken into account in the previous chapter, as well as a disincentive effect (distortion relative to leisure) resulting from the consumption tax aspect of the tariff. Furthermore, it seems reasonable to argue that the order in the hierarchy will probably be unaffected. Policies at the top of the hierarchy are those which are directed precisely to the point of the divergence; relevant subsidies required will then cost rather little, less than when the subsidies are less discriminating. Hence the welfare gains from choosing a policy high up in the hierarchy as compared with one lower down will probably be even greater than before.

IV
SUBSIDY DISBURSEMENT COSTS

The fiscal transaction of raising revenue and then using it for subsidization has two transaction costs, the collection cost of raising the revenue, and the cost of disbursing the subsidy. So far we have only allowed for the former.

In a country with a developed taxation system embracing all sections of the community, subsidies can be provided through reductions or exemptions in taxes already payable. Net subsidies could be paid out through the same system. The more the subsidies overlap with an existing category of taxpayers the lower the disbursement costs. There are always costs in complicating any system, or in setting up a new system, but in general, in developed countries subsidy disbursement is not likely to be very costly.

On the other hand, subsidy disbursement can be costly in a less-developed country, and sometimes impossible. Just as there are untaxable sectors, so there are unsubsidizable sectors. How would one provide a production subsidy to handicraft manufacturers? An all-embracing subsidy on employment in manufacturing could also be administratively very difficult. If one wished to replace the existing protection of manufacturing industry provided by tariffs and import restrictions with a system of direct subsidization one would, in many countries, have to disburse very high subsidies; it would not be sufficient to exempt industries from existing excise taxes, since these tend to be much lower and to be confined to a few products. In practice, the simplest form of subsidization is often through the provision of low-cost facilities by the state, or subsidization of some inputs that are produced by a few domestic firms or that are imported. This gives rise, of course, to by-product distortion in factor-use.

Thus subsidy disbursement costs increase the by-product costs of intervention, may make some forms of intervention impossibly costly, and may alter the order in the hierarchy of policies. They may make it costly to intervene *precisely* at the point of a marginal divergence because of the costs of defining and identifying the relevant recipients. Indeed, absolute precision in this regard is almost impossible, so that some by-product distortion costs resulting from an intervention would be inevitable even if there were no by-product costs of raising the revenue needed to finance the subsidy.

One form of protection has no disbursement costs. The restriction of imports through a tariff or a quota causes the domestic price level of competing products to rise, so that automatically domestic consumers are taxed and producers subsidized. The simplicity of this device when compared with taxing numerous domestic consumers and then disbursing the revenue to numerous producers is obvious. Exactly the same effect can be obtained by an export tax if it is desired to subsidize some consumers. The tax lowers the price-level at which the product is exchanged domestically and hence it taxes producers and subsidizes domestic consumers in one action.[6]

An export subsidy may also be preferred to a production subsidy for this reason. It is often easier to disburse a subsidy

[6] The tariff or the export tax will, or may, raise customs revenue. It might be argued that this will need to be disbursed and hence disbursement costs will be incurred. But this is not so. It can go into the general funds of the government, permitting other distorting taxes to be reduced.

at the port, when a product leaves the country or when the foreign exchange transaction takes place, than at the point of production. Furthermore, the amount to be disbursed is less than with a production subsidy if a given degree of protection is to be provided: the export subsidy causes the price of the exportable product sold to domestic consumers to rise (for otherwise producers would export all their production); hence there is no need to subsidize that part of production which is sold at home.

We have thus an argument for protection through taxing or subsidizing trade rather than domestic production, consumption or factor-use. It becomes conceivable that a tariff may be first-best on these grounds. This new argument must be set against all our earlier arguments on the other side. The force of the argument depends, of course, completely on the empirical question of whether subsidy disbursement costs are high. In developed countries they are unlikely to be high, nor in less-developed countries when it is a matter of protecting manufacturing industries in the advanced sector of the economy.

V
INCOME DISTRIBUTION EFFECTS AND THE THEORY OF DOMESTIC DIVERGENCES

Many economic theorists have tended to be either blind or nihilistic about the income distribution implications of economic policies. A vast amount of normative economic theory simply ignores income distribution by concentrating on 'Pareto-optimality'. The *theory of domestic distortions* on which Chapter 2 is based comes into this category. Some theorists, on the other hand, consider that every policy will have some effect on income distribution, that such effects cannot be ignored, that economists can make no inter-personal comparisons of utility, that therefore they cannot make any judgement about the desirability of any policy at all, and hence they cannot say anything useful at the normative level. On this basis the discussion in Chapter 2, and no doubt most *a priori* economic theory, would be irrelevant for practical issues.

We shall be discussing these matters further in Chapter 5.

Here we are concerned solely with how income distribution effects bear on the theory of domestic divergences. It will be argued that it is necessary neither to be blind nor to be nihilistic.

In Chapter 2 we assumed that the raising of revenue involved no by-product distortion or collection costs and the disbursement of subsidies no disbursement costs. This meant that it would also be possible to alter income distribution by taxing some people and subsidizing others without any costs of this kind. Hence the assumption that income redistribution is costless would be consistent with this model. Let us make this assumption now. It will be removed in the next section.

In any given situation, with any given set of marginal divergences, whether corrected or not, there will be (a) an *actual* income distribution resulting from that situation in the absence of any deliberate redistribution, and (b) a *socially desired* income distribution. We need not discuss the latter concept at this stage. Let us just accept that there is some kind of social ordering of possible income distributions. If the *actual* distribution differs from the *socially desired* one then there is a *domestic divergence*. We can call it an *income distribution divergence*. This is very similar to the divergences we have discussed earlier. First-best policy calls for an intervention as close as possible to the point of the divergence, namely appropriate subsidies and taxes on incomes. If some other device were used to redistribute incomes—for example a subsidy on production of some industries or a tariff—by-product distortions of the usual type would be generated, so that these policies would be second-best, third-best, and so on. Taxes and subsidies on trade, or for that matter on production or consumption, will never (given our assumptions) be first-best to deal with income distribution divergences.

Now let us introduce the concept of the *by-product income distribution distortion*. Suppose that a marginal divergence caused by an externality is corrected by a first-best subsidy. This alters income distribution. Suppose, further, that the income distribution initially was the socially desired one and that the income distribution that results when the externality-divergence has been corrected is not desired. The correction of the externality-divergence has yielded a by-product *income distribution distortion.* It is then necessary to correct this with an appropriate income redistribution policy. Given our present assumption of costless income redistribution, it is possible to do this without creating any other by-product distortions.

Hence the first-best package of policies involves the use of two instruments, (a) the first-best subsidy (say, a production subsidy) that corrects the initial divergence created by the externality, and (b) the income tax-subsidy that corrects the income distribution distortion created by the first subsidy. The two policies are thus directed towards two targets, (i) correction of the initial marginal divergence, and (ii) avoidance of an income distribution distortion. With two costless instruments both targets can be attained. This approach assumes, it must be repeated, that the raising of revenue involves no by-product distortion or collection costs and the disbursement of subsidies no disbursement costs.

If we assume that income distribution distortions created by various policies are always corrected by appropriate first-best—and hence costless—income taxes and subsidies we can ignore the income distribution consequences of different policies. The hierarchy of policies to deal with a wage-rate divergence, for example, will remain exactly as set out in Chapter 2.

VI
INCOME REDISTRIBUTION NOT COSTLESS

The inevitable by-product distortion costs of raising revenue must now be reintroduced; hence income redistribution can no longer be costless. This requires the preceding discussion to be modified in three ways. In each case we suppose that there is initially the familiar marginal divergence in production owing to an external economy created by an import-competing industry.

Firstly, optimal subsidization policy must now be influenced by the income distribution effects. *Trading-off* is required. The subsidy to the industry concerned will increase the incomes of (a) the producers of the protected products and (b) the persons who benefit from the external effect generated by the industry. If this redistribution is not in a desired direction some *re*-distributive policies are required. These will tax the beneficiaries in some way that does not reverse the initial effect of the subsidy on production, but will inevitably impose by-product distortion costs.

It follows that optimization requires a trading-off process. The subsidy should not be so high as to correct the initial marginal divergence as far as it would have been corrected in the absence of an undesirable income distribution effect. Similarly, the income redistribution policy should not be fully corrective: some of the income distribution effect of the subsidy should *stick*.

To assess the desirability of the optimum subsidy and how far it should go it is thus necessary to take the income distribution consequences into account. One can no longer confine oneself to 'Pareto-optimality'.

Secondly, a subsidy to correct the marginal divergence remains preferable to a tariff. So the main message of the theory of domestic divergences stands (in the absence of subsidy disbursement costs). Indeed, the argument in favour of a subsidy is strengthened. There are two reasons for this.

(1) A subsidy redistributes incomes from the general taxpayers—the community choosing to distribute the tax burden as it wishes—to the producers of the particular product. Income redistribution policy will try, at a cost, to offset, partially, the latter effect. By contrast, a tariff redistributes incomes from particular consumers to particular producers. Income redistribution policy has then a bigger job to do—to reverse, partially, not only the effect on the particular producers, but also the effect on the particular consumers. It may be that the income distribution was initially not optimal, and some redistribution against these consumers was desirable; it is still unlikely that the tariff would be the best way of achieving this.[7]

(2) The tariff raises customs revenue, which transfers income from particular consumers to the general taxpayers. But they may not be the particular consumers whom it is optimal to tax for income distribution reasons; it would be pure chance if this were so. Hence, in principle, some redistributive policies are needed to offset this. The tariff has

[7] This argument will be qualified in Chapter 5, pp. 109–10.

thus a bigger income distribution effect than the subsidy, and if this effect is generally not in a desired direction the required correction gives rise to more by-product costs.

The fact that income redistribution is not costless has a third effect. Suppose the country is constrained to use a tariff, rather than a subsidy, for protection, so that we have the second-best situation described in the previous chapter, pp. 18–20. Suppose, further, that the income distribution consequences of the tariff do not happen, by chance, to be desirable. It must now be remembered that, even though income redistribution through income taxes and subsidies should accompany the tariff, it should not go so far as to restore the original distribution; because of the by-product costs of redistribution some of the distributive effects of the tariff should stick. It follows that the height of the second-best optimum tariff will be less than otherwise, and the less desirable the income distribution consequences of the tariff, the lower it should be.

VII
ILLUSIONS AND AN IMPERFECT WORLD

How can subsidies *always* be preferable to trade interventions for correcting a domestic divergence, at least aside from possibly minor considerations of subsidy-disbursement costs and of particularly low collection costs of protective tariffs relative to revenue tariffs? Instinctively one rebels against so sweeping a generalization. Furthermore, the whole 'subsidy-biased' approach seems to have a great air of unreality about it. How does it come about that tariffs and import restrictions are used all round the world when their use for what are often the professed purposes of the restrictions can be so broadly condemned? Surely one can hardly conceive of a country such as India, for example, replacing its regime of tariffs and import controls with a mass of subsidies to producers.

What then have we left out?

One cannot necessarily assume that the revenue to finance subsidies is always raised in a minimum-cost way. Of course it *could* be, in principle, but it just may not be. One has then the typical second-best problem of an imperfect world: one must compare a tariff, which is certainly second-best or worse, with a subsidy financed in a second-best way. In many countries it is always difficult politically to increase tax-rates, though there is no similar resistance to tariffs. An imbalance between private spending and public facilities may result: we start then with a marginal—indeed, much more than marginal—domestic divergence. A subsidy may well be financed not by increasing taxes but by forgoing other public expenditures and so increasing the existing imbalance.

Similarly, one cannot necessarily assume that first-best income distribution policies are being followed, so that we do not start with an 'undistorted' income distribution situation.

The protected products may, broadly, be 'luxury' goods consumed mainly by the middle and upper-classes and the income distribution consequences of protection may therefore be desirable. If subsidies were used, the required revenue might be obtained by general sales taxes regressive in their effect, or perhaps by forgoing government expenditures on social services.

This situation means that there is an inconsistency. If the people that control the government are willing to accept higher prices for imported luxury goods, why cannot the subsidy be financed by a sales tax on luxury goods, or an income tax on the sort of people who consume these goods? The answer is that people and governments are often inconsistent, or *apparently* so. Perhaps one interest group controls the tariff policy and another is able to influence tax policies. More important, there may be an element of *illusion*. This is a matter of great importance for tariff and import quota policies, and needs to be looked at more closely.

Explicit taxation imposes psychic costs which implicit taxation does not. These are costs in the minds of the taxed public and hence costs to the government in loss of popularity. One cannot ignore an element of irrationality and illusion in popular attitudes to taxation. The payment of personal income tax is a clear payment and is just as clearly resented, explicit taxes on production are noticed and resented by producers, while consumption (sales) taxes may be less obvious but are often brought to public notice by producers or retailers, though they are probably more acceptable to the public than income taxes. But tariff and import quotas, provided they are not mainly on inputs into domestic production, and especially if they are prohibitive, are another matter.

It may be understood that on the imports that remain, tariffs are equivalent to sales taxes, but it is often not understood that the *subsidy-equivalent* of the tariff is like a subsidy to producers financed by a sales tax. So a fiscally hard-pressed government's reluctance to replace protective tariffs with subsidies is understandable. This is the *cosmetic* attraction of tariffs.

A similar cosmetic attraction attaches to export taxes in countries where a high proportion of exportables is consumed domestically. It may be realised that the actual export tax revenue is collected, in effect, from producers. But it is not widely understood that the lower domestic price created by the export tax is a way of taxing producers of exportables for the benefit of consumers or other industries.

A subsidy financed by tax revenue and hence going through the budget makes it obvious that an industry is protected. A clear sum of money stands witness to the cost, even though it is not, strictly, a measure of welfare cost. By contrast, a tariff or quota hides the reality. Quite elaborate research may be needed to calculate the subsidy-equivalent of the tariff or quota. For the same reason, if subsidies are chosen, the beneficiaries often prefer indirect subsidies given to factors of production or through the provision of public facilities below cost rather than direct output subsidies.

Protection unnoticed is protection more secure. And, even when it is noticed, tariff legislation is not normally annually renewable, while by contrast, in democracies budgets are annual and their contents are often scrutinized with embarrassing thoroughness. For this very reason—the obscurantist aspect of tariffs and quotas—free-trade-minded economists preferred subsidies to tariffs long before the theory of domestic distortions was developed. The chances of sustained protection are certainly less with subsidies than with tariffs, so that it is in the interests of exporters and relevant consumers that explicit subsidies rather than tariffs are used as protective devices. At the same time, it must be noted that many countries do use subsidies to sustain very expensive activities that most economists would regard as uneconomic—as witness Britain's aircraft extravaganzas—but this is usually when tariffs could not achieve the protective purpose, as in the case of export industries.

It is thus not difficult to understand the widespread use of tariffs and quotas in preference to subsidies, and perhaps in some cases, even to defend such use. But insofar as economists have a bias against protection on the grounds that generally it results from special interests overruling general interests, and insofar as they have a bias in favour of explicitness and rationality—a belief not only that the 'cost-benefit' technique is their special skill, but also that its practice is a *good thing*— then one would expect them to prefer subsidies to tariffs. This is quite apart from the likelihood that the use of subsidies will minimize the by-product distortion costs of protection.

4

TRADE TAXES AS SOURCES OF GOVERNMENT REVENUE

TAXES on foreign trade are substantial sources of revenue for governments of almost all less-developed countries. In at least twenty less-developed countries trade taxes have in the late nineteen fifties accounted for one-quarter or more of total government revenue.[1] In a number of countries—including Indonesia, Burma, Ceylon, Malaysia, Thailand, Nigeria, Ghana, and Colombia—the proportion has been over 40 per cent. The main type of trade tax has been the tariff, but in addition there have been the profits of multiple exchange rate systems, export taxes, and profits from export marketing boards, the latter being really forms of export taxes.

Export taxes, including the profits of marketing authorities, were important sources of revenue in the early nineteen-fifties, though they have declined in importance since then. Export marketing boards were established in a number of countries during or immediately after World War II to stabilize internal prices and improve marketing facilities, and rather incidentally yielded large revenue surpluses until the mid-1950s. Ordinary export taxes became important as a result of the Korean raw materials boom 1950-51, when rates of export tax were increased in many countries. By the early nineteen-sixties there were 12 countries where export taxes (including marketing board profits) accounted for more than 10 per cent of government revenue, and four (Uganda, Malaysia, Sudan, Ceylon) where the proportion was more than 20 per cent.[2]

[1] Stephen R. Lewis, Jr., 'Government Revenue from Foreign Trade: An International Comparison', *The Manchester School of Economic and Social Studies*, 31, Jan. 1963, pp. 39–46. (Indonesia has been added to his list here.) This is the only available comprehensive statistical study of this subject. There is no strong reason to believe that trade taxes have become significantly less important in the last ten years.

[2] Richard Goode, George E. Lent and P. D. Ojha, 'Role of Export Taxes in Developing Countries', *I.M.F Staff Papers*, 13, Nov. 1966, pp. 453–503.

For developed countries, by contrast, taxes on trade are generally not now significant sources of revenue, and usually account for less than 10 per cent of central government revenue, and sometimes much less. But this was not always so. In the early histories of the now-developed countries trade taxes have often been very important, and the principal purpose of tariffs has been to raise revenue. In the United States, customs duties accounted for over 25 per cent of revenues at all levels of government in 1890, though only 0·8 per cent in 1960. In Germany they accounted for 16 per cent of revenue in 1914, though only 4 per cent in 1960.[3] Furthermore, in the E.E.C. countries revenue from customs duties which supplement taxes on domestic production, such as excise taxes or value-added taxes, are still of some significance. Such tariffs are 'border-tax adjustments' which convert taxes on production of import-competing goods into taxes on domestic sales.

In view of the role that trade taxes play currently in less-developed countries as sources of government revenue, and that they once played in the now advanced countries, it is clearly necessary to consider trade taxes not only from a protective but also from a revenue point of view. The aim of this chapter is primarily to consider two issues. Firstly, what role, if any, should taxes on trade play in an optimum revenue-tax package, bearing in mind the availability of other taxes, notably income, sales and excise taxes? Secondly, given that trade taxes are used to raise revenue, what considerations determine the optimal structure of trade taxes for this purpose?

I

SOME SIMPLE PRINCIPLES OF
OPTIMUM TAXATION

The problem of the revenue role of trade taxes will be posed here in a limited way: what is the optimum way of raising a given amount of revenue, and especially what is the role of trade taxes in such an optimum system? We shall assume here that domestic divergences have all been corrected (though,

---

[3] Richard A. Musgrave, *Fiscal Systems*, Yale University Press, New Haven, 1969, pp. 138–139.

in fact, it will normally not be optimal to make full correc-
tions), that the country cannot affect its terms of trade, and
that the exchange rate maintains external balance and
monetary policy internal balance. The revenue raised from
taxes may be used to finance public goods, to finance subsidies
to correct domestic divergences, to finance social services, or to
finance capital accumulation. Illusion or 'cosmetic' aspects
of various taxes will be ignored. We shall also assume throughout this chapter that savings are unaffected by the choice of
tax, since any possible effects on savings are peripheral to the
main issues here.

Three considerations will then determine the optimal tax
mix, namely income distribution effects, distortion effects and
collection costs. Collection costs will turn out to be crucial,
for without them trade taxes cannot be part of a first-best
tax package. Even if there were collection costs for income
taxes—perhaps even so high that income taxes were completely ruled out so that all revenue had to be raised with
commodity taxes—it would remain true that trade taxes as
such should not be used. One has to introduce collection costs
not only for income taxes but also for commodity taxes to get
a first-best case for revenue tariffs or export taxes.

The argument can be developed in a number of stages.

Suppose, to begin with, that there are no collection costs for
any type of tax, and furthermore that the elasticity of
substitution of effort for leisure for all potential taxpayers is
zero, so that taxation has no disincentive effects. All revenue
should then be raised by taxes on persons—on their wealth,
incomes or expenditures—and not on commodities. The issue
of the choice between wealth, income and expenditure taxes
hinges on the effect on savings, which we are ignoring, and it
will be simplest to assume an income tax. The desired income
distribution effects can be brought about in a straightforward
manner through the income tax. One can apply any desired
principle for distributing the burden of taxation—ability to
pay, benefit principle, class interest, and so on. To avoid any
distortion the tax must not be related in any way to the type
of activity or industry in which the taxpayer is engaged or to
the way in which he spends his income.

When we allow for the effects of taxation on incentives,

complications are introduced which have long preoccupied economists. There are then inevitable distortion costs because taxation brings about substitution of the untaxable good, leisure, for effort, the reward of the latter being taxed income. Given that there are inevitable distortion costs, these can be reduced by modifying the simple income tax system based on income distribution considerations alone. Firstly, progressiveness can be reduced. Secondly, taxes can be raised relatively on those members of the community who have relatively less opportunity to vary the supplies of their factor services. Finally, an element of differential commodity taxation can be superimposed on the income tax system, so that prices of goods that are poor substitutes for leisure, or actually complements with leisure, rise relatively to the prices of goods that are close substitutes for leisure.[4]

These adjustments will reduce somewhat the distorting substitution towards leisure that taxation is likely to bring about, though the resultant gains must be traded off against the failure to attain the desired income distribution objectives and against the new distortion in the consumption pattern between the various goods concerned that is introduced. The last adjustment, concerned with various goods being substitutes for or complements with leisure to different degrees, is probably unimportant, and it will henceforth be assumed that all goods are equal substitutes for leisure.

The next step is to suppose that collection costs for income tax are so high that all revenue must be raised from taxes on commodities—sales taxes, excise taxes (taxes on domestic production) and trade taxes. We might, alternatively, suppose that some income tax is being collected, but increases in income tax are not possible. In the absence of disincentive effects, distortion costs would now be avoided by a uniform *ad valorem* tax either on domestic production or on domestic sales. If there are disincentive effects but if all goods are equal substitutes for leisure, a uniform tax would minimize, rather than avoid, distortion costs. But a uniform tax does not make

---

[4] W. J. Corlett and D. C. Hague, 'Complementarity and the Excess Burden of Taxation', *Review of Economic Studies*, 21, 1953–54, pp. 21–30; *Trade and Welfare*, pp. 112–118; also, *Trade and Welfare: Mathematical Supplement*, pp. 29–30.

it possible to take into account income distribution considerations since it is equivalent to a proportional income tax. If it is desired to impose the tax burden mainly on the rich, it becomes appropriate to have a system consisting of a differential sales tax combined with a differential excise tax. The sales tax structure would consist of higher taxes on 'luxury' goods consumed by the rich and lower taxes on 'essentials' consumed by the poor. The excise tax structure would consist of higher taxes on those products which are intensive in factors earning high incomes, and low or zero taxes on products intensive in low-income factors.

While there is still no role for trade taxes as such, a trade tax could be a component of either a sales tax or an excise tax. A sales tax on an import-competing good could be levied in the form of a tariff combined with a tax on domestic production at the same rate (with an allowance for wholesale and retail trading margins). Similarly, a production tax on an exportable good could be levied in the form of an export tax combined with a tax at the same rate on sales of the exportable product to domestic consumers.

Finally, we can narrow down the range of taxes even more by supposing that high collection costs compel the exclusion not only of income tax but of all kinds of taxes except two, namely (a) tariffs and (b) excise taxes on domestic production of import-competing goods. We can also suppose that a sales tax on importables is possible since this is more or less equivalent to a combination of tariff and excise tax. The assumptions are somewhat extreme, but in many less-developed countries income tax can be levied only in the advanced sector of the economy, and this sector produces mainly import-competing goods, so that the income tax is essentially like a production or excise tax on import-competing production. Since export taxes are (broadly) symmetrical with tariffs, one could also introduce export taxes without altering the analysis significantly.

The problem then arises of making up an optimum mix of taxes. This case is expounded fully in Appendix 1 to this chapter (which neglects income distribution considerations). Essentially the choice is between a tax on import-competing *production* and a tax on import-competing *consumption*, the latter tax being obtained by combining a tariff with an excise tax at the same rate. Each form of tax creates a distortion, and the two distortions must be traded off against each other. In addition, income distribution must be taken into account: a shift from production tax to consumption tax would shift the burden of taxation away from import-competing producers towards exporters, and towards those who are especially heavy consumers of importables.

There is still no first-best role for a tariff, unaccompanied by an excise tax on domestic production. This point can be made

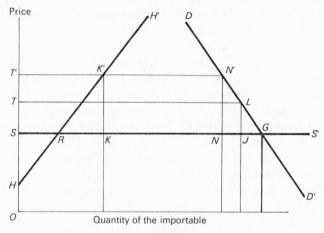

Fig. 4.1

clearly if we imagine, realistically, that it is desired to raise a given amount of revenue from commodity taxes on import-competing goods, the burden of the tax to be borne mainly by the richer members of the community. It can be shown that a tariff on 'luxury' goods on its own—unaccompanied by an excise tax—would be second-best in this case; it would involve a greater distortion cost for given revenue raised than a sales tax, and in addition would involve by-product income distribution effects in favour of domestic import-competing producers which would be in a desired direction only by pure coincidence.

This comparison between a sales tax and a tariff is made in Figure 4.1. It refers to an importable good consumed mainly by rich people. DD' is the domestic demand curve, HH' the import-competing supply curve, and SS' the foreign import supply curve.

A sales tax of ST would raise revenue of $STLJ$ and inflict a consumption-distortion cost of JLG. It would be the equivalent of a tariff of ST combined with an excise tax on domestic production, also of ST.

A tariff on its own to raise the same revenue would have to be ST', raising revenue of $KK'N'N$ ($= STLJ$). It would inflict the production-distortion cost $RK'K$ and the consumption-distortion cost $NN'G$. Thus, not only would it add a production-distortion cost, but the consumption-distortion cost would be greater than in the case of the sales tax because the rate of tax would have to be greater to raise the same revenue. In addition, it would bring about a redistribution of income not only from consumers of the product to the Treasury equal to the revenue, but also (in partial equilibrium terms) a redistribution of $ST'K'R$ from these consumers to import-competing producers; and the latter redistribution might not be socially desired, for example, if the producers are even richer than the particular consumers, or if the producers are foreigners. It follows that only relatively high collection costs of sales or excise taxes can justify a preference for tariff over sales tax.

II
THE IMPORTANCE OF COLLECTION COSTS

Economists have tended to ignore collection costs of taxes, and in general have neglected the administrative aspects of fiscal systems.[5] Any particular economist can be excused from analysing these matters in detail when they happen to be

[5] There are exceptions. See especially John F. Due, *Indirect Taxation in Developing Economies*, Johns Hopkins Press, Baltimore, 1970, which is the best reference on the subject. Also C. S. Shoup et al., *The Fiscal System of Venezuela*, Johns Hopkins Press, Baltimore, 1959; and Richard A. Musgrave, *Fiscal Systems*, Yale University Press, New Haven, 1969. Johnson, in 'Optimal Trade Intervention ...' dismisses administrative costs as 'of practical rather than theoretical consequence', and similarly, these matters find no place in either of the two modern classic theoretical studies, *Trade and Welfare*, and R. A. Musgrave, *The Theory of Public Finance*, McGraw-Hill, New York, 1959.

outside his personal area of competence; for this reason they are not discussed in detail in this book. But one cannot excuse a failure to stress their importance, and to build them into theoretical models, when these models are meant to be some guide to the choice between different types of tax, or between complicated and simple versions of any particular type of tax.

The costs with which we are concerned are really of three types.

Firstly, there are the costs of tax administration incurred by the tax-collecting authority; an important element of this is the labour-cost of ensuring compliance.

Secondly, there are the resource costs incurred by taxpayers to fulfil their tax obligations while minimizing payment, or perhaps to ensure successful evasion; these costs can be quite considerable when complicated taxes are involved or when taxpayers actually choose to break the law. Bribery of tax officials should not be included in these costs, since bribes as such are simply income-redistributive.

Thirdly, there are the distortion-costs which result from taxpayers rearranging their affairs as part of an avoidance or evasion effort. In principle these are no different from the rearrangements of production and consumption patterns, or of the choice between work and leisure, which have already been allowed for in our analysis. But tax systems are often designed to minimize such distortions, and the less effectively a tax system is administered, the more distortions one can expect.

The term *collection costs* will now be used to embrace all three types of cost, even though a part of the second cost and the whole of the third cost are attributable to taxpayers' attempts to reduce collections, and we will mainly have in mind the first type of cost, namely the cost of tax administration.

The central point is that collection costs for trade taxes are generally much lower than for other taxes. This is mainly because foreign trade usually flows through a few ports or *bottlenecks*, and even when it does not, it is easier to police a border and collect taxes on goods passing across it than to seek out a large number of individual taxpayers, whether persons or firms, or to ensure that they produce accurate tax returns. The point is really quite obvious.

In less-developed countries collection costs of export taxes are sometimes especially low, lower indeed than costs of import taxes, both because exports are often more homogeneous products than most imports, and because, in many cases, there are only a few main products and—even more conveniently—a few main producers.

Once we allow for collection costs, tariffs and export taxes may form part of a first-best tax package. The tariffs need not necessarily be accompanied by excise taxes on domestic production at the same rate. This is indeed the reason why taxes on trade are so significant in less-developed countries. In determining the optimum tax mix, say, between tariffs and excise taxes, differential collection costs may weight the scales heavily in favour of tariffs in spite of the production distortion cost that may result from the protective effects of tariffs.

Collection costs also present an argument for indirect taxes collected on domestic production (like an excise or a value-added tax) or on wholesale or retail sales, in preference to a comprehensive personal income tax, since the indirect taxes require contact with far fewer taxpayers. In some countries a case for corporate income tax might also be based on collection costs. In most less-developed countries income tax is collected only from very few persons and from large companies, partly because of collection costs, and hence reliance has to be placed on indirect taxes. In an economy with a high ratio of imports to total domestic sales, the ease of collecting tariffs on imports is a strong argument for the use of sales taxes, or taxes having effects similar to sales taxes, since a substantial part of these can be collected at the point of importation.

In comparing tariffs with more general indirect taxes on domestic sales, one must compare the costs of collecting taxes on imports with the costs of taxing domestic production. Most less-developed countries do have taxes on domestic production, whether described as excises or sales taxes, but they usually apply to only a limited number of products produced or sold by large establishments. The problem is the difficulty of collecting taxes from small-scale establishments.

For this reason, a retail sales tax is usually not practicable. It may be technically *possible* to collect taxes from small firms, so that they are not literally 'untaxable', but the costs in relation to tax yield would be very high because of economies of scale in tax collection from any particular firm.[6]

III
SMUGGLING AND UNDER-INVOICING

Collection and evasion problems also arise in the case of taxes on trade, though generally they are less critical than for other taxes. Essentially there are three problems: smuggling, under-invoicing, and misclassification.[7]

(1) *Smuggling* is the entry of unrecorded and hence untaxable imports, or the exit of unrecorded exports. It is a particular problem for island nations. Its prevention can use up substantial resources of a customs administration. Today smuggling is a serious problem for Indonesia and the Philippines, two multi-island states. Smuggling into and out of these countries is eased by the proximity of Singapore and Hong Kong, both great commerical centers with which traders in Indonesia and the Philippines have close business connections.

---

[6] It may be interesting to look at a particular case here. In Pakistan export duties and import duties are supplemented by excise taxes on domestic production as well as by sales taxes. The sales tax revenue can be split up according to whether it is collected on imports or on domestic production. In the early nineteen sixties taxes collected on domestic production, including relevant sales tax revenue, accounted for only 25% of total revenue from indirect taxes while 75% was collected on trade. Of the production taxes 62% were collected on three products, textiles, tobacco and petroleum. There were only twenty-nine excisable goods, though a somewhat larger number were subject to sales tax. In the earlier post-war years the proportion collected from domestic production was even less. See S. R. Lewis and S. K. Qureshi, 'The Structure of Revenue from Indirect Taxes in Pakistan', *Pakistan Development Review*, 4, Autumn 1964, pp. 491–526; and G. M. Radhu, 'The Rate Structure of Indirect Taxes in Pakistan', *Pakistan Development Review*, 4, Autumn 1964, pp. 527–551.

[7] See Due, *Indirect Taxation in Developing Economies*, pp. 38–53; J. Bhagwati, 'Fiscal Policies, the Faking of Foreign Trade Declarations, and the Balance of Payments', in his *Trade, Tariffs and Growth*, Weidenfeld and Nicolson, London, 1969; C. G. F. Simkin, 'Indonesia's Unrecorded Trade', and H. V. Richter, 'Problems of Assessing Unrecorded Trade', in *Bulletin of Indonesian Economic Studies*, 6, March 1970, pp. 17–60. I have also benefited from seeing an unpublished paper on the Indonesian tariff system by Richard Cooper.

Comparison of prices in Djakarta markets with Singapore and Hong Kong prices of comparable goods indicates that in many cases importers cannot have paid the legal tariffs.

Smuggling is a form of tax evasion which involves resource costs for the smugglers, as well as risks of discovery and prosecution to which some monetary value could be attached, and smugglers are likely to balance these against the tariff or export tax payments avoided. Hence the lower the *ad valorem* rates of tariff or export tax and the greater the difficulty of smuggling—which depends, in particular, on the size and weight of the smuggled goods—the less smuggling there is likely to be. Wrist watches, on which a high *ad valorem* duty used to be payable, have long been the favourite smuggled item in Britain.

When all imports are legal and hence taxable there is one tariff rate which maximizes the customs revenue for any product, the level of this maximum revenue tariff being higher the lower the elasticity of the import demand curve for that product.[8] Once we allow for smuggled imports we can conceive of another demand curve, this time only for legal imports, which will be more elastic than the demand curve for imports of the product as a whole because its elasticity will take into account the falling ratio of legal to smuggled imports as the tariff rate rises. The maximum revenue tariff rate will then be lower than if there were no smuggling. A similar argument applies on the export side: smuggling will lower the maximum revenue export tax rate.

(2) *Under-invoicing* means that the values of imported or exported goods are invoiced by traders below the actual values paid or received by them. Hence, when tariff rates are fixed on an *ad valorem* basis (related to *value* of imports), there is a loss of revenue. The collaboration of foreign exporters may be required for this, and if they obtain some reward for their collaboration the prices paid by importers will be higher than otherwise; so under-invoicing might worsen the terms of trade. Similarly, under-invoicing of exports leads to a loss of export tax revenue and some of the gains may go to foreign importers.

The problem of obtaining correct values of imported goods

---

[8] See pp. 72 below.

for customs purposes is a very important aspect of customs administration. It is avoided when customs duties are *specific* rather than *ad valorem*, but specific duties have many disadvantages and inequities, and are not suitable for highly differentiated manufactured goods. Normally customs authorities deal with the problem by having lists of foreign trade prices from principal supplying countries, using manufacturers' catalogues, and so on.

Evidence that there is *mal*-invoicing (under- or over-invoicing) of imports can sometimes be obtained by comparing the import statistics of the country concerned with figures of exports from its supplying countries (since there may be no incentive to misrepresent export values in the supplying country). Similarly, evidence of under-invoicing of exports can sometimes be obtained by comparing recorded export figures with import statistics of customer countries.

Under-invoicing of imports, like smuggling, means that the actual payment for imports made by importers is greater than that recorded in the import statistics; assuming that there is no under-invoicing of exports at the same time, under-invoicing of imports must show up somewhere else in the balance of payments as some form of unrecorded payment, remittance or capital outflow, or perhaps just as a residual debit item. But if there is under-invoicing also on the export side this may not be so: the foreign exchange unrecorded and retained by Indonesian exporters in Singapore may be sold directly to Indonesian importers, not passing through the Indonesian Central Bank. Hence both export and import tax payments are reduced and export and import values may be understated to an equal extent.

In many countries restrictions on certain imports and on capital outflow create an opposing incentive for *over*-invoicing of imports. The purpose of over-invoicing is to obtain foreign exchange from the central bank nominally to buy legal imports, but in fact to sell part of this foreign exchange on the black market, where a high rate is sustained by restrictions on certain imports which can only be obtained with black market foreign exchange. Alternatively, the excess foreign exchange may be stowed away in Hong Kong or Swiss bank accounts. For the same reason there may be *under*-invoicing of exports.

(3) Finally, another major problem of customs administration is that of *misclassification* of goods. This problem only arises because tariffs are not generally uniform *ad valorem*. If there is any doubt about the appropriate classification for a good, importers will have an incentive to make it appear to fit into a low-duty category. In a world of continually changing and differentiated products, classification systems need to be continually revised. The greater the costs of administration and the less developed the customs administration, the stronger the case for uniform *ad valorem* nominal tariffs, or at least for having only a small number of categories. Distinctions made on the basis of end-use present particular problems: if a motor-car for business use is taxed at a lower rate than a car for private use, how is the customs inspector to be sure that the 'business' car will not be resold for private use, or that business and private use will not be combined?

Having said all this, it must be pointed out that customs administration is highly developed in most advanced countries. In most less-developed countries collection costs are significant and evasion of taxes on trade do present some problems (even in a country with such a highly developed administrative system as India), but the problem is only severe in a limited number of countries—notably the island states of Indonesia and the Philippines, and some African countries. Collection costs and evasion problems of trade taxes seem to be generally much less than for most non-trade taxes.

IV
THE OPTIMUM TARIFF STRUCTURE

If a given total revenue is to be raised from tariffs alone, what tariff structure will minimize the distortion costs of raising this revenue? This question is quite distinct from the question asked so far, namely, whether tariffs, rather than other taxes, should be used to raise revenue. We now take the decision to raise revenue through trade taxes as given. It will be assumed that there are many importable goods, and hence many possible tariff rates, and no domestic divergences. The small country assumption applies, as usual. Income distribution effects are disregarded. Collection costs per dollar of revenue raised are assumed to be the same for all tariffs, and hence can be disregarded.

(1) *The Marginal Cost of Raising Revenue*

If (a) the elasticity of supply of exportables and of domestic demand for exportables were zero and if (b) taxation had no disincentive effects, with the elasticity of supply of effort zero, the answer would be simple. Tariffs would not distort the production or consumption pattern relative to exportables or leisure, and the only distortion possible would be in the pattern of production and consumption of importables. The optimum tariff structure would be a uniform tariff. A uniform nominal tariff would avoid distortion of the consumption pattern, and a uniform effective tariff would avoid distortion of the production pattern. If exportables were never inputs in importable production, a uniform nominal tariff would automatically lead to a uniform effective tariff; otherwise, precise uniformity in both could not be attained, and some distortion would be inevitable.[9]

If substitution effects relative to leisure and to exportables have to be allowed for, the answer is different. The optimum-revenue structure should then be non-uniform, based on the principle familiar from public finance theory that, to minimize deadweight costs, taxes on low-elasticity goods should be higher than on high-elasticity goods. The conclusion applies in a general equilibrium model, but a simple partial equilibrium exposition will be used here. We shall ignore the leisure substitution and focus on substitution relative to exportables. Suppose that for every importable product we can draw a demand and a supply curve which shows how quantities consumed and produced change as the price of the relevant product changes relative to the fixed prices of exportables; each of these curves is assumed to be independent of all the other curves, cross-elasticities being zero. For each product the supply curve is then subtracted from the demand curve, and a demand curve for imports is obtained.

The optimum revenue-tariff structure will now involve high tariffs on goods where the elasticity of import demand is low— so that little distortion is caused by a tariff—and low tariffs

[9] See *The Theory of Protection*, pp. 188–9.

on goods where the elasticity of import demand is high. It has to be remembered that the elasticity of import demand is a derived elasticity: it could be high because the good is a close substitute domestically for exportables either on the demand or the supply side.

A useful concept is the *marginal cost of raising revenue.* This must be distinguished from ordinary marginal revenue.

Figure 4.2 shows the import demand curve, DD', for one

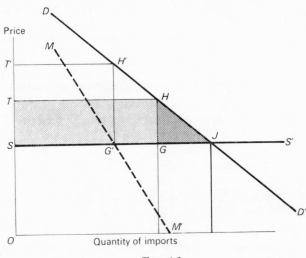

Fig. 4.2

product, while SS' is the import supply curve. A tariff of ST raises revenue of $STHG$ at a total distortion cost of GHJ; the latter is the sum of the *production-distortion cost* and the *consumption-distortion cost.* As the tariff is increased, the revenue increases up to the maximum-revenue tariff ST'. The revenue $ST'H'G'$ is the maximum attainable. Note that when revenue is at its maximum, ordinary marginal revenue is equal to the import price (MM' is the ordinary marginal revenue curve, and crosses SS' at G').

As the tariff rate is increased up to the maximum revenue tariff, both revenue and the distortion cost increase. Hence there is a relationship between the distortion cost of raising the revenue and the actual revenue raised. This then yields a *marginal cost of raising revenue.* At the maximum-revenue tariff rate this marginal cost will be infinite.[10] If the DD′ curve becomes steeper (so that the elasticity of the curve falls at J), a given tariff rate will both raise more revenue and impose a lower distortion cost.

This is illustrated in Figure 4.3 for the case where there are two imports and the two import demand curves are linear. The marginal cost of raising revenue is shown on the vertical axis, and the tariff rate on the horizontal. The curve relating the two for product 1 is $O\alpha_1$. It rises at an increasing rate. The maximum revenue tariff for that product is OR_1. A similar curve is drawn for product 2. The latter product has a lower import demand elasticity at the free trade point, so that its curve, $O\alpha_2$, is to the right of product 1's curve: for any given tariff the marginal cost of raising revenue is less, and the maximum revenue tariff rate is higher.

Minimizing total costs of raising given revenue requires the marginal costs of raising revenue to be equalized. Thus in Figure 4.3, a low tariff of OL_1 on the high-elasticity product 1 combined with a high tariff of OL_2 on the low-elasticity product 2 will yield an identical marginal cost of raising revenue of OT^*. If more revenue is needed, both tariffs must rise, subject to the constraint that neither tariff should be above its maximum revenue level. The tariff on the low-elasticity product should always be higher than that on the high-elasticity product.

If the demand curves are not linear the basic approach and conclusions remain the same. It must always be true that if the marginal costs of raising revenue are to be equalized, the rate

[10] Let P = the domestic pre-tariff price, $\mathrm{d}P$ = increase in domestic price owing to tariff, Q = quantity imported pre-tariff, $\mathrm{d}Q$ = change in quantity owing to tariff, t = tariff rate, e = import demand elasticity at point J (defined as positive number), C = cost of protection (triangle *GHJ*, *assuming HJ is a straight line*), and R = revenue raised.

Then $t = \mathrm{d}P/P$; $e = -(\mathrm{d}Q/Q)/(\mathrm{d}P/P)$; $C = -\frac{1}{2}\mathrm{d}P \cdot \mathrm{d}Q$; $R = \mathrm{d}P(Q+\mathrm{d}Q)$; and from these

$$\frac{\mathrm{d}C}{\mathrm{d}R} = M(t) = \frac{t.e}{1-2.t.e}$$

where M(t) is the *marginal cost of raising revenue.*

Furthermore

$$\frac{\mathrm{d}M}{\mathrm{d}t} = \frac{e}{(1-2te)^2} > 0$$

and when $\mathrm{d}R = 0$, $M(t) = \infty$.

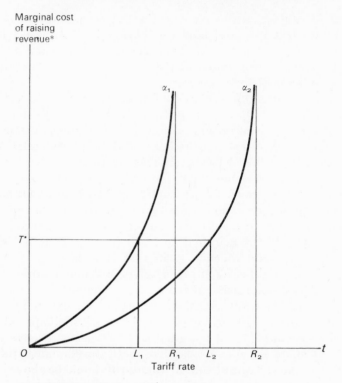

* = dC/dR where C is distortion-cost and R is revenue

FIG. 4.3

of tariff must be relatively higher on the good with the relatively lower elasticity. Along a linear demand curve the elasticity rises as the price rises, and we have distinguished the linear curves on the basis of their elasticities at the free trade points. In the case of non-linear curves it is necessary to distinguish them on the basis of their elasticities for given marginal costs of raising revenue. It could be shown that in general (and not just with linear curves) the marginal cost of raising revenue must increase with the rate of tariff provided the elasticity rises as the price rises. It is possible that locally the elasticity falls as the price rises, so that the marginal-cost-of-raising-revenue curve might slope negatively over a range. But if the demand curve eventually cuts the vertical axis the elasticity must eventually increase. Hence the marginal cost of raising revenue must eventually increase as the tariff is increased and must become infinite at the maximum revenue point.[11]

(2) *Tariffs on Inputs*

Should an optimum revenue tariff structure include tariffs on imported inputs, whether inputs into importable or exportable goods? This is an interesting question because it is sometimes suggested that revenue tariffs on inputs should be avoided.[12] Indeed, countries tend generally to avoid tariffs on inputs because they want to avoid reducing effective protection for final goods, or even imposing negative effective protection on them.

---

[11] Let P = domestic pre-tariff price, T = initial tariff per unit, $\mathrm{d}T$ = *small* increase in tariff per unit, Q = quantity imported with tariff T, $\mathrm{d}Q$ = change in quantity owing to increase in tariff, t = rate of tariff, $e(t)$ = import demand elasticity *at tariff point* (positive number), C = cost of protection, and R = revenue raised.

Then $\dfrac{\mathrm{d}C}{\mathrm{d}T} = -T\dfrac{\mathrm{d}Q}{\mathrm{d}T}$ and $\dfrac{\mathrm{d}R}{\mathrm{d}T} = Q + T\dfrac{\mathrm{d}Q}{\mathrm{d}T}$. Since $t = T/P$ and $e(t) = -\dfrac{\mathrm{d}Q}{Q}\dfrac{T+P}{\mathrm{d}T}$ it follows that

$$\frac{\mathrm{d}C}{\mathrm{d}T} = \frac{e(t)tQ}{t+1} \text{ and } \frac{\mathrm{d}R}{\mathrm{d}T} = Q - \frac{e(t)tQ}{t+1}.$$

Hence

$$\frac{\mathrm{d}C}{\mathrm{d}R} = M(t) = \frac{e(t)t}{t+1-e(t)t}.$$

[The denominator must be positive provided the tariff is less then the maximum revenue rate, where $t = 1/(e(t)-1)$.]

Furthermore

$$\frac{\mathrm{d}M}{\mathrm{d}t} = \frac{e't(1+t)+e(t)}{(1+t-e(t)t)^2}$$

where $e' = \mathrm{d}e/\mathrm{d}t,$

and this is positive provided e' is non-negative.

In addition, if $\mathrm{d}M = 0$ then $\mathrm{d}t = -\mathrm{d}e\,\dfrac{t^2+t}{e}$. [The higher the elasticity the lower the tariff for a given M.]

[12] Little et al. in *Industry and Trade in Some Developing Countries*, p. 137, make the point, though undogmatically. One may also get the impression from V. K. Ramaswami and T. N. Srinivasan, 'Optimal Subsidies and Taxes When Some Factors are Traded', *Journal of Political Economy*, 76, Aug. 1968, pp. 569–82. They argue that when trade taxes are levied for revenue, there should be no tariffs on inputs to make exports. But they assume that there is no domestic consumption of the export and so rule out the possibility of consumption-distortion resulting from levying the trade tax on the final good.

The brief answer is that an optimum revenue tariff structure is likely to include tariffs on inputs. A tariff on a final good will impose the familiar consumption- and production-distortion costs. A tariff on an input will avoid the consumption-distortion cost but, if imposed on its own, will impose two production-distortion costs: firstly, the cost resulting from the protection provided for domestic production of the input, and secondly the cost resulting from the anti-protection (negative effective protection) imposed on the final good. Furthermore, if the taxed input can be substituted for untaxed inputs in the production function of the final good, the input tariff will distort factor proportions. If a tariff on a final good is supplemented by tariffs on its inputs so as to reduce or even eliminate effective protection for the final good, the production cost created by the final good tariff may be reduced by the input tariff, though a new production cost may be created through the protection provided for domestic production of the input.

If a given amount of revenue is to be raised from a tariff on a final good and on its input, there is likely to be some optimum mix of the two tariffs. If the consumption-distortion effects were very small, owing to a low demand (substitution) elasticity for the final good, the optimum mix might contain very little input tariff. On the other hand, initial imports of the final good might be very low and imports of the input high. A final good tariff would then have to be at a very high rate to raise a given amount of revenue, and indeed it might not be able to raise it. This, then, would increase the weight of the input tariff in the optimum mix.

V
EXPORT TAXES

The role of export taxes in less-developed countries and their relatively low collection costs have already been noted. The main point about export taxes is their essential 'symmetry' with tariffs. An export tax lowers the domestic after-tax price of exportables, and a tariff raises the domestic after-tax price of importables. Thus both lower the prices of exportables relative to the prices of importables facing domestic producers and consumers. This argument is developed more fully, taking into account the multiplicity of export and import goods, in *The Theory of Protection*.[13] An export tax will cause domestic resources to shift out of export industries into import-competing industries, just like a tariff, and similarly it will, like a tariff, cause the consumption pattern to shift towards exportables away from importables. Of course, an import tariff will in the first instance improve the balance of payments and so require a rise in domestic factor prices or an exchange rate appreciation to restore equilibrium, while an export tax will (with given world prices) worsen the balance of payments, and so require a depreciation. The symmetry principle applies if we compare equilibrium situations, assuming appropriate balance of payments adjustment.

It follows that, in the absence of collection costs for income and sales taxes, an export tax would be no more optimal than a tariff, and the reasons are the same and need not be repeated. But given that there are collection costs, and that these are sometimes especially low for export taxes compared with other taxes, a set of export taxes may well form part of a first-best tax package. The question of what structure of export taxes will raise a given amount of revenue at minimum distortion-cost is thus of interest. The issue is the same as in the case of tariffs: the *marginal cost of raising revenue* from different export taxes must be equalized, and this will involve high rates of tax when the domestic elasticity of supply of exports is low, and low rates when the elasticity is high.

VI
THE DECLINING IMPORTANCE OF TRADE-TAX REVENUE IN THE PROCESS OF ECONOMIC DEVELOPMENT

Finally, let us return to the point stressed at the beginning of this chapter, namely the importance for government revenue of taxes on trade. There is some evidence that the importance of taxes on trade declines as *per capita* income increases, at least once the latter passes a certain threshold.

We have already noted that trade taxes are more important for less-developed countries than for developed countries. A

---

[13] *The Theory of Protection*, pp. 119–120.

comparative cross-country study by Lewis,[14] based on average figures for the years 1954–60, defined the relationship between the fiscal importance of trade taxes and *per capita* incomes more precisely. In the group of the 23 richer countries included in his survey—including many developed countries as well as some of the better-off less-developed countries—the share of trade taxes in total revenue fell as *per capita* income increased. On the other hand, within the group of 18 poorer countries there was no such noticeable relationship. For both groups the fiscal importance of trade taxes varied with the importance of the foreign trade sector of the economy, as indicated by the ratio of trade to gross national product. For all the countries together the latter ratio seems thus capable of explaining the differing shares of trade taxes in total revenue better than *per capita* incomes, but clearly *per capita* incomes are of some relevance, above all in explaining the overall difference between less-developed and advanced countries.

A number of considerations would lead one to expect trade taxes to be proportionately less important as sources of revenue in advanced countries than in less-developed countries.

(1) As countries become more developed, collection costs of non-trade taxes, and notably of the income tax, fall; this is a form of 'technical progress'. Collection costs of trade taxes are also likely to decline, but these costs are already low, so that the gap between the two costs narrows, and the optimal tax pattern shifts towards non-trade taxes.

(2) The capacity to produce manufactured import-competing goods in response to tariff inducements increases so that a given tariff has an increasing protective, and a lower revenue, effect. This means that the maximum revenue obtainable from tariffs decreases, at least in relation to gross national product; more important, even if tariff rates remain below maximum revenue levels, the production-distortion costs of raising given revenue increase, so that the optimal rate of trade tax relative to other taxes declines.

(3) In very underdeveloped economies imported goods are purchased mainly by the rich, and in many less-developed

---

[14] Stephen R. Lewis, Jr., 'Government Revenue from Foreign Trade: An International Comparison', 1963.

countries a limited group of imported goods can be fairly clearly specified as being mainly consumed by the well-to-do. Tariffs on these goods are then quite convenient *progressive* taxes. But with the process of development imported goods become more prominent in the consumption patterns of the general population, or enter as inputs into domestically produced goods that are widely consumed, so that it becomes more and more difficult to devise a tariff structure that has an assured progressive impact. A broad system of revenue tariffs is quite likely to become more like a proportional, or even a regressive, tax. Thus tariffs become less efficient instruments for distributing the tax burden equitably.

(4) The ratio of government expenditure to gross national product tends to rise in the process of growth for a variety of reasons: government spending is 'income elastic in the social welfare function'. Taking this as given, it becomes almost inevitable that the share of trade taxes in total revenue will decline. The ratio of maximum trade-tax revenue to imports is likely to fall for reason (2) above, so that, for a given trade-GNP ratio, there will be a decline in the ratio of maximum trade-tax revenue, and also of optimum trade-tax revenue, to GNP. Hence taxes on trade are likely to be income-*in*elastic while total government spending is income-elastic.

It should be noted here, incidentally, that empirical evidence does not suggest that the trade-GNP ratio changes in any consistent way as *per capita* income rises.[15] Since the importance to the government budget of taxes on trade depends, above all, on this trade-GNP ratio, if such a relationship between *per capita* income and the trade-GNP ratio did exist, we would have an additional link between the fiscal importance of taxes on trade and the process of development.

We have been concerned here with explaining the share of trade taxes in total revenue, taking the ratio of government expenditure to GNP at each level of development as given, and have suggested that changes in this ratio may influence the trade-tax share in revenue. But the relationship may also go the other way: the ability to collect taxes on trade may affect the ratio of government expenditure to GNP for a given *per capita* income. Cross-country comparisons by

[15] Musgrave, *Fiscal Systems*, pp. 119, 151.

Hinrichs suggest that such a relationship may exist for poorer less-developed countries. He has found a relationship between a country's *openness*—that is, the trade-GNP ratio—and the ratio of government revenue to GNP. For the poorer less-developed countries, openness appears to be a better indicator of the government-revenue-share than *per capita* income.[16]

(5) As countries become more developed, the pattern of imports tends to shift away from manufactured consumer goods towards intermediate goods and capital-goods. Apart from income distribution considerations (see point (3) above), and *provided there is no protective motive*, tariffs on intermediate goods and capital-goods might then be expected to replace tariffs on final goods as sources of revenue.

In practice it is usually desired to protect import-competing manufacturing industries, or at least not to *anti*-protect them. Tariffs on final goods provide such protection; on the other hand, tariffs on intermediate and capital-goods *lower* effective protection for final goods, and if there is little scope for domestic production of the intermediates, this anti-protective effect will not be compensated by the effects of positive protection of intermediates. The approach in this chapter has been to assume absence of domestic divergences, and hence no reason for protection, but the introduction of a protective motive would thus yield another reason why the share of tariffs in an optimum-tax package can be expected to decline in the process of development.

APPENDIX
THE OPTIMUM EXCISE TAX AND TARIFF MIX

Assume that a given amount of revenue is to be raised and only two tax instruments are available, a tariff and an excise tax on

[16] Harley H. Hinrichs, 'Determinants of Government Revenue Shares Among Less-Developed Countries', *Economic Journal*, 75, Sept. 1965, pp. 546–556. See also V. Tanzi and C. McCuistion, 'Determinants of Government Revenue Shares among Less-Developed Countries: A Comment', *Economic Journal*, 77, June 1967, pp. 403–5, who show that on the basis of Hinrichs' figures, his generalization does not apply to the whole group of less-developed countries, but only to the poorest group (*per capita* income below $150).

import-competing production.[17] The rate of excise tax will be expressed in relation to the price of domestic production excluding tax. This puts it on the same basis as the tariff rate. When the tariff and the excise tax on a particular product are at the same rate (so expressed) then the two together make up a consumption tax (or sales tax).

(1) *Assumptions*

There is only one importable product on which excise tax and tariff are levied, with no taxes elsewhere (other than, possibly, non-distorting income taxes). The one taxed product can be thought of as either the sole importable in the two-good general equilibrium trade model or as one among many importables. Initially there are no domestic divergences between private and social costs, other than those created by the two taxes; the revenue which is being raised finances offsetting subsidies that counteract completely any divergences that would otherwise exist.

A desired income distribution is being continuously maintained by non-distorting taxes and subsidies which are outside the model. Alternatively one could assume that (i), the existing income distribution is the desired one, (ii) factor-intensities are identical between the taxed product and other products as a group, so that income distribution is not affected by production shifts induced by the tax structure, and (iii) all persons have identical homothetic indifference maps, so that income distribution is also unaffected by changes in relative prices facing consumers.

Given all these assumptions, free trade without any excise tax or subsidy would be optimal, were it not for the need to raise revenue.

The disregard of income distribution effects is the most serious limitation of the approach here. Assumptions (i) to (iii) are grossly unrealistic, while it is hardly reasonable to suppose that a desired income distribution is being maintained independently of the model when the whole concern of the model is with the raising of revenue, which must be one part of such an income distribution policy. It follows that the conclusions must be supplemented by allowing for income distribution effects.

---

[17] One could justify this either by supposing that high collection costs rule out income tax (or further increases in income tax), or that non-economic considerations do so. The problem is analysed in *Trade and Welfare*, pp. 190–196, and in *Trade and Welfare: A Mathematical Supplement*, Ch. V. Meade is concerned with the world optimum, not the national optimum, but this distinction does not in any case arise in the analysis here since the terms of trade are held constant.

(2) *The Model*

In Figure 4.4 the vertical axis shows the rate of excise tax, e, and the horizontal axis the rate of tariff, t. Both tax and tariff can be negative. Along the 45° line NON' the two rates are equal, so this line traces out consumption taxes or (in the south-western quadrant) consumption subsidies.

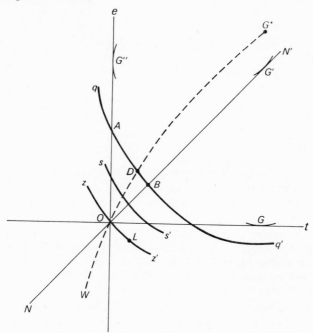

Fig. 4.4

The curve qq' traces out all the combinations of excise tax and tariff that yield a given amount of revenue. If less revenue were desired there would be a lower curve, such as ss'. If no net gain or loss in revenue were desired the curve would be zz', which goes through the origin. The curve qq', as drawn, assumes that elasticity conditions are such that it is possible to raise the desired amount of revenue by an excise tax alone or a tariff alone.

What is the right mix of instruments?

If the whole of the revenue were raised by an excise tax (everywhere along the vertical axis) the price to consumers would remain the free trade price and there would be no consumption-distortion cost. But the domestic producer of the importable product would be anti-protected in relation to other producers, so there would be a production-distortion cost. If the two rates were equal (everywhere along NON'), so that there is a consumption tax on the product, the result would be a consumption-distortion cost but no production-distortion cost. These are the two limiting cases. (The two distortion costs include distortion relative to leisure). The optimum will be in between, yielding less production cost than the first case and less consumption cost than the second. The essential point is that the marginal distortion cost per unit of revenue raised rises from zero both as the excise tax is raised from zero, and as the consumption tax is raised.

In Figure 4.4, with revenue indicated by curve qq', the point of zero consumption cost is A, of zero production cost is B, and the optimum is D, somewhere between A and B. The optimum involves a positive tariff and a positive excise tax, with the excise tax greater than the tariff.[18] The curve $WODG^*$ traces out optimum mixes of the two instruments for various revenue levels.[19]

The amount of extra revenue that can be raised by steadily increasing the tariff and excise tax rates is limited. In Figure 4.4 these maximum revenue effects are expressed by the constant revenue curves acquiring positive slopes at high rates of tariff or excise tax. In fact the curves are concentric contours, with the maximum revenue point G^* the peak. The maximum revenue tariff with a zero excise tax is at G, the maximum revenue excise tax with a zero tariff is at G'', and the maximum revenue consumption tax is at G'. The maximum revenue point G^* requires the same tariff rate as G' but a higher excise tax rate. The relationships between, G, G', G'', and G^* are further spelt out in Section (5) of this Appendix.

(3) *Introducing Protection*

Suppose that it is desired to protect the import-competing industry with a given percentage margin to obtain a target increase

---

[18] This result may seem surprising, but will not be unexpected to readers of *Trade and Welfare*. Meade in *Trade and Welfare*, pp. 194–196, and *Trade and Welfare: A Mathematical Supplement*, Ch. V, shows that the rate of tax should be higher on home production than on imports when the elasticity of supply of home production is smaller than the elasticity of foreign supply (which is the case in our model) and that, from a *world* efficiency point of view, if the home supply elasticity is greater than the foreign supply elasticity the production tax should be less than the tariff.

[19] One could add the extra constraint that the policies could not be mixed, but that the choice is between tariff alone, excise tax alone, or the two at equal rates, yielding a consumption tax alone. There may be an indivisibility in tax administration which makes this a sensible way of formulating the problem. See P. M. Mieskowski, 'The Comparative Efficiency of Tariffs and Other Tax-Subsidy Schemes as a Means of Obtaining Revenue or Protecting Domestic Production', *Journal of Political Economy*, 74, Dec. 1966, pp. 587–99.

in its output. This protective margin could be obtained by various combinations of tariff and excise tax, positive or negative, traced out by the straight line QTQ' in Figure 4.5. The point T indicates the case where protection is attained by a tariff alone. At all points in the north-eastern quadrant, tariff and excise tax rates are both positive but the tariff rate is always appropriately greater.[20]

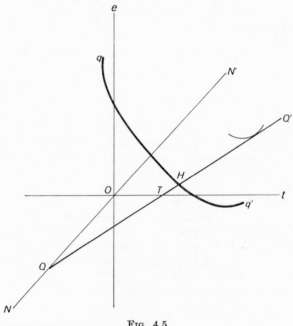

FIG. 4.5

Hence we have two instruments and two targets. If the revenue target is that represented by the curve qq', both targets will be satisfied at the point H. It cannot be said that protection as such yields or costs revenue. Given the desired protection it is possible to attain any revenue target (including zero revenue), subject to a maximum (given by the tangency of a revenue contour to QTQ').

[20] If P = free trade price per unit, E = excise tax per unit, T = tariff per unit, and R = protective margin per unit, so that $R = T-E$, then we define tariff rate $t = T/P$, excise tax rate $e = E/(P+R)$, and protective rate $r = R/P$. Hence $r = (t-e)/(1+e)$ or $t = e+r(1+e)$.

The slope of the line QTQ' in Figure 4.5 is $dt/de = (1+r)$. Further, if $t = e$, then $r = 0$; if $e = 0$, then $r = t$. If $e = -1$ then $t = -1$; this is true for any value of r, and is represented by the point Q in Figure 4.5. An increase in r means that the line QTQ' pivots on Q to the right.

(4) *Collection Costs*

Now introduce collection costs on the taxes under consideration.

Collection costs might be expressed as a proportion of gross revenue raised; we can imagine that part of revenue is used up in transit, rather like transport costs in international trade. To high-light the point that in general collection costs in proportion to revenue raised are lower for tariffs than for excise taxes, it will be assumed here that collection costs of the tariff are zero and of the excise tax positive.

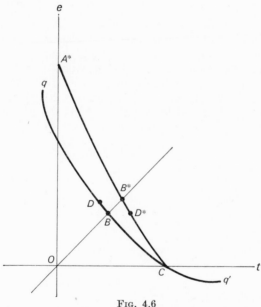

Fig. 4.6

In the absence of collection costs, the revenue target is given by curve qq' in Figure 4.6. With collection costs, the same *net* revenue target will require more *gross* revenue, and the higher the ratio of excise tax to tariff, the greater the gross revenue will have to be. We thus obtain a curve $A*B*C$ showing various combinations of excise tax and tariff that yield the same net revenue as the combina-tions on qq' yielded in the absence of collection costs.

At $A*$ there is no consumption-distortion cost, at $B*$ no production distortion cost and at C no collection cost. The optimum point, $D*$, will involve trading-off three costs at the margin. Compared with the optimum, D, on qq', $D*$ will involve a higher tariff rate because more gross revenue is to be raised to finance collection costs (income effect) and because, for a given ratio between tariff and excise tax rate, the marginal cost of raising revenue from the tariff relative to the marginal cost of raising it from the excise tax has been reduced (substitution effect). On the other hand, it may involve either a higher or a lower excise tax rate.

(5) *Maximum Revenue Tariff and Excise Tax*

Let us now retrace our steps and examine the relationships between maximum revenue tariff, maximum revenue consumption tax and maximum revenue excise tax, (the points G, G', G'' and G^* in Figure 4.4).

We now use Figure 4.7. The vertical axis shows the price of the importable in terms of the price of the exportable (or in terms of the average price of all other goods, in a multi-product model), and the horizontal axis shows quantities of the importable.

SS' is the import-competing supply curve, DD' is the domestic demand curve for the product, and NN' is the import demand curve (SS' horizontally deducted from DD'). PP' is the foreign import supply curve.

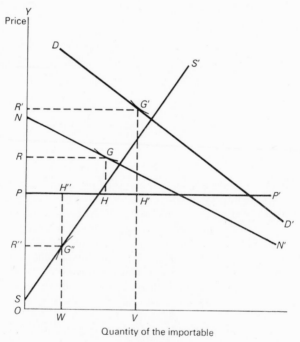

Fig. 4.7

Imagine a map of rectangular hyperbolae in the YPP' space with origin at P. At G' a curve is tangential to DD': this gives us the maximum revenue consumption tax (t'_g), the rate of tax being PR'/OP and the revenue raised $PR'G'H'$. At G a curve is tangential to NN'; this gives us the maximum revenue tariff (t_g), the rate of tax being PR/OP and the revenue raised $PRGH$. It is clear that t'_g must be greater than t_g.

Next, imagine a map of rectangular hyperbolae in the OPP' space, with origin at P. At G'' a curve is tangential to SS'; this gives us the maximum revenue excise tax, the rate of tax being $R''P/OR''$ and the revenue raised $R''PH''G''$.

What combination of taxes will maximize revenue? It can be shown that a tariff of PR' per unit combined with an excise tax of $R''R'$ per unit will do so. This combination will bring consumption to OV—which would also result from the maximum revenue consumption tax alone—while it will bring production to OW—which would also result from the maximum revenue excise tax alone. Revenue has then two parts, (i) the consumption tax element, $PR'G'H'$, which is at a maximum with that rate of consumption tax, and (ii) the additional excise tax element, $R''PH''G''$, which is at a maximum when the additional excise tax (above the amount needed to turn the tariff into a consumption tax) is PR''.

It is clear that the rate of tariff in the maximum revenue *set* of tariff and excise tax is equal to the maximum revenue consumption tax rate (t'_g). To obtain the rate of excise tax required for the maximum revenue *set* one might first impose the maximum revenue consumption tax and then add an extra excise tax which will maximize the extra revenue.

5

PROTECTION AND INCOME DISTRIBUTION

HISTORICALLY, one of the main reasons for the imposition of protective tariffs in the now-advanced countries has been to prevent changes in internal income distribution that would otherwise have taken place as a result of market forces. The aim has been to maintain sectional incomes, often in response to sectional pressures.

Typically, a war cuts off or reduces supplies of foreign goods and so provides protection for import-competing industries. If market forces were allowed full play when the war has ended, imports would flood in and destroy some of the infant industries or at least severely lower the incomes derived in import-competing industries. The affected producers exert political influence and—most important—the wider community is persuaded of the injustice that the market is likely to do to them, and so protection is provided to shelter the shorn lamb.

Thus the Napoleonic Wars cut off supplies of corn (wheat) to Britain from the Continent and led to expansion of acreage and higher prices of corn. When the war was over the Corn Law of 1815 was designed to maintain prices, with an import prohibition as long as the domestic price was below a certain level. In the United States the War of 1812 had a similar effect in protecting manufacturing industries, and the beginnings of American protectionism can really be traced to the period immediately following this war. Significant tariff protection of manufacturing industry in Australia began in a similar way in the nineteen-twenties, World War I having provided the infant protection. The second big boost to Australian protection came immediately after the second World War. This war also had such an effect in Latin America, notably Brazil and Argentina.

Depressions, whether sectional or general, can also explain a good deal of protectionism. Thus the Great Depression led to the imposition of tariffs and quotas all around the world. But this cannot really be described in income distribution terms, though the aim was certainly income maintenance— essentially through exporting unemployment by running balance of payments surpluses, not an aim that many countries could successfully achieve at the same time. Sectional depressions or declines of industry are another matter. One version of the income maintenance motive for protection is what has been called (though not by its advocates) the 'senescent industry argument'. The two outstanding cases are the protection of continental European agriculture since the late nineteenth century and the world-wide protection of the textile industry in recent years.

European agricultural protection, which takes its contemporary form in the Common Agricultural Policy of the European Economic Community, originated in the depression of 1873–79 and in the more prolonged fall in agricultural prices produced by new supplies of produce from Russia and the New World. The response was not uniformly protective: notably Britain, where the free trade-protection battle had been fought so vigorously earlier in the century and had been won by the free-traders, allowed its agriculture to decline. But agricultural protection in Germany, France, and some smaller countries has not looked back since.

With the inflow of cheap textile manufactures in the nineteen-thirties from Japan and in the nineteen-fifties and -sixties from Hong Kong and India, the response was characteristic in all the Western countries, though less in Britain than elsewhere. Protection of textiles and a few other labour-intensive goods is now one of the main forms of non-agricultural protection in most advanced countries, and the objective is straightforward income maintenance. The revival of protectionism in the United States during the last few years, with strong support from the trade unions, is also an obvious sectional response to increased imports from Japan.

One can dignify these responses and motivations in various ways, and a new concept will be suggested later in this chapter. The present author, when trying to explain (but not defend)

tariff policy in Australia in the nineteen-fifties and early
-sixties put forward the 'existence principle'. The implication
of policy at the time was that any industry that has in one way
or another come into existence should be protected. There
could still be debate about how much protection an industry
really "needs" and whether further industries or activities
should be brought into existence, but the existence of an
activity appeared to set a constraint on policy. Literally, the
role of tariffs was *protection*. No doubt these attitudes could
be found in many other countries.

In many countries 'arguments for protection' have been
advanced at the time by policy-makers or subsequently by
economists which have provided rationalizations for tariffs—
suggesting that in some way or another the nation benefits,
and not just a section. There may indeed have been such
beneficial 'efficiency' effects. But if one is looking for explanations of what actually happened, one need go no further than
look at income distribution or income maintenance.

Any history book will show that these income distribution
effects are the very stuff of politics. The great free trade versus
protection controversies of the ninteenth century in Great
Britain and the United States brought out the conflicting
interests of different sections of the community. It was the
debate about the effects of the Corn Laws which really stimulated the beginnings of the modern theory of international
trade. In Australia it has been recognized that a principal
effect and possibly a purpose of tariffs on imports of manufactures has been to raise real wages at the expense of rural
rents—a method of 'painless extraction' from the rural
exporting sector. It has already been stressed in Chapter 3
that tariffs and other trade interventions are unlikely to be
first-best devices for altering income distribution. Nevertheless,
tariffs are ofter preferred, if only because of the illusion they
create that some citizens are not being taxed for the benefit of
other citizens. Thus the income distribution effects of protection call for close attention in this book.[1]

---

[1] The various remarks about tariff history here are rather casual and, of
course, over-simplifications, though they have drawn on a wide range of
sources. No single reference can be given as there is no systematic study of
protectionist history covering many countries and looking for any consistencies

I
THE EFFECTS OF PROTECTION ON INCOME DISTRIBUTION

A tariff or an export tax has two types of income distribution effect. Firstly, the revenue raised represents redistribution from particular consumers to the government and hence indirectly to those sections of the community which benefit from extra public expenditures or from the reduction of other taxes. This has been discussed in the previous chapter. Secondly, the changed domestic price structure brought about by the trade taxes or other trade interventions will have redistributive effects, and on this we focus in the present chapter.

The effects of a protective structure on income distribution in a multi-commodity model can be analysed in the same way as one analyses the effects on resource allocation and the consumption pattern. We are concerned here with comparing the income distribution resulting from a system of protection— a protective structure consisting possibly of tariffs, quotas, export subsidies, taxes and so on—with an alternative situation without any trade taxes or subsidies (and with other taxes given), the exchange rate and fiscal and monetary policies being adjusted so as to maintain the overall level of employment and external balance. Production and consumption taxes and subsidies on traded goods might be included in the protective structure. One can also analyse the effects of

---

of behaviour. I am particularly indebted here to discussion with Stanley Engerman. For the United States, the classic is F. W. Taussig, *The Tariff History of the United States*, 7th ed., New York, Putnams and Sons, 1923. See also Sidney Ratner, *The Tariff in American History*, Van Nostrand Company, New York 1972. Motives and attitudes affecting more recent U.S. policies are discussed in R. A. Bauer, I. S. Pool, and L. A. Dexter, *American Business and Public Policy: The Politics of Foreign Trade*, Aldine-Atherton, Chicago, 1972. Any economic history of Britain will discuss the Corn Laws; see especially J. D. Chambers and G. E. Mingay, *The Agricultural Revolution 1750–1880*, Batsford, London 1966, Chapter 5. The responses of different European countries to the agricultural slump of the 1870s are compared in C. P. Kindleberger, 'Group Behaviour and International Trade', *Journal of Political Economy*, 59, February 1951, pp. 30–46. The post-war history of Latin American protectionism is described in S. Macario, 'Protectionism and Industrialization in Latin America', *Economic Bulletin for Latin America*, Vol. 9, March 1964, pp. 61–101. On Australia, see W. M. Corden, 'The Tariff' in A. Hunter (ed.), *The Economics of Australian Industry*, Melbourne University Press, 1963.

changes in the structure, hence comparing one structure with another, rather than using the free trade situation as the reference point.

The analysis of the resource allocation effects of a protective structure was the main theme of *The Theory of Protection* and can be readily extended to the analysis of income distribution effects. Broadly speaking, just as the *scale of effective protective rates* is likely to give some indication of resource allocation effects, so it will also give some indication of the nature of the income distribution effects on the production side. Factors of production specific to activities high in the scale of effective rates—the outputs of which are likely to increase—will probably gain in real incomes, and factors of production specific to activities low in the scale of effective rates are likely to lose, compared with free trade. But incomes of mobile factors will also be affected, the relative 'factor-intensities' of the various activities (value-added products) being relevant. If the effective protective rate on a product is increased one would normally expect factors of production 'intensive' in the activity concerned to gain in real income.

In addition, effects on people in their capacity as consumers will be indicated, broadly, by the scale of nominal rates, provided the proportions of expenditure on different goods vary among consumers—which will be either (i) because their tastes are different, so that they choose to divide up their spending differently even when they have the same incomes, or (ii) because their incomes differ and income elasticities of demand are not unity.

Of course there are limiting cases. When factors of production are quite immobile, and hence *specific* to particular activities, a change in a protective structure may have a big effect on income distribution but little or none on resource allocation. This is often likely to be the case in the short-run, especially when agriculture is being protected.

At the other extreme, resource allocation may be affected, but not income distribution. Conceivably, a tariff may expand capital-intensive relative to labour-intensive industries, and so increase the demand for capital relative to labour, but corporate share ownership may be well dispersed through the community, so that for most individuals, increases in dividend receipts compensate for reductions in wages. More realistically, the output patterns within firms or farms may shift but their factor proportions may remain broadly unchanged. This may help to explain the acquiescence with which tariffs on manufactured goods were eliminated within the E.E.C. (though another possible explanation will emerge later). Freer trade within the Community led to expansion of intra-trade in manufactures, many firms expanding their export sales at the same time as they were facing more import competition.

All this is a little general and vague, though perhaps it is all that can be said about income distribution effects without giving a false impression of precision. But international trade theorists have attempted to analyse these effects more rigorously and, to do so, have made severely limiting assumptions. We turn now to a review of this work.

II
THE STOLPER-SAMUELSON MODEL

A celebrated article by Stolper and Samuelson provides the basis for the rigorous analysis of the income distribution effects of a tariff in a general equilibrium model.[2] The authors asked whether a tariff could raise the real income of a substantial section of the community, say the whole body of wage-earners, even though in the potential sense, it lowers the real income of the community as a whole. By a fall in potential community real income is meant that gainers could not fully compensate losers without suffering a loss themselves as a result.

[2] W. F. Stolper and P. A. Samuelson, 'Protection and Real Wages', *Review of Economic Studies*, 9, 1941 pp. 58–73 (reprinted in H. S. Ellis and L. A. Metzler (eds.), *Readings in the Theory of International Trade*, Blakiston Co. Philadelphia, 1949; and also in J. Bhagwati (ed.), *International Trade*, Penguin Modern Economics, 1969).

An excellent discussion of the Stolper-Samuelson theory, and also of the relevant literature that preceded it, is in John S. Chipman, 'A Survey of the Theory of International Trade: Part 3, The Modern Theory', *Econometrica*, 34. Jan. 1966, pp. 35–40. The basic argument, with its principal conclusion, can be found in a book published in 1906, A. C. Pigou, *Protective and Preferential Import Duties*, Macmillan, London, 1906 (London School of Economics Reprint, 1935), pp. 58–9.

A thorough exposition and discussion of many implications is in *Trade and Welfare*, Ch. XVII. See also, R. E. Caves, *Trade and Economic Structure*, Harvard University Press, Cambridge, 1960, pp. 68–76.

Previous to this article many writers on protection had doubted that the real income of any large and mobile factor of production could be raised by protection when the latter lowers real income for the country as a whole. It had of course rarely been questioned that a factor which is quite specific to a protected industry could gain from protection, or that a relatively minor factor could gain. It was also accepted that a large factor could gain *relatively*. It might gain a larger *proportion* of the cake, but the common argument was that, when the total cake was being reduced in size because of protection, it is improbable that the *absolute* size of the slice obtained by a large factor could increase.

It had also been the general view that whether a factor gains or loses, relatively or absolutely, must depend to an extent on its spending pattern. A factor is more likely to lose if protected goods—the prices of which have gone up—play a large part in its expenditure pattern. With the domestic prices of some goods rising, while the prices of others remain unchanged, there would then be an index number problem in determining whether a factor gains or loses.

(1) *The Simple Model: Assumptions*

Stolper and Samuelson set up a simple model with two industries and two homogeneous factors of production. There is some production of both products even in the absence of a tariff; hence consistent *non-specialization* is assumed. The factors are each in fixed supply to the economy, and each factor is always employed in both industries; the factors are perfectly mobile between the two industries. Let us call the industries the exportable and the importable industry, and the factors, labour and land. Since the factors are homogeneous and mobile, their rewards—the *real wage* and the *real rent*— must be uniform throughout the economy.

In each industry there is (i) marginal productivity factor pricing, and (ii) constant returns to scale. These assumptions are crucial.

Assumption (i) means that in each industry the real factor price will be equal to the marginal product of the factor in that industry, the real factor price being expressed in terms of that product. Thus, the real wage in terms of importables

must be equal to the marginal physical product of labour in
the importable industry. In terms of exportables it must be
equal to the marginal physical product of exportables. Bearing
in mind that wage-earners are likely to consume some of each
product, the *true* real wage should be expressed partly in
terms of importables and partly in terms of exportables—in
other words, as a weighted average of the two. If the marginal
physical product of labour goes up in *both* industries then the
real wage in terms of both importables and exportables must
go up, and hence the *true* real wage must go up. This point
will be central to the argument to follow.

Assumption (ii) means that in each industry the marginal
product of a factor depends purely on the ratio in which the
factors are employed, and not at all on the scale of output.
A fall in the labour-land ratio in an industry will involve a
rise in the marginal physical product of labour and hence a
rise in the real wage in terms of the product of that industry.

Next, let us introduce the concept of *factor-intensities*.
Given the production functions of the two industries, one
industry is likely to employ a higher ratio of labour to land
for any given ratio of the factor prices than the other industry.
In each industry the ratios will of course change if the factor-
price ratios change. If the wage-rent ratio goes up the labour-
land ratio will go down as the industries adjust to the higher
cost of labour in terms of land. But we assume that for any
factor-price ratio uniform for the economy, one of the
industries always employs a higher ratio of labour to land
than the other. This is the *labour-intensive industry*. Let us
assume here that importables are labour-intensive, while
exportables are land-intensive.

Stolper and Samuelson also assume that the country
concerned faces given world prices, so that there are no terms
of trade effects. This makes it inevitable that when a tariff is
imposed the relative *domestic* price of importables rises. It is
quite likely to do so even if there were terms of trade effects,
but we put this possibility aside for the moment.

(2) *The Simple Model: the Paradox*

A tariff raises the relative domestic price of importables.
Hence labour and land move out of exportables into

importables. Output of importables goes up and of exportables goes down. Now we come to the apparent Stolper-Samuelson Paradox. What happens to the ratio of labour to land in each industry? For the economy as a whole it is fixed. *And yet in each industry separately it will fall.*

Since the labour-land ratio falls in each industry, the real wage in terms of each product must rise; this follows from assumptions (i) and (ii) above. Hence the *true* real wage must rise. Similarly, with a lower labour-land ratio in each industry, the true real rent must fall. Thus the real rewards to the two factors must move in opposite directions. The tariff has led to an *absolute* rise in the real wage, and not just to a relative rise.

Why must the labour-land ratio fall in each industry even though it is fixed for the economy as a whole? The point is simply that the economy-wide ratio is a weighted average of the ratios in each of the two industries. When output of the labour-intensive industry expands and of the other industry declines as a result of factors moving into the former out of the latter industry, the weights alter. It is then necessary for the two industry ratios to adjust to keep the economy-wide ratio constant. The change in the ratios is brought about by the changing factor price ratio, and a change in this will affect the ratios in both industries in the same direction: so, if one ratio falls, the other must also.[3]

If N_x = labour in the exportable industry, L_x = land in the exportable industry, N_m and L_m are the factors in the importable industry, while N and L are the total factor stocks, so that $N_x + N_m = N$ and $L_x + L_m = L$, we have

$$\frac{\dfrac{N_x}{L_x} L_x + \dfrac{N_m}{L_m} L_m}{L_x + L_m} = \frac{N}{L}$$

and in the particular Stolper-Samuelson case,

$$\frac{N_m}{L_m} > \frac{N_x}{L_x}$$

[3] Stolper and Samuelson have expounded the argument in this paragraph with the aid of the Edgeworth-Bowley box diagram, which many students of this subject have found very helpful. As their article has been twice reprinted and there is nothing to add here to their geometric exposition, the diagram is not reproduced here.

Protection raises L_m and lowers L_x. If the two industry labour-land ratios did not change, N/L would then rise. But N/L has to stay constant, so one or both of the industry ratios must fall, and since they both depend on the same factor-price ratio they must fall together.

We note then that in this model the real return to a large and mobile factor is increased by protection, and that this qualitative result does not depend on the spending pattern of wage-earners. But their spending pattern still influences by *how much* the true real wage rises. The greater the share of exportables in their spending pattern, the greater the rise in the true wage will be.

(3) *The Magnification Effect*

The rise in the real wage will be proportionately greater than the rise in the domestic price of importables induced by the tariff, except in a limiting case. Hence the effect of the product price change is *magnified* in its incidence on the relevant factor-price.[4]

Consider first the limiting case where labour is used only to produce importables and land to produce exportables. This is the case where the factor-intensity difference is at a maximum. We hold the price of the exportable in terms of a numeraire constant while the domestic price of the importable goes up, owing to a tariff, say by 10 per cent. The wage rate will then go up by 10 per cent since wage income is identical with income from importables. Hence there is *no* magnification effect. Rent income will stay unchanged.

Next, allow both factors to be used in both industries, as in the Stolper-Samuelson model. We know that rent must go down in terms of both goods, so that, with the price of exportables constant, the rate of rent must go down absolutely. As for the wage, it must go up by more than 10 per cent. This can be explained as follows: if the price of importables rises by 10 per cent, and the reward of one of the factors used in producing importables, namely land, actually goes down, then the reward of the other factor must go up by more than

---

[4] The concept 'magnification effect' comes from Ronald W. Jones, 'The Structure of Simple General Equilibrium Models', *Journal of Political Economy*, 73, Dec. 1965, pp. 561–2. Meade, in *Trade and Welfare: Mathematical Supplement*, pp. 106–108, pointed out the relevance of differences in factor-intensities.

10 per cent for the cost of production of the importable to rise, on balance, by 10 per cent.

In the limiting case above, the factor-intensity difference was at a maximum and the magnification effect was zero. The less the factor-intensities differ the greater the magnification effect will be. If the factor-intensities differ very little, a given change in the factor-price ratio will change the relative costs of the two goods, and hence the product price ratio, only by a small amount. Putting this the other way around, a given change in the product-price ratio will be associated with a large change in the factor-price ratio.

The term 'magnification effect' refers to the effect of a given change in commodity prices on factor prices. But if income distribution is a policy target one may prefer to reverse the question, and ask what tariff rate is required to achieve a given increase in the real wage. The tariff will need to be less, the less the difference in factor-intensities.

(4) *The Distribution of Tariff Revenue*

We have taken no account so far of the revenue that is raised by the tariff, and what is done with it. It would be simplest to assume that up to this point in our discussion the tariff is prohibitive, so that no revenue is raised.

If revenue is raised and redistributed to the factors as income supplements, whether directly, or indirectly through benefits obtained from public spending, we must distinguish factor *rewards* from factor *incomes*. The analysis so far has referred to factor *rewards*. But factor *incomes* depend additionally on the tariff revenue. The way the revenue is shared out between the two factors will help to determine what finally happens to their incomes.[5] If the revenue is all distributed to labour, the real income of wage-earners will go up even more than when only the rise in the real wage was taken into account; if the revenue is distributed wholly to landowners, their real incomes will not fall by as much as their factor rewards.

The main conclusion of the Stolper-Samuelson analysis is that the real income of wage-earners must go up as a result of a tariff when importables are labour-intensive. This must

[5] See *Trade and Welfare: Mathematical Supplement*, Ch. XVIII.

remain true when the distribution of the tariff revenue is taken into account. But is it now possible that the real incomes of land-owners also go up? This is not possible. With given terms of trade, in this simple model the tariff causes real income for the country as a whole to fall. Hence, if the real income of wage-earners rises, that of the other factors must, on balance, fall.

(5) *The Assumptions Reviewed*

How crucial are the assumptions of the model? One might accept perfect competition, or at least some fairly consistent relationship between factor prices and marginal products. Constant returns to scale may also be an acceptable first approximation.[6] Here let us review the four assumptions of (a) non-specialization, (b) constant terms of trade, (c) two commodities only, and (d) two factors only.

(a) *Specialization.* Specialization in one of the products breaks the crucial link between the domestic product price ratio and the factor-price ratio. It is this link which is central to the Stolper-Samuelson mechanism.

Suppose that in free trade production is specialized in the exportable. Furthermore, the country is not just on the margin of starting to produce the importable if its price rose a little. Rather, a 20 per cent tariff would be required to initiate any production. Up to this level a tariff has no production effect. All factors will stay in exportable production, and each factor's income in terms of the exportable is fixed.

The tariff raises the domestic price of the importable and so lowers both the real wage and the real rent. Allowing for tariff revenue distribution, the real income of one of the factors might increase, though the real income of the other factor must fall since (because of the consumption distortion created by the tariff) real income for the country as a whole will have fallen. The more the spending pattern of a factor is biased

---

[6] There is a literature exploring the effects of increasing returns to scale on the model. See M. C. Kemp, *The Pure Theory of International Trade and Investment*, Prentice-Hall, 1969, pp. 161–4; N. Minabe, 'The Stolper-Samuelson Theorem under Conditions of Variable Returns to Scale', *Oxford Economic Papers*, 18, July 1966, pp. 204–12; R. W. Jones, 'Variable Returns to Scale in General Equilibrium Theory', *International Economic Review*, 9, Oct. 1968, pp. 261–72; and R. Batra, 'Protection and Real Wages under Conditions of Variable Returns to Scale'', *Oxford Economic Papers*, 20, Nov. 1968, pp. 351–60

towards the importable, the more its real income is likely to fall. This is a result which was intuitively accepted before the Stolper-Samuelson model.

Now imagine the tariff to be raised above 20 per cent. The Stolper-Samuelson mechanism will come into play again. Hence, if the importable is labour-intensive, from then on the real wage rises and the real rent falls further. If one compares the results of a 30 per cent tariff with the free trade situation one may then, on balance, find that the real wage is either higher or lower.[7] The net result will depend, among other things, on the spending pattern of wage-earners.

(b) *Terms of trade effects.* Let us now abandon the small country assumption and allow the terms of trade to improve as a result of a tariff. Two distinct complications then arise.

The first complication is the possibility of the *Metzler paradox*.[8] The tariff lowers the duty-free price of the importable and raises the domestic price of the importable above its duty-free price. On balance, in certain rather special conditions, the domestic price of the importable might actually fall, so that the tariff, rather surprisingly, protects the exportable and not the importable. If importables are labour-intensive the real wage will then fall, not rise.

This, rather improbable, possibility does not really affect the main Stolper-Samuelson conclusion. It is that 'protection (prohibitive or otherwise) will raise, reduce or leave unchanged the real wage of the factor intensively employed in the production of a good according as protection raises, lowers or leaves unchanged the internal relative price of that good'.[9] There is a unique relationship between the domestic commodity price change and the factor price change, the sign of the relationship depending on the factor-intensity condition.

[7] See John Black, 'Foreign Trade and Real Wages', *Economic Journal*, 79, March 1969, pp. 184–5, for a geometric exposition of this argument. (He compares no-trade with free trade, so the tariff-revenue distribution question does not arise.)

[8] L. A. Metzler, 'Tariffs, the Terms of Trade and the Distribution of National Income', *Journal of Political Economy*, 57, Feb. 1949, reprinted in R. E. Caves and H. G. Johnson (eds.), *Readings in International Economics*, Richard Irwin, Homewood, 1968. For an exposition, see *Trade and Welfare*, pp. 292–5, and also *The Theory of Protection*, pp. 250–52.

[9] Jagdish Bhagwati, 'Protection, Real Wages and Real Incomes', *Economic Journal*, 69, Dec. 1959, p. 743.

The second complication is that a tariff may now raise the real *incomes* of both factors even though the *reward* of one of the factors must decline.[10] When there are terms of trade effects total potential real income of the country may rise, not fall.[11] Suppose that it does rise, and that, as usual, importables are labour-intensive. Consider the normal non-Metzler case where importables are protected and hence the real wage rises and the real rent falls. When tariff revenue is distributed it becomes possible that the income of landlords has gone up, even though rents have declined. The higher the proportion of the revenue that goes to landlords, the more likely this is. If the Metzler paradox applies at the same time the real wage will have gone down but, allowing for the tariff revenue distribution, it becomes then possible for the incomes of wage earners to rise as a result of the tariff in spite of the adverse effect on wages.

Again, the main Stolper-Samuelson conclusion is not really affected. The tariff must raise the real income of that factor which is intensive in the production of that product the domestic price of which has risen. Only a subsidiary Stolper-Samuelson conclusion, not really crucial to the main argument, has to be modified: it is no longer certain that the real income of the other factor will be reduced.

(c) *Many commodities.* If there are many commodities but only two homogeneous mobile factors of production, the Stolper-Samuelson theorem can be readily generalized. This was correctly pointed out by Stolper and Samuelson. An industry would be defined as labour-intensive if its labour-land ratio were higher than the country's overall labour-land endowment ratio, and the real wage would be raised if a tariff were imposed on such an industry. If a tariff structure is imposed there is again no difficulty if the protected importables are all labour-intensive, so defined. But if some are labour-intensive and some land-intensive, the net effect will

---

[10] *Trade and Welfare*, p. 297 (footnote). See also Bhagwati, loc. cit.; Harry Johnson, *Aspects of the Theory of Tariffs*, Allen and Unwin, London, 1971, pp. 36–39; and V. S. Rao, 'Tariffs and Welfare of Factor Owners: A Normative Extension of the Stolper-Samuelson Theorem', *Journal of International Economics*, 1, Nov. 1971, pp. 401–15.

[11] See Chapter 7.

depend on whether, on balance, the demand for labour is raised relative to the demand for land.

It follows that one can relate the Stolper-Samuelson approach to the multi-commodity theory of protection. Indeed, one can think in terms of *activities* rather than commodities and *effective* tariff rates rather than nominal rates. The question is whether particular activities are labour-intensive or not, and whether they expand or contract as the result of the protective structure.

Once one allows for many commodities it is almost inevitable that there will be some specialization: at any given time not all commodities will be produced in the country. Strictly, the Stolper-Samuelson argument requires that at all times all commodities are being produced. There may be no production of some under free trade and production may start when tariffs are imposed. Production of others may cease as a result of protection because activities that compete for resources with them receive high protection. Obviously one can disregard those commodities that are neither produced nor consumed in the country either in free trade or in the protection situation. Difficulties can probably also be avoided by assuming that in those cases where there is no domestic production under free trade no tariffs are imposed; hence they are also unlikely to be produced under protection.

(d) *Many factors*. The theoretical difficulties really begin once one allows for more than two factors, as Stolper and Samuelson were well aware. If the number of factors exceeds the number of commodities it is difficult to obtain clear-cut results because one cannot determine factor rewards from the marginal conditions alone. Meade has provided the fullest exposition of such a case, namely one with just two commodities and three factors. He shows that it might not be possible by a manipulation of the domestic commodity price ratio to obtain a desired income distribution between the three factors. Generalizing from this case, he concludes that

the greater is the number of traded commodities relatively to the number of factors of production used in their production and the more diverse are the ratios in which the various factors are used in the production of the various commodities, the more probable it is that any desired redistribution of income could be obtained among the various factors by a combination of trade policies which produced an appropriate shift in the demand for the various products.[12]

The essential problem is that there is no longer a unique association through an intensity condition between a particular commodity and a particular factor; some commodities may, in a sense, have to do double duty as being 'intensive' in several factors.[13]

(6) *Conclusions*

What significant conclusions about the real world are indicated by the Stolper-Samuelson model?

In a multi-commodity model the output changes induced by a protective structure will inevitably be complex, the outputs of some industries with positive effective rates possibly falling even in the absence of the Metzler paradox. But the general argument might be broadly accepted that from the signs of the output changes we should be able to obtain some indication of the signs of the factor price changes given that we know, in a rough way, which factors are specific to or intensive in which activities.

Furthermore, it can be concluded that it is at least possible, and quite plausible in some circumstances, that a large mobile factor gains absolutely from protection. What is obvious for a minor factor or a specific factor may also then apply to a large mobile factor, though, in the absence of some of the crucial Stolper-Samuelson assumptions, the result is no longer certain. But it cannot be concluded that spending patterns do

[12] *Trade and Welfare*, p. 310.

[13] Mathematical economists have analysed the special case where the number of factors is equal to the number of commodities, but is greater than two. Broadly, the Stolper-Samuelson theorem can be extended to this case. Uekawa has established conditions for a weak form of the theorem: 'There exists an association of goods and factors such that a rise in the price of a good will bring about a more than proportionate increase in the price of the corresponding factor; while some other factor prices may increase too, their increase rate must not be larger than that of the intensive factor'. (Yusuo Uekawa, 'Generalization of the Stolper-Samuelson Theorem'. *Econometrica*, 39, March 1971, pp. 197–217.) See also M. C. Kemp and L. L. Wegge, 'Generalizations of the Stolper-Samuelson and Samuelson-Rybczynski Theorems', in M. C. Kemp, *The Pure Theory of International Trade and Investment*, Prentice-Hall, 1969; and John S. Chipman, 'Factor Price Equalization and the Stolper-Samuelson Theorem', *International Economic Review*, 10, Oct. 1969, pp. 399–406.

not matter. Even in the Stolper-Samuelson model the *extent* of the gain depends on the spending pattern, and once the assumptions are removed, notably the assumption of non-specialization, the spending pattern may influence not only the size but also the sign of the effect.

III
INCOME DISTRIBUTION CRITERIA

When a policy of protection alters the distribution of income, how should one assess the welfare implications? There are, broadly, three kinds of approach: the *Pareto-optimal* approach, the *agnostic* approach, and the *social welfare function* approach.

(1) *Three Approaches*

The *Pareto-optimal* approach says that a policy is desirable if it increases *potential* real income—if it *could* make at least one person better off without anyone being worse off.[14] Since 'could' is not 'will' this really ignores income distribution. To emphasize that it may not be 'optimal' to pursue a policy with undesirable income distribution effects it is better described as the pursuit of *Pareto-efficiency*.

There is clearly no virtue in the pursuit of Pareto-efficiency on its own. To make sense of this approach one must assume that there is an independent income distribution policy through the fiscal system which brings about whatever income distribution is desired. In using this approach the central issue then is whether that assumption is reasonable. It is arguable that for some countries, at least, it may not be wholly unreasonable. They have progressive income taxes which syphon off large parts of income gains and modify income losses, they have social security systems, and so on. The devices for income redistribution are available, to some extent are used, and perhaps are not very costly (in terms of distortion and collection costs).

It is common for international trade theorists to focus on

[14] An increase in potential real income must strictly be defined in the Samuelson sense: for *every possible initial distribution of income* (and not just the one actually existing initially) it must be possible to make at least one person better off without making anyone worse off. This allows for the possibility that the overall demand pattern will affect the conditions determining whether there is an improvement. See P. A. Samuelson, 'Evaluation of Real National Income', *Oxford Economic Papers*, January 1960, pp. 1–29.

Pareto-efficiency, and so implicitly to make the assumption. This will also be done in some parts of this book. The assumption on these occasions is that a separate policy will turn *potential* gains into *actual* ones or, alternatively, that recommendations must subsequently be qualified to take into account income distribution effects.[15]

The *agnostic* approach comes in two versions, the *destructive* and the *constructive* version. The *destructive* version says that if a policy makes one person better off and another worse off, an economist can say nothing about whether or not it is a good policy. Since most economic policies make at least one person worse off, strictly nothing can be said about aggregative welfare effects.[16]

The *constructive* version is the explicit use of welfare or 'distributional' weights, as in Meade's *Trade and Welfare*,[17] hence combining considerations of income distribution and of efficiency explicitly. For example, a policy that gives $100 to Smith while taking $40 from Jones raises the weighted combined income of the two, and hence is desirable, if the marginal welfare weight attached to an income by Smith is more than 40 per cent of that attached to a decrease in income by Jones. This is still an agnostic approach, since it says nothing about what is the correct weighting system. It is a technique for analyzing problems, not for prefabricating answers.

In practice weighting must usually take place in money terms. But a convincing argument can be made that distributive judgments are generally made in terms of the *means* to welfare (i.e. money) and not welfare itself, and that welfare weights should be attached to 'changes in money incomes (with some eye to prices) of different groups'.[18]

---

[15] This approach is used by some cost-benefit analysts. See further discussion on pp. 409–10 below. In the author's view one should not, in general, ignore income distribution effects.

[16] The *locus classicus* of the destructive approach is J. de V. Graaff, *Theoretical Welfare Economics*, Cambridge University Press, 1957, though the main argument that interpersonal comparisons of utility are not permissible to economists *qua* economists goes back to Lionel Robbins, *An Essay on the Nature and Significance of Economic Science*, Macmillan, London, 1942.

[17] See especially *Trade and Welfare: Mathematical Supplement*, Ch. 1.

[18] S. K. Nath, *A Reappraisal of Welfare Economics*, Routledge and Kegan Paul, London, 1969, pp. 142–145.

The *social welfare function* approach means that one takes a consistent welfare weighting system and works out its implications for policy.[19] An economist could base it on his own preferences. He could choose the social welfare function to which governments or sections of the community, or perhaps just some individuals, appear to subscribe. He could assess policies in terms of various alternative functions. One could also view such a social welfare function as essentially an instrument of positive economic analysis, helping one to understand the rationale of actual policies being followed.

There are a number of plausible social welfare functions of this latter kind. One is the *egalitarian social welfare function*.[20] It rests on the idea that the marginal utility of money falls as income rises and that utility curves are similar for different individuals. Marginal welfare weights are then higher for the poor than the rich. This proposition is the basis of the progressive tax principle. It may help to explain why tariffs on luxury goods (consumed by the rich) are often higher than tariffs on necessities, though it cannot be said in general that protection policies have egalitarian consequences.

Sometimes policy can be explained in terms of a *group*

[19] The concept of the social welfare function originated with Bergson, who defined it in an extremely general way. See A. Bergson, 'A Reformulation of Certain Aspects of Welfare Economics', *Quarterly Journal of Economics*, 52, Feb. 1938, pp. 310–34. Scitovsky describes it as 'a kind of collective utility function, which expresses everybody's preferences relating not only to his personal satisfaction but also to the state of the entire community and to the distribution of welfare among the members of the community' ('The State of Welfare Economics', in Tibor Scitovsky, *Papers on Welfare and Growth*, Stanford University Press, 1964, p. 184.) He points out that individual preferences must be weighted somehow to yield a social preference function. Here we are concerned only with what the function has to say about income distribution and it need not necessarily describe 'everybody's' (weighted) preferences.

Paul Samuelson has been the great advocate of the social welfare function concept, and its widespread use is probably owed to him. He has written 'At some point welfare economics must introduce ethical welfare functions from outside of economics. Which set of ends is relevant is decidedly *not* a scientific question of economics. Any prescribed set of ends is grist for the economist's unpretentious deductive mill ... The social welfare function is as broad and empty as language itself and as necessary'. (*The Collected Scientific Papers of Paul A. Samuelson. Vol.* II, M.I.T. Press, Cambridge, p. 1103).

See also Nath, *A Reappraisal of Welfare Economics*; J. Rothenberg, *The Measurement of Social Welfare*, Prentice-Hall, 1961; and A. K. Sen, *Collective Choice and Social Welfare*, Holden-Day, San Francisco, 1970, Ch. 3.

[20] A. C. Pigou, *The Economics of Welfare*, Macmillan, London, 4th ed. 1932.

social welfare function. The aim of policy may be to maximize the real income of a group within a nation; in an extreme form equal positive weights are attached to all incomes within the group and zero weights to all incomes outside. The approach of maximizing the real income of the nation as a whole is a special case of such a group function since the nation is a subset of the world. One could imagine different groups, all with different group social welfare functions, alternating in control of government and pursuing group interests, but unable to reverse completely or instantaneously the policies of predecessors. At any point in time the laws and institutional arrangements—including for example the tariff schedule— may encrust consequences of the group social welfare functions of many past periods.

(2) *Conservative Social Welfare Function*

Let us now introduce the *conservative social welfare function*, a concept which seems particularly helpful for understanding actual trade policies of many countries. Put in its simplest form it includes the following income distribution target: any significant absolute reductions in real incomes of any significant section of the community should be avoided. This is not quite the same as setting up the existing income distribution as the best, but comes close to it, and so can indeed be described as 'conservative'. In terms of welfare weights, increases in income are given relatively low weights and decreases very high weights.

The conservative social welfare function helps to explain the *income maintenance* motivation of so many tariffs in the past and the reluctance to reduce income maintenance tariffs even when it has become clear that the need for them is more than temporary. It can be regarded as expressing a number of ideas.

Firstly, it is 'unfair' to allow anyone's real income to be reduced significantly—and especially if this is the result of deliberate policy decisions—unless there are very good reasons for this and it is more or less unavoidable.

Secondly, insofar as people are risk averters, everyone's real income is increased when it is known that a government will generally intervene to prevent sudden or large and unexpected income losses. The conservative social welfare function is part of a social insurance system.

Thirdly, social peace requires that no significant group's income shall fall if that of others is rising. Social peace might be regarded as a social good in itself or as a basis for political stability and hence perhaps economic development. And even if social peace does not depend on the maintenance of the incomes of the major classes in the community, the survival of a government may.

Finally, if a policy is directed at a certain target, such as protection of an industry or improving the balance of payments, most governments want to minimize the adverse by-product effects on sectional incomes so as not to be involved in political battles incidental to their main purpose.

The conservative social welfare function is an essentially dynamic concept. One might have all the data at one point in time to tell one the income distribution consequences of free trade and of various protective structures, and the details of a social welfare function might be specified. Yet this could not define the optimum structure since one would need to know also what the real incomes of various sections of the population were at earlier times.

Furthermore, it may call only for temporary protection, since the aim may be only to avoid sudden income losses and hence to slow up the consequences of change rather than to avoid them altogether. Hence the marginal welfare weight attached to a (potential) losing income group will leap up when the change, such as a fall in import prices, takes place, and will then gradually fall until eventually it has returned to its original level.

The conservative social welfare function also explains why it is possible to bring about tariff reductions in a context of general growth in per capita incomes (as in the European Economic Community) when it would not be possible in a more stagnant economy. While a tariff reduction may reduce the incomes of some sections below what they would be otherwise, these incomes may nevertheless not fall absolutely relative to a previous period—especially if tariff reduction is gradual.

The conservative social welfare function bears a superficial resemblance to some of the conclusions of modern *agnostic* welfare economics in its *destructive* version. Both are agreed on two propositions: First, a policy is desirable if it makes no one worse off and at least someone better off. Secondly, if the policy is accompanied by compensation, so that no-one is finally made worse off, the package deal of policy-plus-compensation is desirable.

The two approaches differ in the following respects. Firstly, the 'agnostics' would say that if a policy on its own would make some people better off and others worse off and no compensation took place nothing could be said about the desirability of the policy; by contrast, if the conservative social welfare function is accepted, the policy is undesirable if it makes a significant number of people significantly worse off. Secondly, the 'agnostics' have nothing to say about the desirability of compensation; it would be desirable only if certainty about the direction of welfare effects were an end in itself. But the conservative social welfare function actually makes compensation desirable.

IV
SENESCENT INDUSTRY PROTECTION AND ADJUSTMENT ASSISTANCE

The conservative social welfare function can be used to interpret two common types of government policy, namely the temporary protection of 'senescent industries', or possibly of industries that are temporarily in trouble rather than facing a permanent decline in demand, and adjustment assistance for such industries.

(1) *A First-best Argument for Tariffs?*

It might be argued that protection to halt change in order to prevent its income distribution effects is certainly not first-best. In line with our earlier analysis, first-best policy would be to permit change and its consequences and then use income taxes and subsidies to deal with income distribution. Yet one can base a first-best argument for tariffs on the conservative social welfare function. Essentially it depends on the costs or difficulties of obtaining information.

Suppose import prices of particular products fall owing to foreign suppliers becoming more competitive for one reason or another. This will redistribute incomes against producers of import-competing goods and in favour of consumers or using industries. It may be difficult or even impossible to bring about a redistribution back to the original situation through taxes and subsidies. Quite apart from the institutional difficulties, and collection and disbursement costs, there is the crucial information problem: precisely *who* gained and lost, and by how much? This is particularly serious if the effects are sudden.

There is only one way of reversing or avoiding the income distribution effects *precisely*, and that is to impose a tariff which will keep the prices facing domestic consumers and producers exactly where they were before import prices fell.

Many countries have at times followed such a policy, or at least aimed at it, though not always with complete success. And it is a policy that protectionists are always advocating. It might be called *senescent industry protection*. Pareto-efficiency is sacrificed to the income distribution objective of the conservative social welfare function. First-best policy may not necessarily call for a completely offsetting tariff, since the income distribution gain of the tariff might be traded off against the efficiency losses.

There are two reasons why the argument may be only for temporary tariffs when import prices fall. Firstly, the conservative social welfare function might be interpreted only as being concerned to avoid sudden changes in income. Secondly, information-gathering as well as collection of special taxes and disbursement of special subsidies, take time, so that there is an inevitable time-lag in bringing direct methods of compensation into operation. The desire to avoid sudden income change may also make a case for dismantling tariffs gradually rather than suddenly, in order to reach a preferred position.

(2) *Adjustment Assistance*

Some countries provide adjustment assistance to industries that are adversely affected by changes of various kinds. Normally this is not tied specifically to the effects of import competition, and often takes the form of regional policies. It includes subsidized retraining programmes.

The European Economic Community has its Social Fund
for this purpose and individual E.E.C. countries have active
regional policies which will eventually be taken over by a
single Community regional policy acting through a Regional
Fund. The United States and Canada are the only countries
at present to have schemes which specifically tie adjustment
assistance to import competition, though this has sometimes
been urged for other countries. The U.S. Trade Expansion Act
of 1962 required firms or groups of workers seeking such
assistance to show that tariff cuts were the main reason for
injury, which was interpreted after 1969 to mean a substantial cause of injury. Up till the end of 1969 not one petition
was granted under this Act, but since then some assistance
has actually been provided.

Britain has provided assistance to her cotton textile industry,
subsidizing the scrapping of machinery, and so making it
possible to allow a large increase in imports of cotton textile
goods from Hong Kong and India. Under the U.K. Cotton
Industry Act of 1959 re-equipment was also subsidized, but
this must be regarded as straightforward protection rather
than just adjustment assistance. In addition, Britain has
government-sponsored retraining schemes.[21]

The idea of adjustment assistance is to encourage rather
than to slow up change, and at the same time to provide some
compensation for the industries or factors of production
concerned. It is a preferable method than protection since it
protects or at least supplements real incomes without em-
balming or protecting patterns of production. Actually it
contains two elements.

Firstly, it is a way of providing income compensation to
factors adversely affected by changes. Hence it expresses the
idea of the conservative social welfare function. Normally this

[21] There is a large literature (rapidly expanding at the time of writing) on
adjustment assistance. See Stanley D. Metzger, 'Injury and Market Disruption
from Imports', and Marvin M. Fooks, 'Trade Adjustment Assistance', both
in Vol. I of the *Williams Committee Report* (President's Commission on
International Trade and Investment Policy) Washington, 1971; T. W. Murray
and M. R. Egmond, 'Full Employment, Trade Expansion, and Adjustment
Assistance', *Southern Economic Journal*, 36, April 1970, pp. 404–4240; Caroline
Miles, *Lancashire Textiles: A Case Study of Industrial Change*, Cambridge
University Press, 1968; Little et al., *Industry and Trade in Some Developing
Countries*, Chapter. 8.

assistance would be temporary, since the main impact is felt over a short period, unless indeed prospective changes had been gradually anticipated. In principle, any amount of temporary adjustment assistance will have a present value (obtained by applying an appropriate discount rate to the assistance to be provided in the different years) and there will be some level of permanent assistance at a constant annual rate that will have the same present value. Hence, from the point of view of income subsidization, one can always compare a scheme of temporary assistance with another scheme, say through tariffs or subsidies, which is intended to provide permanent assistance.

Secondly, adjustment assistance is a way of fostering the transfer of resources by subsidising costs of transfer of those factors of production, usually labour, that actually move. This aspect is concerned with Pareto-efficiency. If people and industries have a natural tendency to under-adjust in response to change, perhaps owing to ignorance or private risk aversion exceeding social risk aversion, or if the private costs of adjustment exceed the social ones, subsidizing adjustment is preferable to providing income compensation. In the absence of such divergences adjustment assistance might lead to over-adjustment. Insofar as the alternative to adjustment assistance is straightforward protection, rather than non-intervention, one can say that adjustment assistance fosters the transfer of resources even when it consists purely of income compensation not tied directly to the transfer of resources.

V

INTERREGIONAL INCOME DISTRIBUTION

Protection will often affect interregional income distribution. Historically this has been very important. For example, in the United States protection of manufacturing in the early nineteenth century probably benefited the North at the expense of the cotton-exporting South.

If the North region contains most of the import-competing manufacturing industry and the South most of the agricultural exporters, tariffs and import restrictions will, in effect, tax the latter for the benefit of the former. Of course the gainers will also include any Southern manufacturers and the losers any Northern farmers, since the essential redistribution is between factors of production, the regional effect being a by-product of the different factor-intensities of the regions.

One can analyse this case in terms of customs union theory. If there were free trade the South region would export agricultural products and import manufactures at world prices. As part of a customs union which has tariffs on manufactures the region's terms of trade will deteriorate since it will have to buy at least some of its manufactured imports from the North at the higher, protected, prices. To the extent that it shifts purchases from outside to the North, there is *trade diversion*, and to the extent that it pays higher prices for goods previously bought from the North even under free trade, there is an income redistribution from South to North.

The North gains from the second part of this terms of trade effect, since it will still be able to buy its agricultural imports at world prices, whether from the South or from outside. It will gain from the trade diversion effect only if more external economies or non-economic benefits attach to manufacturing than to agricultural production. In addition there will be a trade contraction effect in the North, domestic production of manufacturing replacing imports from outside. This will inflict a loss on the North if there are no domestic divergences and foreign prices remain unaffected, but may yield a gain if, for example, external economies or non-economic benefits attach to manufacturing.

It might be argued that only income distribution effects on *persons* are significant. But in fact the regional implications of protection may be important.

(1) The region may be the unit the income of which should be maximized in the minds of its citizens, and it may have the institutions to give expression to these ideas.

(2) Each region may have a preference in its spending pattern for goods and services produced within the region rather than for goods produced in other regions. Hence the initial income redistribution effect between the regions may have secondary effects shifting the overall domestic demand pattern for the country as a whole towards non-traded goods

5

and services produced within the region that has gained the
real income. Thus the income redistribution effect filters right
through the region to encompass sectors that have nothing
directly to do with traded goods.

(3) The region may be the unit within which any compensation through the fiscal and public expenditure system takes
place. If the incomes of farmers within a region fall while
manufacturers in the same region gain it will be possible to
maintain regional tax collections and so continue to provide
regional public services, perhaps even at an improved standard
if compensated regional income has risen. But if the beneficiaries are manufacturers in the other region, and there is no
interregional compensation, tax collections will fall and there
will be no element of automatic compensation. This effect will
operate the more public services are financed regionally rather
than nationally, and in the absence of adequate systems of
regional fiscal adjustment.

(4) The changed pattern of factor prices may cause labour
and capital to move from South to North. The question will
then arise whether one judges the differential income effect in
terms of aggregate income of the region (regional mercantilism),
income per head of the original population (including those
who have chosen to migrate), income per head of the remaining population, or perhaps just population size. If one
accepts an individualistic social welfare function one would be
concerned with income per head of some appropriate group
irrespective of where its members finally choose to live.
Labour mobility would then modify any adverse effects of
policies such as protection. But if one places some value on
cultural homogeneity or continuity, or upon 'place prosperity'
as such as distinct from 'people prosperity' then anything which causes people to move out of a region may need
to be scored negatively quite apart from any income
effects.

VI
TARIFFS AND THE BALANCE OF PAYMENTS

The income distribution effects of tariffs and of import
restrictions are closely related in two ways to their role in
improving the balance of payments.

(1) *Tariffs on Luxury Imports to Improve the Balance of Payments*

A country with a balance of payments deficit and with its output at full capacity or full employment needs to reduce the sum of consumption and investment spending (*absorption*) if it is to eliminate the deficit. This must be accompanied by measures to switch the pattern of domestic and foreign spending away from foreign goods on to domestic goods, sufficient to make up for the fall in demand for domestic goods brought about by the decline in spending. Normally one would argue that the first-best switching device is exchange rate adjustment since it is neutral between import-replacement and export promotion and makes maximum use of the price mechanism. The reduction in expenditure, insofar as it is not brought about by the devaluation itself, would be brought about by fiscal and monetary policies. Fiscal measures would, as part of a first-best package, take into account income distribution considerations.

In practice countries often take into account income distribution considerations when selecting their switching policies. Tariffs and quantitative restrictions, which can discriminate between different domestic consumers and producers, are preferred to the essentially neutral price device of exchange rate adjustment. This is the logic behind the common policy of imposing tariffs or import restrictions on so-called 'luxury' imports—goods purchased by the well-off—when there is a balance of payments deficit, while not applying such restrictions on other imports, usually called 'essentials'. The logic is that of the progressive tax principle: when real expenditure is to be reduced, partly through higher domestic prices of imports or even complete deprivation of some imports, the burden should be borne mainly by the well-off. It is, of course, not first-best, since excess protection will be provided for domestic production of luxury goods, a consumption distortion will be created, and taxing *goods* cannot attain income distribution objectives as efficiently as taxing *incomes*.

(2) *Wage Rigidity and the Balance of Payments*

We now come to another way in which balance of payments

policy and income distribution effects are related. Conceivably a system of tariffs might succeed in improving the balance of payments of a country where a devaluation would fail to do so. At least, that is the logic of an argument that is very common, especially in less-developed countries—and perhaps notably in Latin America—but that is rarely made explicit.

It is argued that if a country devalued, money wages would rise so as to avoid the decline in real wages that would otherwise be brought about by the rise in the internal prices of importables and exportables. This rise in money wages would offset the effects of the devaluation, and a wage-price-devaluation spiral would be set up. On the other hand, if discriminatory devices, such as carefully-chosen tariffs, were used the initial rise in the cost of living of wage-earners could be avoided, or could be offset by increased money wages made possible by the higher domestic prices that import-competing industries could charge.

This argument hinges on an alleged rigidity of real wages. Assuming full employment or full capacity initially, the elimination of the balance of payments deficit requires *absorption* to fall. This absorption might be divided into expenditure by wage-earners and expenditure by others. If the expenditure of both sections were fixed in money terms a devaluation would reduce both in real terms. But real wages are assumed to be rigid; with full employment this means that the real wage bill is rigid, and given wage-earners' propensity to save, real expenditure by wage-earners must then be rigid. Hence the reduction in absorption must be focused on the non-wage sector and some device, such as a tariff, must be introduced which redistributes income relatively towards labour if the desired real wage is to be made compatible both with full employment and balance of payments equilibrium. Tariffs which are imposed on luxury goods rather than on wage-goods, and on labour-intensive goods rather than on land- or capital-intensive goods, will have the necessary redistributive effect.

An alternative way of attaining the necessary balance of payments improvement while maintaining the desired real wage would be through sufficient deflation, which in turn would cause unemployment. Total expenditure by wage-earners would then fall. Thus the tariff might be regarded as a way of attaining balance of payments equilibrium while avoiding deflation-induced unemployment. In popular Latin American language, unemployment and stagnation would be the costs of forgoing tariffs or quotas, that is, the costs of free trade. One might then go on to compare the two solutions to the balance of payments problem, the first combining a tariff with full, or maximum, employment, and the second, free trade with unemployment. Looking at the effects on the economy as a whole, one has to set the distortion costs of the tariff against the loss of output and disutility resulting from unemployment.

The income distribution argument for protection seems then to have been transmuted into a balance of payments (or an employment) argument. In the income distribution argument the level of the real wage is a policy target; in the balance of payments argument it is a constraint. Furthermore, crucially it is implied that the incomes of at least some of the factors of production other than labour are flexible downwards in real terms. It is obviously necessary that tariffs *can* significantly affect income distribution. It must be possible to distinguish traded goods which are clearly wage-goods, or alternatively, there must be significant differences in factor-intensities between different tradeable goods.

There are two important weaknesses in the whole approach. Firstly, it is simply not true that real wages are generally so rigid that devaluations fail to reduce them. An investigation by Richard Cooper of twenty-four devaluations occurring over the period 1953–66 and including most of the major devaluations by less-developed countries in the early 1960's, supplemented by a study of some later devaluations, has shown that generally, within a year or two following devaluation, real wages *have* fallen.[22] Of course, normally devaluations lead to *some* increases in money wages and hence *some* inflation, and the real wage finally does not fall by as much as in the immediate aftermath of the devaluation.

[22] Richard N. Cooper in Chapter 13 of G. Ranis (ed.), *Government and Economic Development*, Yale University Press, New Haven, 1971; and also Cooper, *Currency Devaluation in Developing Countries*, Essay in International Finance No. 86, June 1971 (Princeton University International Finance Section), pp. 24–28.

The second weakness is that discriminatory trade restrictions are unlikely to be first-best devices to deal with the problem. First-best policy would be to increase income tax on non-wage factors, devalue so as to bring about the necessary switching, and use part of the extra income tax collections to compensate wage-earners for the rises in the prices of importable and exportable wage-goods resulting from the devaluation. If collection costs rule out income tax the revenue could be obtained instead by consumption taxes on non-wage goods. Alternatively, consumption of the principal wage-goods might be subsidized; this is indeed often part of a devaluation package.

Finally, there is nothing equitable in allowing some members of the community to maintain their real incomes— normally through trade union action—while others have to put up with decreases in their incomes. In many less-developed countries urban wage-earners have real incomes well above the national average. If they do not accept some of the burden of disabsorption required by a balance of payments improvement, the extra burden may be borne not just by rich traders, industrialists or landlords, but by much poorer peasants.[23]

[23] More could be said at a formal level about this wage rigidity argument.

Consider the simple Stolper-Samuelson model, with importables labour-intensive. Balance of trade equilibrium is always assumed. Suppose the free trade real wage is below the desired (constrained) one. The latter may be attained by a tariff, full employment being maintained. (But there is a limit: the maximum attainable real wage results when all imports have ceased, and this maximum real wage might still be below the desired one.)

Alternatively, free trade could be preserved and the desired real wage attained through reduction of employment. This is equivalent to a reduction of the labour supply, leading to a biased shrinking of the production-possibility curve. With given terms of trade, output of the exportable will rise and of the importable will fall until the country is specialized in the exportable. Up to this point the real wage and rental remain unchanged, being determined by the externally-given price ratio. Employment must fall further beyond this specialization point if the real wage is actually to rise to the desired level.

The wage-rigidity argument has really been anticipated by G. Haberler in 'Some Problems in the Pure Theory of International Trade', *Economic Journal*, 60, June 1950, pp. 223–40, who shows that in a two-sector real trade model the opening of an economy to trade may cause unemployment if there is real-wage rigidity. Of course, he does not put it in balance-of-payments terms. See also Johnson, 'Optimal Trade Intervention in the Presence of Domestic Distortions'.

6

EMPLOYMENT AND INDUSTRIALIZATION

PROTECTION of manufacturing industry in less-developed countries is usually justified on the grounds that it increases employment, that it moves labour out of low-productivity agriculture into higher-productivity modern manufacturing, and that there are various special advantages of manufacturing relative to agriculture. These may not always be the true reasons for protection: the income maintenance motive can explain some protectionist history in less-developed countries as well as in developed countries. But they are oft-used arguments, and some are rightly taken seriously by economists. All suggest that there is some kind of domestic divergence in the economy, so that tariffs would rarely be the appropriate first-best devices, other than to help finance any necessary subsidies. We discuss some of these reasons now. Others, notably the infant industry argument, will be dealt with in later chapters.

I
PROTECTION TO FOSTER EMPLOYMENT

Practical men in many countries and, no doubt, many centuries have pointed out that tariffs increase employment in the protected industries. Economists, also in many countries, and for at least two centuries, have pointed out that the extra employment in the protected industries may be at the expense of employment in other industries, and hence does not necessarily provide an argument for protection. The crude practical men's arguments can obviously be dismissed as naive. Nevertheless, it is possible to construct plausible 'employment-creation' arguments for protection.

The most familiar argument was at one time espoused by Keynes, and hinges on both exchange rate rigidity and downward rigidity of the general level of money wages.[1] General unemployment can be reduced or eliminated by an increase in aggregate domestic expenditures brought about by fiscal and monetary policies. But, on its own, this will worsen the balance of payments, and hence must be accompanied by some kind of policy that switches the pattern of domestic and foreign demand away from foreign on to domestic goods. The first-best 'switching' device is a devaluation of the exchange rate. Hence, in the presence of money-wage rigidity, exchange rate flexibility is implicit in the free trader's approach to this type of problem.

If the exchange rate cannot be altered, then tariffs or import restrictions can be used to avoid the adverse balance of payments consequences of the fiscal and monetary expansion. In the absence of tariffs or import restrictions, or some other 'switching' device (such as export subsidies), and given that an adverse balance of payments cannot be allowed to continue for any length of time, it would not be possible to eliminate the unemployment. Hence tariffs make possible full employment.

This argument has great similarities with the wage-rigidity argument advanced in the previous chapter. But it is not identical with it. In both cases a tariff can be described as making external and internal balance compatible and can be presented either as a balance of payments or an employment-creating device. Furthermore, in both cases tariffs impose distortions: a uniform effective tariff will distort the production pattern by favouring import-competing relative to export industries, and a non-uniform effective tariff structure will also distort the production pattern within the import-competing sector. In addition, there will be distortions on the consumption side depending on nominal tariffs. The costs of these distortions must be set against the losses of output and of utility imposed by unemployment. One could think of these costs as being 'traded-off' and conceive of an optimum tariff structure which imposes some distortions and allows some unemployment to remain.

[1] See R. F. Harrod, *Life of John Maynard Keynes*, London, Macmillan, 1952, pp. 424 ff., and for a general review of Keynes' views, and other references, S. J. Wells, *International Economics*, London, Allen and Unwin, 1969, pp. 118–120.

The difference between the two cases is that in the wage-rigidity case *real* wages were rigid while in the present case only *money* wages are rigid. In the real wage-rigidity case, a devaluation could not do the job that tariffs did, since an income redistribution was required to make full employment and balance of payments equilibrium compatible. In the present case a devaluation could do the job and would be the first-best device. Thus the distortion costs imposed by the tariff are really the costs of *not* using the exchange rate. The optimum tariff structure that results from trading-off these distortion costs against the costs of the remaining unemployment clearly represents a *second-best* optimum.

The sectional employment argument of the practical men may sometimes be very close to the real-wage rigidity argument; and if one wishes to be sympathetic to the practical men, one could interpret their argument in these terms.

Suppose that it is proposed to remove a tariff that protects an existing industry. The neo-classical economist will suggest that labour is likely to move out of that industry into other industries since he will be assuming labour mobility, appropriate Keynesian full employment policies, and exchange rate flexibility. But it is possible that the workers in the industry are quite immobile, at least in the short-run. We have then a limiting case of factor-intensities, the particular workers being specific to the industry concerned. If they allowed their money, and hence real wages to fall, there would still be no unemployment problem. The industry's costs would fall, so that it could survive without a tariff. Furthermore, to some extent labour might be substituted for capital in the industry. An employment problem only arises if wages do not fall, or do not fall sufficiently. There may be some small decrease in wages as various marginal extra payments, overtime payments, and so on, are reduced, but normally money wages and hence real wages would indeed not fall sufficiently and unemployment would result.

We have then a special case of the more general real wage-rigidity argument advanced in the previous chapter. It is the limiting case in which that factor of production that needs to accept a reduced real wage if full employment and non-intervention are to be maintained, is specific to an industry.

The argument is relevant mainly in the short-run, or when a protected industry is the main employer in a region and there is considerable interregional labour immobility. Given the real wage constraint, a direct subsidy on labour employed in the industry would be first-best. A subsidy to the industry would be second-best since, unlike the labour subsidy, it would also encourage capital to stay in the industry although capital may be potentially mobile and some should leave the industry. A tariff would be third-best because it would impose, in addition, a consumption distortion. But there is an argument here for temporary subsidization or protection.

To return to the earlier argument for tariffs based on exchange rate rigidity, many examples can be produced for its relevance. Until recently a conventional attachment to exchange rate fixity has inhibited many countries from altering their exchange rates appropriately. This is not the place to go into the reasons for this: there have been elements of pure ignorance, of fear of the unknown, of a concern for the effects on capital movements, and of an exchange rate fetishism, the latter powerful, for example, in pre-1967 Britain. All this has changed in recent years. If one goes more deeply into the reasons why some countries have preferred to use tariffs or import restrictions rather than to devalue, one is often likely to bring income distribution considerations to light. Similarly, the reluctance of Germany and Japan to appreciate may also be explained by a desire to protect incomes of exporting interests.

In 1964 the United Kingdom imposed general import surcharges for balance of payments reasons because of a reluctance to devalue. Until the devaluation of the rupee in 1966, India had never altered her exchange rate relative to sterling; and the actual devaluation, like Britain's a year later, was a most reluctant one. Until 1966 one could certainly justify some degree of trade intervention in India on balance of payments, and hence indirectly on employment, grounds. In the post-war period until 1967, Australian policy-makers would not contemplate any exchange rate alteration relative to sterling mainly because of fears that it would disturb or inhibit capital inflow, but possibly also for income distribution reasons. All discussions of tariff reforms in Australia had to bear this in mind, and it was argued—for example by the present writer—that a general case for tariffs could be based on balance of payments and employment grounds.

Many less-developed countries, and others such as France, have never been reluctant to alter their exchange rates, but recently attitudes to exchange rate flexibility (if not always to floating) have generally become more sympathetic. Hence the exchange rate rigidity argument for tariffs may be more relevant for explaining the use of tariffs in the past than for justifying it in the future.

II
DISGUISED UNEMPLOYMENT

We now come to an extremely influential argument for fostering labour-intensive industrialization in less-developed countries. It is generally associated with the names of Nurkse and Lewis, and is so well-known that one hardly needs to elaborate it in any detail.[2] Nevertheless, its full implications and limitations are not always appreciated.

(1) *The Surplus Labour Model*

One may visualize a dual economy consisting of an 'advanced' sector and a 'subsistence' sector. In the advanced sector profit maximization rules and labour is paid the value of its marginal product. If the wage rate facing the advanced sector falls, employment in this sector will increase. Thus there is nothing special about the advanced sector. On the other hand, the principle of *income-sharing* operates in the subsistence sector. Peasants in this sector receive their average products, not their marginal products. Since there will be diminishing returns, a peasant will thus receive more than his marginal product. An extended peasant family will share out the fruits of its labours among all its members even though the withdrawal of labour by particular members might have very

---

[2] R. Nurkse, *Problems of Capital Formation in Underdeveloped Countries*, Oxford, Blackwell, 1953; W. A. Lewis, 'Economic Development with Unlimited Supplies of Labour,' *Manchester School of Economic and Social Studies*, 22, May 1954, pp. 139–191. See also, Gustav Ranis and John Fei, *Development of the Labor Surplus Economy: Theory and Policy*, Homewood, Irwin, 1964. For a critical discussion, see Hla Myint, *The Economics of the Developing Countries*, New York, Praeger, 1965, pp. 86–90.

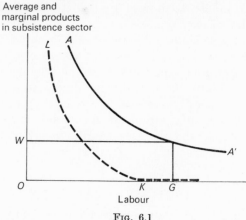

FIG. 6.1

little effect on the family's total product. In the limiting case—that of *surplus labour*—the marginal product is actually zero, so that a reduction in the labour-force would have no effect on total product.

The model is represented in Figure 6.1. The horizontal axis shows the quantity of labour in the subsistence sector, and the vertical axis average and marginal products. AA' is the average product curve and LKG the marginal product curve. At K the marginal product begins to be zero. Initially the

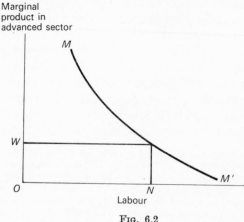

FIG. 6.2

quantity of labour in the subsistence sector is OG. Hence the marginal product is zero and the average product is OW.

Figure 6.2 shows the quantity of labour in the advanced sector on the horizontal axis, and the value of the marginal product in that sector, as well as the wage-rate, on the vertical axis. Units are so chosen that one unit of the advanced sector's product is equivalent to, or can buy—perhaps through foreign trade—one unit of product produced in the subsistence sector. To avoid unnecessary complications, the terms of trade are held constant here.[3]

According to this approach, the wage that the advanced sector has to pay is determined by the average product in the subsistence sector. In practice the two may not be equal since there may be a 'wage-differential.' Here we shall assume that they are equal; the 'wage-differential' will be introduced in the next section. Hence the wage in the advanced sector is OW, employment is ON, and total labour in the economy is OG (Figure 6.1) plus ON.

As drawn in Figure 6.1, some part of labour in the subsistence sector, namely KG, is *surplus labour*. If it were induced to move into the advanced sector, there would be an increase in output in the latter without any decline in output of the subsistence sector. This surplus labour represents 'disguised unemployment.' Some persons may be literally inactive, and hence unemployed, but they do obtain an income even in the absence of any publicly-provided unemployment benefits; thus they are not *openly* unemployed. The unemployment is 'disguised' principally because there is likely to be *under*-employment; everybody on the farm does some work, but is involuntarily unemployed some of the time. Surplus labour is thus made up of part of the potential working-time of the whole labour-force; and when some of the labour-force leaves

---

[3] These diagrams are not partial equilibrium. We assume that in the subsistence sector there is a specific fixed factor, say land, and in the advanced sector, another specific fixed factor, say capital, and that each of these sectors uses mobile labour. For the economy as a whole there will be a production possibility curve with a diminishing marginal rate of transformation showing how full employment output varies as labour is transferred from one sector to the other. The point of tangency of the terms of trade line (its slope indicating the externally-given price ratio) with the production possibility curve will be the point of maximum aggregate output, where the values of the marginal products in the two sectors are equal.

the subsistence sector, the remaining people each work sufficiently longer hours to keep the total product constant.

It needs to be stressed that even if there were no surplus labour, the allocation of labour between the sectors would not be optimal since the values of the marginal product of labour in the two sectors would not be equal. An optimal reallocation would require labour to be moved out of the subsistence sector until the marginal product in that sector were equal to the marginal product in the advanced sector, and hence would need to go beyond the point where disguised unemployment is eliminated.

The practical implication of this model for the theory of optimal trade policy is that the advanced sector has to pay a wage-rate which is higher than the opportunity cost of labour; when there is surplus labour, the latter will be zero, this being the limiting case. First-best policy is then to subsidize the use of labour in the advanced sector, possibly financing the subsidy by taxing the subsistence sector, if that is practical. There is a *marginal divergence* in labour cost facing the advanced sector, and this implies a hierarchy of policies as explained in Chapter 2, a tariff for the import-competing products of the advanced sector being third- or fourth-best.

(2) *Qualifications to the Surplus Labour Model*

This model contains some grains of truth and insight. Otherwise it would not have had so powerful an intellectual hold for several years, nor would it be worth discussing at length here. But it has some limitations, the simple exposition just provided needs to be qualified, and it is not as widely applicable as may once have been thought.

Firstly, empirical research in several countries seems to suggest that if labour moves out of peasant agriculture, the total product is likely to fall.[4] Many people may be unoccupied for large parts of the year, but at harvest time all hands are needed. Thus the marginal product is not zero, and strictly, there is no surplus labour. Conceivably one can imagine that,

[4] C. H. C. Kao, K. R. Anschell and C. K. Eicher, 'Disguised Unemployment in Agriculture: A Survey', in C. K. Eicher and C. W. Witt, *Agriculture in Economic Development*, New York, McGraw-Hill, 1964; David Turnham and Ingelies Jaeger, *The Employment Problem in Less Developed Countries*, O.E.C.D. Development Centre, Paris, 1971, pp. 64–68.

when labour is withdrawn, there might be significant re-organization representing a productivity or production function improvement, so that on balance output stays constant, but one cannot assume that the withdrawal of labour would automatically bring this about.

The surplus labour assumption has been important in some growth models constructed on the foundations of this basic model, but it is not really crucial here. The main point is that labour receives its average product rather than its marginal product, and this follows from the *income-sharing* assumption. Whether the marginal product of labour is very low but positive or is actually zero does not really matter. The insight that the income-sharing assumption originally gave was to fit surplus labour into a coherent income-distribution model. Casual empiricism had suggested that there was surplus labour and so provided the stimulus for the income-sharing model. But while the surplus labour assumption requires the income-sharing assumption, the latter does not need the former, and, as it happens, the latter assumption is probably more firmly based in reality, as well as being the crucial assumption for our purposes.

Secondly, the simple version of the model assumes that peasants place zero marginal value on leisure, at least over the relevant ranges of their utility functions. This is implied in the assumption that when some of them leave the sector, the remainder will work appropriately longer hours so as to maintain the same total output and that the extra work does not represent a social cost. With this assumption the oppor-tunity cost of the labour which moves into the advanced sector is zero. In practice the benefits of relaxation, especially in an enervating climate, cannot be ignored.

One could revise the model by supposing that the peasant family trades off the fruits from work against the joys of relaxation. When some members of the family leave the farm, the productivity of extra work per head for the remaining members goes up; hence each member may work longer hours, substituting work for leisure by reason of the substitution effect. On the other hand, there is also an income effect, inducing him to take out some of his gains in extra leisure. On balance he may still work longer hours, but only with very special assumptions would each remaining member of the family work so much more as to keep the total hours worked for the family as a whole constant. In any case, the costs of any leisure forgone must be introduced into the analysis. All this could be analysed more formally, but the qualification does not affect the crucial income-sharing assumption: each peasant will still receive an income in excess of his marginal product.

Thirdly, the model as presented so far assumes that peasants own their own land and so obtain the whole product. In the case of a landless peasantry, the family income may be determined at some sort of subsistence level, landlords obtaining the surplus. But this amendment is not inconsistent with the family income-sharing assumption: each member of the family may still receive an income in excess of his marginal product.

Fourthly, if there is a significant number of hired agricultural labourers—as there is in many less-developed countries— then the model ceases to be appropriate. Such labourers are likely to be paid their marginal products. It is true that, at the same time, there may be extended peasant families working on an income-sharing and work-spreading basis, but as long as hired labourers are available in the subsistence sector who are free to move into the advanced sector, employers in the latter need only pay a wage equal to the marginal product in the subsistence sector. This is an important consideration that limits the wide applicability of the model. Indeed, it is arguable that hired agricultural labour is so prevalent in the less-developed world—in India some 25 per cent of the labour-force consists of hired workers—that the model is clearly inappropriate.

Finally, the model assumes that the extended family is not a consistent product- or utility-maximizing group. As long as a member of the family is on the farm, he is allowed to share in the work and the income. The family as a group may indeed organize its affairs so as to maximize its returns, given its knowledge of possible methods and taking into account the utility of leisure. But when a member of the family leaves the farm, he is also assumed to leave the maximizing group. He will no longer share in the family income or contribute to it, and hence will leave the farm only if the income he can

obtain by his own work in the advanced sector is at least equal
to his share of income in the subsistence sector. Yet there is
some evidence that this assumption may be further from reality
than an alternative assumption.

One could assume that the family as a constant group seeks
to maximize its real income, and that the possibility of
sending some of its members to work in the advanced sector is
one option it can choose in its maximizing process. It will then
seek to equate the value of the marginal product of its labour
in all activities, whether in the subsistence or the advanced
sector. Some sons will work in the mines, some in the factories,
and some will stay on the farm; and it will pay to send them to
the mines and the factories even when the net returns are no
greater than the rather low marginal product on the farm.
The advanced sector will then need to pay a wage rate no
greater than the social opportunity cost of labour; there will be
no marginal divergence.

These are matters which are difficult to test. One can
measure the marginal physical and value products of labour in
different uses, but it is more difficult to place a value on
leisure—an important matter when the rhythm of work on the
farm is very different from mine and factory—and there are
other considerations on which it is difficult to place a monetary
value and which are likely to raise the wage required in the
advanced sector. More and more evidence seems to suggest
that peasants tend to behave like fairly rational economic
agents, and it certainly seems more rational for a family to
equate the value of its marginal product in various potential
uses than to follow the somewhat irrational rules of the
standard model. But the matter must be left open.

III
WAGE DIFFERENTIAL: SIMPLE MODEL

So far we have assumed that the wage that has to be paid in
the advanced sector is equal to a worker's income in the
subsistence sector but that the latter is greater than his
marginal product in that sector. We now come to another
model which has also been very influential in the study of
less-developed countries and appears, on first sight, to yield

policy implications which are almost identical to those produced by the preceding model.

This time we assume that labour is paid its marginal product not only in the advanced sector but also in the subsistence sector. Thus there is no *income-sharing*. But the wage is not identical in the two sectors. There is a *wage-differential*, the wage payable being higher in the advanced sector. The latter sector should be thought of as consisting of large-scale manufacturing, mining and plantations, while the 'subsistence sector' includes not only subsistence agriculture

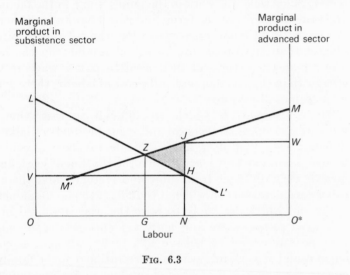

Fig. 6.3

but also small-scale agriculture producing for the market, as well as small-scale manufacturing.

This case is represented in Figure 6.3. This diagram combines a representation of the subsistence sector and the advanced sector.[5] The total quantity of labour in the economy is indicated by the length of the horizontal axis, the quantity in the subsistence sector being read from the origin on the left, and the quantity in the advanced sector from the origin on the right. The vertical axes show, as in the earlier diagrams, the

[5] Note that this is a general equilibrium diagram. The remarks in footnote 3 fully apply.

values of marginal products and the wage-rate. The curve LL' shows the value of the marginal product of labour in the subsistence sector. The quantity of labour in that sector is initially ON, and hence the wage is OV. The curve MM' shows the value of the marginal product of labour in the advanced sector. The quantity of labour in that sector is initially O*N and the wage O*W. Hence the wage-differential is HJ, and the marginal products of labour are not equal in the two sectors.

An optimal allocation of labour from a Pareto-efficiency point of view requires labour to move from the subsistence sector to the advanced sector until the marginal products are equalized. This assumes no other domestic divergences, and also the small country assumption, so that prices of traded goods are given. In the diagram, the optimum is given by the point Z. It requires GN of labour to move from the subsistence into the advanced sector. The value of subsistence output foregone will be GZHN and the value of advanced output gained will be GZJN. The excess of the gain in advanced-sector output over the loss of subsistence sector output is represented by the shaded area ZJH.

The optimum result could be attained by subsidizing the use of labour in the advanced sector appropriately. The extent of the subsidy required will depend on (a) the way the subsidy is financed and (b) the precise behaviour of the wage-differential in response to the subsidy. In the more theoretical literature, the unrealistic assumptions are usually made that (a) the subsidy is financed by lump-sum non-distorting taxes with no collection costs and (b) the wage-differential is fixed proportionately or absolutely. The assumption of lump-sum taxes means, among other things, that even if the taxes were paid mainly in the agricultural sector they would not affect employment in that sector for a given wage.

This excessively simple case is represented in Figure 6.4. When labour moves out of the subsistence sector into the advanced sector to the optimal extent, the wage in the subsistence sector rises from OV to OR. To maintain the original proportional wage differential HJ/NH, the wage in the advanced sector must then rise to GQ (ZQ/GZ = HJ/NH). Hence the subsidy per man required is ZQ. If

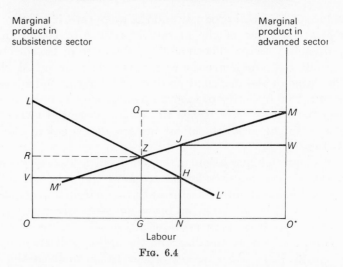

Fig. 6.4

the wage-differential were fixed absolutely rather than proportionately, the result would be much the same with ZQ = HJ.

One should really remove the two very unrealistic assumptions. Let us first remove assumption (b) and make the slightly more plausible assumption that there is a fixed minimum wage of O*W in the advanced sector. This is represented in Figure 6.5. The rise in the marginal product in the subsistence sector owing to labour moving out of that sector automatically

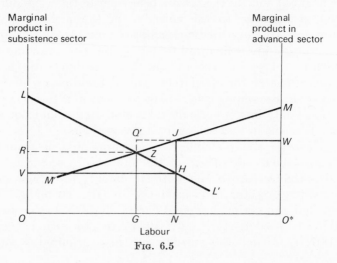

Fig. 6.5

causes the differential to be squeezed from HJ to ZQ'. The subsidy per man required is only ZQ'.

Removing assumption (a) may make things a little more complicated. Let us suppose that the subsidy is financed by a tax on labour in the subsistence sector, at least to some extent. We will continue to ignore collection and various by-product distortion costs. Let us assume also, for the moment, that the wage differential is fixed proportionately or absolutely rather than that there is a fixed minimum wage in the advanced sector. The tax on labour in the subsistence sector will drive

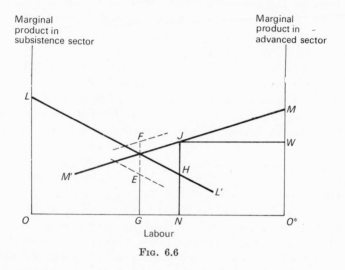

Fig. 6.6

labour out of that sector into the advanced sector, thus reinforcing the effect of the subsidy on labour in the advanced sector. Indeed, if the tax were used to finance some other activity, or if the revenue from it were distributed in a lump-sum way, the tax could do the job of shifting labour into the advanced sector on its own. In any case, the rate of subsidy required will now be less than before.

This is represented in Figure 6.6. The dotted curves are the MM' curve with the subsidy receipt added and the LL' curve with the tax payment subtracted. The advanced sector wage is GF and the subsistence-sector wage GE. If the differential is fixed proportionately, $EF/GE = HJ/NH$.

The story is slightly different when the wage differential is not fixed, but there is instead a fixed minimum wage in the advanced sector. It can easily be shown that the rate of subsidy required will be unaffected by the manner in which it is financed (still assuming no by-product distortion and collection costs); it will simply be determined by the excess of the minimum wage over the value of the marginal product of labour at the optimum allocation point. On the other hand, the wage-differential *will* be affected. Since the tax in the subsistence sector will force down the wage there, it will increase the differential compared with] what it would have been if the subsidy had been financed in some other way. Compared with the situation before any intervention, the differential might increase or decrease.

If the advanced sector produces import-competing products, tariffs can achieve some of the results of the wage subsidy. There will, as usual, be a hierarchy of policies, and a tariff is likely to be third- or fourth-best. It is particularly worth noting that labour in the 'subsistence' sector is likely to consume some imported goods, so that a tariff would be somewhat equivalent to a subsidy on labour use or production in the advanced sector which is financed by a tax on labour in the subsistence sector.

This is the simple model. It is very similar to the income-sharing model because both yield the result that in the absence of intervention the marginal product of labour in the advanced sector will exceed the marginal product in the subsistence sector. The principal policy implications are thus also identical. It is common to combine the two models into one, so that there are *two* reasons why there is a *marginal product differential* and why the use of labour in the advanced sector should be subsidized, or a shadow wage lower than the actual wage should be used in planning public investment in the advanced sector. But it seems important to keep the two elements separate, if only because they are subject to quite distinct criticisms and one model might well survive without the other.

A large theoretical literature has explored the wage-differential model in the traditional two-product two-factor general equilibrium context, using the box diagram, transformation curve, and so on.[6] This yields essentially the same results as the exposition here.

The exposition has disregarded the possible effects of the subsidy on the terms of trade between the subsistence and the advanced sector. Changes in internal terms of trade play an important part in some of the best-known dual-economy models, but these generally do not allow for international trade. Here we are assuming that the prices of the products of both sectors are given from outside the country, apart from the effects of tariffs or other indirect taxes. Allowing for terms of trade effects would not affect the main argument, though it would complicate the geometry.

The argument for protection that emerges is sometimes described as the *Manoilesco* argument, though this gives much more credit to a former Rumanian Minister of Finance than he deserved. In a rather confused book published in 1929[7] he noted the generally low productivity of agriculture compared with industry, and suggested that protection which draws resources from the low-productivity to the high-productivity sector would bring about a gain. His argument was clarified, and the element of truth in it brought out, in reviews by Viner[8] and by Ohlin[9]. The latter review, in particular, foreshadowed many ideas that have subsequently been rigorously analysed and stressed, notably the idea that in the presence of a wage differential a subsidy on labour-use is preferable to a tariff.

Lewis' celebrated dual economy model[10] included a wage differential as an ingredient, as well as income-sharing, and the whole matter was revived and expounded in modern two-sector general equilibrium terms by Hagen in 1958.[11] It is the Hagen

[6] See Stephen P. Magee, 'Factor Market Distortions, Production and Trade: A Survey', *Oxford Economic Papers*, 25, March 1973, pp. 1–43, and many articles cited therein.

[7] Mihail Manoilesco, *The Theory of Protection and International Trade*, London, P. S. King, 1931.

[8] Jacob Viner, *International Economics*, The Free Press, Glencoe, 1951, Ch. 7.

[9] Bertil Ohlin, 'Protection and Non-Competing Groups', *Weltwirtschaftliches Archiv*, 33, Heft 1, 1931, pp. 30–45.

[10] W. A. Lewis, 'Economic Development with Unlimited Supplies of Labour'.

[11] Everett E. Hagen, 'An Economic Justification of Protectionism', *Quarterly Journal of Economics*, 72, Nov. 1958, pp. 496–514.

article which initiated the recent rigorous literature on the subject.

One can certainly observe in many less-developed countries that the wage obtainable in the advanced sector, whether in manufacturing or in mining, is far above incomes earned in peasant agriculture.[12] The crucial question is why this is so and whether the differential represents a genuine marginal divergence.

It may not do so. Labour in the advanced sector may really be a package of subsistence labour plus human capital, and hence may not be identical in quality with subsistence labour. The higher real wage may thus be a return on investment in training and experience, and in the costs of movement from one sector to another. It may be a necessary payment to obtain a permanent, stable labour-force, living with family near the place of work, rather than the casual type of labour generally available in the traditional sector.[13] Even if the quality of the labour is identical, the monetary wage payments may not give a correct indication of the true wages, taking into account the availability of non-market food supplies in the subsistence sector, the higher urban cost of living, and the possible disutility of uprooting oneself to go into the city or the mines. It is thus by no means certain that every apparent wage differential represents a true marginal divergence requiring correction by subsidies or tariffs.

Nevertheless, when all this is allowed for, it is plausible that in many countries a genuine differential remains. The main reason for it is likely to be trade union organization in the advanced sector. Generally one should include government administration in the advanced sector here, while small-scale manufacturing and urban services of various kinds would be included in the subsistence sector, along with peasant agriculture. Trade unions often manage to create a small privileged

[12] Turnham and Jaeger, *The Employment Problem in Less Developed Countries*, Ch. IV.

[13] Hla Myint, 'Dualism and the Internal Integration of the Underdeveloped Economies', *Banca Nazionale del Lavoro Quarterly Review*, No. 93, June 1970, pp. 3–31, reprinted in H. Myint, *Economic Theory and the Underdeveloped Countries*, Oxford University Press, New York, 1971; also Dipak Mazumdar, 'Underemployment in Agriculture and the Industrial Wage Rate', *Economica*, 26, November 1959, pp. 328–340.

elite of urban workers, government clerks and miners; and their position is usually upheld by governments. Alternatively, the wage differential may be created by minimum wage laws applying, inevitably, only to the advanced sector.

Foreign corporations are likely to pay wages well above those needed to obtain labour from the subsistence sector. Subsistence sector wages would appear as starvation wages in comparison with the levels of wages that the corporation's other subsidiaries are paying in the developed countries, and company officials are not immune to humanitarian or guilt feelings. In addition, strong moral and political pressures may be applied on the corporations to raise their wage payments. The evidence that these sorts of factors operate is very widespread, so that it is by no means difficult to explain widespread wage differentials. Hagen has also provided a dynamic explanation discussed below.

In some countries one can observe two distinct wage differentials. One is the excess of the wage paid in the small-scale urban sector, whether manufacturing or services, over the wage obtainable in agriculture. This may not represent a true marginal divergence but may measure the costs of movement from the rural areas to the city, the higher cost of living in the city, and so on. The second wage differential is the excess of the wage paid in the advanced sector—notably large-scale urban manufacturing—over the wage paid in the urban small-scale or 'subsistence' sector. This is the genuine differential that represents a marginal divergence and may call for some social intervention. It follows that wage-rates paid in the urban small-scale sector may be the best indicators of the social opportunity cost of the labour that moves into the advanced urban sector.[14]

IV

WAGE DIFFERENTIAL: SOME QUALIFICATIONS

It is probable that genuine wage differentials do exist and have non-optimal labour allocation effects, though any precise measurement is likely to be difficult, if not impossible. The real doubts concern the appropriate policies.

[14] A. C. Harberger, 'On Measuring the Social Opportunity Cost of Labour', *International Labour Review*, 103, June 1971, pp. 559–579.

(1) *Differential in Cost of Capital*

Gross distortion in the allocation of capital is a common characteristic of less-developed countries. Here there is also a factor-price differential between sectors, capital being much cheaper or more readily available for firms in the advanced sector than for peasants or for small-scale industry.[15]

In this case, also, one must distinguish that part of the differential that represents a true marginal divergence. The risks in lending to peasants or to small businesses are undoubtedly much greater than the risks in lending to large firms in the advanced sector. In addition, there are economies of scale in loan administration. Hence some capital cost differential would be justified. But it is arguable that the differential is often much greater than this, so that a misallocation of resources results from the over-supply of capital to the advanced sector relative to the subsistence sector, or more generally, relative to small-scale enterprises of all kinds.

Some students of less-developed countries, notably Hla Myint, convincingly argue that capital market imperfections, and especially discrimination against small-scale labour-intensive activities, whether agricultural, manufacturing or commercial, are caused by unwise government policies which make credit to large-scale urban enterprises unduly cheap and so allow them to preempt one of the scarcest of national resources.[16] Together with a similar discrimination in the allocation of import licenses for cheap capital-goods and in the provision of public facilities, notably transport, this inhibits

[15] To allow for the potential mobility of capital between sectors, one must depart from the assumption required for our earlier diagrams specified in footnote 3, namely, that factors other than labour are specific to each sector. One could use the two-sector two-mobile-factor diagrams of orthodox trade theory to represent the case in which there are *two* factor-market divergences, one for labour and one for capital. See the literature surveyed in Magee, 'Factor Market Distortions, Production and Trade: A Survey', 1973.

[16] Myint, 'Dualism and the Internal Integration of the Underdeveloped Economies', and *The Economics of the Developing Countries*, Ch. 5; also Little et al., *Industry and Trade in Some Developing Countries*, pp. 86–92. See also Ronald I. McKinnon, *Money and Capital in Economic Development*, The Brookings Institution, Washington 1973. McKinnon stresses the many adverse effects of fragmented capital markets (and especially of the need for self-financing) in less-developed countries and underlines that a first-best fiscal policy (or trade policy) solution cannot be found for what is essentially a financial problem.

indigenous small-scale enterprise, discourages the extension of the exchange economy and, through misallocation, reduces the productivity of capital.

With two factor-market imperfections, first-best policy requires *two* interventions. Attempts to improve the capital market, or alternatively to subsidize the supply of capital to the subsistence sector (or end the existing implicit subsidy to the advanced sector), should be combined with attempts to remove the labour market imperfection or subsidize the use of labour in the advanced sector. Indeed, it is not so much that intervention is required for the first time to improve the capital market, but that the nature of existing intervention needs to be changed to create a freer, more unified market, rather than to fragment it. One cannot say whether, on balance, output of the advanced sector should expand. It is clear only that subsistence or small-scale production should become more capital-intensive and advanced-sector production more labour-intensive. The marked difference in factor proportions between the two sectors is sometimes described as 'technological dualism', but it is more likely to be the consequence of distorted factor prices.

General protection of manufacturing industry is likely to be a most inadequate policy now, and conceivably any degree of protection may reduce real income. The main point is that it is no longer obvious that manufacturing as such, as distinct from labour-intensive manufacturing, should expand. Indeed, it may well be that the distortion in the capital market has led to an *over*-expansion of the protected large-scale manufacturing sector in spite of the opposite handicap that the wage-differential has imposed.

(2) *Flexible Wage Differential*

If there is a minimum wage in the advanced sector which is fixed by law, by convention or on the basis of wage levels in advanced countries then a wage subsidy will have the effects on employment in the advanced sector described earlier. Since the wage in the subsistence sector will rise when labour moves out of that sector, the differential will decline, so that the required subsidy per worker will be less than the initial differential.

If the wage differential has been brought about by trade union action, the matter is more difficult. There is no reason why the original wage in the advanced sector should stay fixed, nor is there any reason at all why the differential should stay constant. One needs to know something about the motivations of the trade unions and about the bargaining situation.

It is at least conceivable that a wage subsidy would lead to a rise in the wage in the advanced sector sufficient to absorb the subsidy completely, hence allowing no increase in employment. The unions might be concerned only with maintaining employment of their existing members and maximizing their existing members' combined incomes. Alternatively they might force an increase in the wage as long as profits increase with the aim of eliminating any rise in profits. Since an increase in profitability would normally be associated with a rise in employment, they will then effectively prevent extra employment. In both cases the wage in the advanced sector will rise sufficiently to fully absorb the wage subsidy, and no level of subsidy can lead to the desired transfer of labour out of the subsistence sector. It will simply lead to an income redistribution from taxpayers to wage-earners. This is the limiting case, but indicates the nature of the problem.

A tariff could have exactly the same effect. In the first instance it might increase profits, but very soon the wage might rise so that any stimulus to output would be prevented. Perhaps in practice there might be some lag in wage adjustment, or the adjustment might be only partial so that some increase in profitability and hence some labour transfer would stick.

(3) *By-product Distortions and Income Distribution Effects*

Financing the wage subsidy will impose deadweight and collection costs; in addition there may be subsidy disbursement costs. If a tariff is used, collection costs may be lower but distortion costs higher. The analysis of Chapter 3 is completely relevant and must be taken into account here, especially as we are concerned with less-developed countries where collection costs can be expected to be high. Hence it will no longer be optimal to have a wage subsidy or a tariff so high that the values of the marginal products of labour in the two sectors are completely equalized.

If there are these costs of revenue collection, one cannot assume that income distribution effects will be, or should be, completely offset by independent tax-subsidy policies. Hence—unlike much of the theoretical literature—one needs to consider whether the income distribution effects of the measures proposed are in an optimal direction or not; this will affect the desirability of the policy, at least at the margin.

The analysis of income distribution effects turns out to be rather complex. A crucial question is who pays the taxes to finance the wage subsidy, or, if a tariff is used, who are the consumers of the protected products. One can obtain the following results with the help of our various earlier diagrams.

In the subsistence sector, if there are landlords and landless peasants, income will be redistributed towards the latter as peasants move out of the sector and the wage rate rises. If peasants own their own land and there is income-sharing, those of them that stay behind will gain since the average product in the sector will rise. But their gains, and the gains of wage-earners in the earlier case, may be reduced or eliminated by the taxes required to finance the subsidy or by the higher prices of imported consumer goods resulting from a tariff.

In the advanced sector the migrants from the subsistence sector will gain in higher wages; these extra wage payments will be partly financed by the productivity rise resulting from their move from a low-productivity to a higher-productivity occupation, and partly by the wage-subsidy. If the wage in the advanced sector has stayed constant in spite of the subsidy, then the pre-existing labour-force will be unaffected, and the whole of the subsidy payment on their employment will raise profits. Of course, the subsidies may be partly or wholly financed by a tax on profits, and, on balance, profits may remain unchanged or even fall. If the wage rises, some of the subsidy payment will go to these wage-earners. If the financing of the subsidy, or the higher prices resulting from a tariff, imposes a burden on wage-earners in the advanced sector, one would have to specify whether the minimum wage or the wage differential is fixed in money or in real terms, and after or before tax.

In any case, the principal redistribution is likely to be from taxpayers or from consumers of imported goods, whoever they are, to wage-earners; but profits could also increase, possibly substantially. It is certainly conceivable that very little net redistribution results. The taxes or higher prices of imported goods might bear on both profits and wages, so that capitalists and workers would both be taxed, the revenue being handed back to them as a wages or production subsidy, or as the subsidy equivalent of the tariff.

It is also possible that a seriously adverse income redistribution is brought about. This is clearest in the limiting case in which an increase in wages absorbs the whole of the subsidy or gain from a tariff, so that no labour transfer results. The only gainers will be the existing privileged group of highly-paid workers. If the wage subsidy is financed from taxes on exports produced in the subsistence (or small-scale agricultural) sector, or from tariff revenue on goods purchased at least in part in this sector, the poor will be subsidizing the better-off (unless the whole burden is borne by rural landlords). This will also be true if tariffs on goods purchased in the subsistence sector provide protection for the products of the advanced sector.

V
WAGE DIFFERENTIAL: DYNAMIC APPROACH

So far the approach has been comparative static. But in 1958 Hagen advanced a *dynamic* explanation of the wage-differential and built on it a case for protection.[17]

He argued that in the process of growth it is continually necessary to transfer labour from agriculture to manufacturing partly because the size of the manufacturing sector grows relatively (because of high income elasticities of demand for manufactures relative to agricultural products) and because the rural birthrate is higher than the urban one. To induce

[17] E. Hagen, 'An Economic Justification of Protectionism', *Quarterly Journal of Economics*, 1958. This article has been criticized for applying comparative static analysis to a dynamic argument; but the analysis can certainly be adapted to bring out the logic of his argument.

the transfer of the necessary labour, the manufacturing wage will have to be continually above the agricultural wage, so that the labour market is continually out of static equilibrium.

Hagen produced, in fact, a version of the infant industry argument, resting on imperfection of the capital market and on ignorance.[18] If the capital market were perfect, so that workers could borrow to finance their costs of movement, and if there were perfect knowledge, so that they had correct expectations about the attractions of factory work and urban life, there would be no need for intervention. But workers must finance their own transfer costs; furthermore, they then expect to be reimbursed at once by employers through a higher wage, even though in fact the benefits from the transfer will be spread over their lifetimes. Thus their first employers have to pay an excessive wage, the gainers being the workers themselves and their later employers. Hence the first employers need to be subsidized or protected. In addition, workers have false expectations about a new way of life.

As an argument for subsidizing labour in manufacturing, and even more, of course, as an argument for trade intervention, it seems rather weak.

(1) If ignorance is the source of the trouble, first-best policy is to spread information; if the difficulty is to finance the transfer of labour and to compensate for temporary inconvenience, first-best policy is to make loans or grants to potential movers or to subsidize elements in the cost of transfer—in other words, adjustment assistance.

(2) A policy of subsidizing those wages which in any case have increased, in the hope that a further increase will induce sufficient labour inflow, would make the nouveau-riche richer. If one generalized this, every factor-price gap that appears in response to changing demand or supply conditions should be deliberately widened through intervention. Only an inversion of the conservative social welfare function could justify the income distribution consequences.

(3) A part of the observed wage differential may reflect genuine costs of transfer, including psychological ones, and

---

[18] This was pointed out by A. Kafka in 'An Economic Justification of Protectionism: Further Comments', *Quarterly Journal of Economics*, 76, Feb. 1962, pp. 163–166.

the genuine uncertainty as to whether the high wage rates will be permanent. Private caution may be justified.

(4) Finally, the willingness and indeed ability of labour to move out of the rural sector may be inhibited by its poverty. Minimum health and energy as well as financial resources are required to make a move. Mobility may then be increased more by subsidizing incomes in agriculture—hence raising the potential migrants' ability to finance investment in movement from their own resources, but incidentally reducing the wage-differential—than by subsidizing manufacturing employment.

VI
URBAN UNEMPLOYMENT

In many less-developed countries there is a great deal of urban unemployment.[19] This coexists with a high minimum wage in manufacturing industries and with a marginal product of labour in subsistence agriculture less than the urban minimum wage but not zero. The urban unemployment is likely to be induced by the wage differential itself. Such induced unemployment must be introduced into our analysis.

The model to be presented here is derived from an important article by Harris and Todaro and has interesting policy implications, turning some of the main conclusions of the orthodox wage-differentials theory upside-down.[20] It seems to suggest that, if anything, countries should protect their agriculture rather than their manufacturing! The approach

[19] Turnham and Jaeger, *The Employment Problem in Less Developed Countries*, Chs. III, V; also *Industry and Trade in Some Developing Countries*, pp. 81–85.

[20] John R. Harris and Michael P. Todaro, 'Migration, Unemployment and Development: A Two-Sector Analysis', *American Economic Review*, 60, March 1970, pp. 126–142. Some of the basic ideas originated in M. P. Todaro, 'A Model of Labor Migration and Urban Unemployment in Less Developed Countries', *American Economic Review*, 59, March 1969, pp. 138–48.

At the time of writing the model is much under discussion, and a large literature, some of it critical, stressing the qualifications and dynamic aspects, is growing up. See Deepak Lal, 'Disutility of Effort, Migration, and the Shadow Wage-Rate', *Oxford Economic Papers*, 25, March 1973, pp. 117–20. I am much indebted to Maurice Scott in understanding these matters. The diagrams used here are not in Harris and Todaro but are adapted from a simple diagram used by Ranis and Fei and others.

will be comparative static. Some of the considerations discussed earlier in this chapter (income-sharing in agriculture,
capital market distortions, labour immobility) will be ignored,
so as to focus on the main points.

(1) *The Expected Urban Wage and the Unemployment Pool*

People migrate into the cities in the hope of getting an
urban job because the wage that they would be paid if their
job hunt succeeded would be so much higher than what they

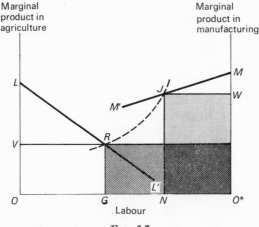

FIG. 6.7

can get on the land. While they are in the unemployment pool,
they are supported at a minimum standard by their relatives,
whether in the agricultural or the manufacturing sector.
It is arguable that, to some extent, the probability that they
succeed in getting a job depends on the size of the unemployment pool in relation to the numbers employed in manufacturing at the given wage. The simple Harris-Todaro approach is to
suppose that the *expected* urban wage is equal to the average
wage of the urban employed and unemployed combined.
This implies no risk aversion and that the manufacturing
labour-force steadily turns over (jobs are not 'tenured').
Modifying these assumptions will modify, but not necessarily
destroy this model.

In Figure 6.7 there are again two sectors, the prices of
agricultural and manufactured goods are constant, being

6

determined in foreign trade, and each sector has a specific factor, the only mobile factor being labour.[21] The length of the horizontal axis shows again the total labour-force available. The marginal product curves are LL' for agriculture and MM' for manufacturing. The fixed minimum wage in manufacturing is $O*W$, so that manufacturing employment is given at $NO*$.

The remainder of the labour-force will distribute itself between the agricultural sector and the urban unemployment pool in the following manner. In equilibrium the *expected* urban wage must be equal to the agricultural wage. Draw a rectangular hyperbola (in the space with origin at $O*$) through J and let it intersect LL' at R. This point R gives the agricultural wage OV and agricultural employment OG. Furthermore, it yields an unemployment pool of GN. If the manufacturing wage-bill $NJWO*$ is spread over the whole urban labour-force, we obtain the *expected* urban wage GR, which is the average of the minimum wage $O*W$ received by the employed and the zero wage received by the unemployed. (The two shaded areas are equal.) Hence the *expected* urban wage is equal to the agricultural wage.

(2) *Subsidizing Employment in Manufacturing*

Suppose that manufacturing employment is subsidized. Assume that the subsidy is financed by a tax on rents or profits which does not affect the supply of either. The effects are shown in Figure 6.8. A wage subsidy of QJ' per man in manufacturing will expand manufacturing output by $N'N$. The value of the extra manufacturing output is the shaded area $N'QJN$. One must then draw a new rectangular hyperbola through J' (in the space with origin $O*$), and the intersection of this new curve with LL' will show the new equilibrium allocation of the remaining labour between the agricultural sector and the unemployment pool. Labour in agriculture declines by $G'G$. The value of agricultural output lost is the shaded area $G'R'RG$.

---

[21] The Harris-Todaro model as presented by them is a closed economy one, and hence has the internal terms of trade determined endogenously, which complicates their analysis considerably, though it does not seem to affect the main results. Opening the model actually simplifies it, provided one makes the small country assumption.

If one is interested in the effect on total output, one must compare the two shaded areas. One cannot say in general whether total output will rise or fall or what will happen to the size of the unemployment pool. It is perfectly possible that output falls. The steeper the slope of the MM' curve and the flatter the slope of the LL' curve, the more likely it is that a

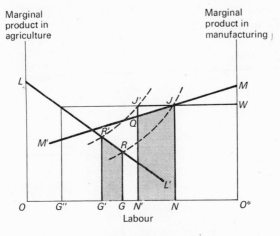

Fig. 6.8

wage subsidy which achieves a given increase in manufacturing employment raises urban unemployment and actually lowers real output. This result also applies to a tariff which protects import-competing manufacturing industry. Quite apart from its usual by-product distortion costs, a tariff may lower real income because of the increased unemployment it creates!

It would be possible to have a wage subsidy that is sufficient to restore full employment. More and more labour would be drawn out of agriculture, and this would steadily raise the marginal product of labour in agriculture until it has reached the level of the fixed minimum wage in manufacturing. At that point unemployment will have disappeared since the only reason for urban unemployment is an excess of the minimum wage over the agricultural wage. In Figure 6.8, agricultural output would be down to OG''. Since the marginal product of labour in manufacturing would be below that in agriculture, the result would not be first-best. It might not even be second-best. A lower wage-subsidy, possibly a very low one, may maximise real output, given that such a wage subsidy in manufacturing is the only available instrument of policy. All this applies to tariffs as well.

If the marginal product of labour in agriculture stayed unchanged when labour moves out of agriculture a wage subsidy would be certain to lower real output. In Figure 6.9

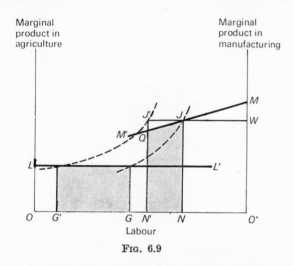

FIG. 6.9

the LL' curve is horizontal. The rise in manufacturing output is shaded, as is the fall in agricultural output. But the latter must be equal to the increased wage-bill in manufacturing $N'J'JN$ (this follows from the properties of rectangular hyperbolae), and the extra manufacturing output falls short of this by the area $QJ'J$. This area thus shows the fall in real output resulting from a wage subsidy. The shadow-wage required for optimal resource allocation decisions in the manufacturing sector will then be equal to the actual wage.[22] There is certainly no case for a tariff then, even as a third- or fourth-best device.

[22] See A. C. Harberger, 'On Measuring the Social Opportunity Cost of Labour', *International Labour Review*, 103, June 1971, pp. 559–579.

(3) *Need to Subsidize Agricultural Employment*

We have yet to come to the really drastic implications of this model. If employment in agriculture, rather than manufacturing, were subsidized, a gain would be certain as long as there is any unemployment pool and provided the minimum wage does not change as a result. A subsidy on the use of labour in agriculture would reduce the wage differential and hence bring some of the urban unemployed back to the land, reducing the unemployment pool. The opportunity cost of

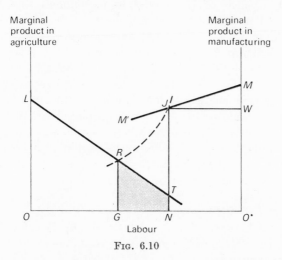

Fig. 6.10

labour to agriculture would be zero. One might not regard a labour subsidy in the subsistence or agricultural sector as feasible (having very high disbursement costs), but similar effects can be brought about by public funds improving the rural infrastructure so as to make life in the villages more convenient and attractive and hence raising the wage in real terms.

In Figure 6.10 the wage subsidy in agriculture is TJ per man. Employment in manufacturing stays at NO^*, but employment in agriculture rises from OG to ON, absorbing all the unemployed. The wage paid by employers in agriculture is NT, while the wage received by wage-earners is NJ, equal to the urban minimum wage. The value of the extra agricultural output is the shaded area $GRTN$ and is a pure gain.

It is labour in agriculture rather than in manufacturing that should apparently be subsidized or protected! If trade policy were to be used, the case would be for an export subsidy rather than a tariff or export tax, assuming that the agricultural sector is also the export sector.

All this is true if one has to choose between subsidizing labour in manufacturing and subsidizing labour in agriculture, and if one does not know the precise shapes of the curves. Subsidizing labour in agriculture is sure to lead to a gain if the general features of the model as expounded so far apply. But it will not lead to the first-best result. If it eliminates unemployment, it will lead to excessive movement of labour into agriculture compared with the first-best solution.

Furthermore, if there were a high discount for risk on the part of potential migrants, the expected urban wage initially would be well below the level indicated in our diagram, so that there would be little urban unemployment. In that case, the gains derived from subsidizing agriculture to the point where unemployment is eliminated would also be low. The gains from using the same amount of revenue to subsidize manufacturing may then be greater.

(4) *First-Best Solution*

The first-best solution requires labour to be allocated such that (a) the values of the marginal products in manufacturing and agriculture are equal and (b) there is no unemployment. This solution is represented by the point Z in Figure 6.11 where the two marginal product curves cross.

As noted by Harris and Todaro, the first-best solution could be attained by a wage subsidy to manufacturing (at the rate of ZZ' per man) combined with a restriction of migration out of agriculture. Alternatively, it could be attained by an equal wage subsidy for both industries of ZZ' per man. To finance this subsidy one would have to find some sectors of the community that are taxable either directly, or indirectly through trade policy, without affecting supplies of the taxed factors as a result. If such factors do not exist, or the collection costs of taxes are significant, first-best tax-subsidy policy could not bring about the resource allocation that wage flexibility would have attained.

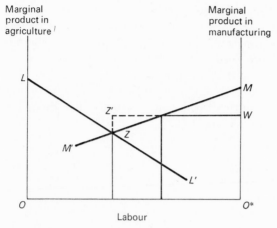

FIG. 6.11

(5) *Who Pays the Taxes?*

It might be thought that a subsidy to labour in manufacturing could be financed by a tax on labour in agriculture, as in the earlier wage-differential model. But it must now be remembered that any tax on labour in agriculture will increase unemployment since it will increase the wage-differential, assuming that potential migrants compare their after-tax wage income in agriculture with the expected urban wage. In Figure 6.8 the taxing of agricultural labour will lower the LL' curve (defined now as a supply curve to the urban sector, not as a social marginal cost curve). This effect strengthens the argument against subsidizing labour in manufacturing, contrary to the popular prescription. The argument also applies to a tariff which protects import-competing manufacturing and which raises the cost-of-living for the rural sector or lowers rural incomes through indirectly lowering the domestic-currency prices of exportable goods.

One can be even more definite in considering a subsidy to labour in agriculture, whether direct or indirect. If this subsidy is financed by taxing labour in manufacturing, the *after-tax* real wage in manufacturing will fall. In the absence of tax illusion, this would lead to a rise in pre-tax wages designed to restore the after-tax wage. This, in turn, would reduce employment and offset the favourable effect of the subsidy in agriculture. Of course, *with* tax illusion the real disposable wage would fall; and since real wage rigidity has been the source of the problem all the time, the latter would indeed have been solved at the source.

It follows that in the absence of tax illusion, a wage-subsidy must be financed primarily or wholly by taxes on profits, on agricultural rents, or both. The financing problem is set out in Figure 6.12.

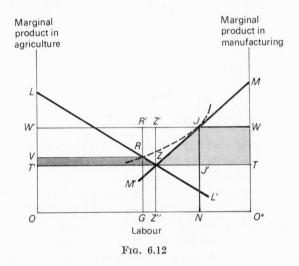

Fig. 6.12

The total cost of the wage-subsidy is $T'W'WT$. It will increase agricultural rents by the shaded area $T'VRZ$. It will increase manufacturing profits by the other shaded area $ZJWT$. If these increases in rents and profits can be recovered through taxation, there remain three areas to be financed: $VW'R'R$ is the portion of the subsidy which has increased the wages of the labour that was originally employed, and remains employed, in agriculture; $RR'Z'Z$ refers to the previously unemployed who are newly employed in agriculture, and is the excess of the wages they receive over the value of their output; similarly, $ZZ'J$ refers to the previously unemployed now working in manufacturing, and is the excess of their wage-bill over the value of their output.

There may be a fixed revenue constraint. If there is not,

one must at least take into account collection costs, disbursement costs, and any by-product distortion costs resulting from the effects of taxes that bear on land use or on capital unevenly or from the effects of taxes on the supply of capital. It will then no longer be first-best to make a full correction with a uniform wage subsidy. A more limited subsidy, perhaps mainly for labour in agriculture—and possibly given through improvement of public infrastructure designed to keep people happily in the villages (and *not* improving mobility into the cities!) may be first-best.

All this rests on a number of assumptions.

(1) Risk aversion is sufficiently low and labour turnover in manufacturing sufficiently high that there is an incentive to be un- or under-employed in the cities because of the wage-differential. The higher risk aversion and the lower labour turnover, the steeper the JR curve in Figure 6.7 (with risk aversion and low turnover it would no longer be a rectangular hyperbola).

(2) The urban after-tax real wage cannot be reduced.

(3) A rise in the agricultural wage resulting from a labour subsidy in agriculture will not cause the urban wage to rise.

(4) The only urban unemployment is of the Harris-Todaro type; in fact, urban unemployment can also have other causes.

(5) Labour in agriculture is paid its marginal product. If there were income-sharing in agriculture a case for subsidizing manufacturing might be restored. In fact, one might combine the Lewis (income-sharing) and the Harris-Todaro (urban unemployment) models; the first focuses on income-sharing in agriculture and the second on *expected* income-sharing in manufacturing.

(6) Investment costs of movement do not inhibit migration into the cities; such investment costs might have a desirable effect in reducing urban unemployment. One might then combine the Hagen and the Harris-Todaro models; one suggests intervention to foster migration, another to inhibit it.

(7) Finally, the wage-differential situation gives rise to *un*employment with zero productivity. In fact it may give rise to urban *under*-employment in a low-productivity high labour-turnover urban services sector. This would present no problems for the analysis: in Figure 6.7 the rectangular

hyperbola would be drawn in a space with origin at a point vertically above O^*, the vertical distance from O^* measuring the wage in the services sector.

VII
IS THERE A CASE FOR FOSTERING INDUSTRIALIZATION?

Neither the disguised unemployment nor the wage-differential argument for protection appears to have stood up very well, when one bears in mind the many qualifications. One must doubt even whether they yield a convincing case for wage-subsidies or for the use of shadow wages less than actual wages in public sector planning. When one allows for urban unemployment one can even generate a case for subsidizing agriculture! But one can hardly arrive at firm general conclusions without considering particular countries in detail. A great deal of information is required about the marginal productivity of labour in agriculture, about the nature and determinants of urban unemployment, and also about the relative marginal productivities of capital, both private and public, in different sectors of the economy.

Myint, on the basis of close study of Asian and African countries, comes down heavily against recent industrialization and import-substitution policies which have been based partly on such arguments. He stresses the importance of 'removing as far as possible the causes of unequal access to the scarce economic resources by the two sectors,'[23] having in mind capital funds, public economic facilities and foreign exchange, as well as the trade union-induced wage differential. Little, Scott and Scitovsky, basing themselves on major studies of seven industrializing countries, conclude that there may be a case for wage subsidies at rates equivalent to subsidies on value added of 5 per cent to 10 per cent, or possibly up to 15 per cent (or effective protective rates of 5–15 per cent).[24] These figures are so low because in most industries of less-developed countries wages represent only a small part of gross value added. The figures should be compared with actual

[23] Myint, 'Dualism and the Internal Integration of the Underdeveloped Economies'.
[24] Little et al., *Industry and Trade in Some Developing Countries*, pp. 146–8.

effective rates in the seven countries which are often over
100 per cent and which do not generally discriminate in
favour of labour-intensive industries.[25]

There are, of course, other arguments for protection to
foster industrialization, and many of these will be discussed
in later chapters. None of them yields first-best arguments
for tariffs or import restrictions, and mostly they are not even
first-best arguments for production subsidies. The most
important by far is the infant-industry argument in its
various forms (Chapter 9); there are various arguments, some
rather confused, based on manufacturing having more economies of scale than agriculture (these require the economies of
scale to be *external* to firms); and there is the argument that
industrialization fosters saving and investment (Chapter 10).

In principle there are numerous possible domestic divergences, and many could yield some second-best, or worse,
arguments for trade intervention. It is hardly worth discussing these in detail since the general method of analysis is
obvious. All sorts of external economies and diseconomies are
possible. Some types of manufacturing activities create undue
pollution. First-best policy would be to deal with this in some
direct way, perhaps taxing or controlling emission of the
polluting agents, while a second-best method might be to tax
production of the relevant product. Reducing or forgoing
protective tariffs that might otherwise be provided, as well as
taxing exports, would only be third-best. The domestic
divergences that are most relevant to traditional industrial
protection arguments will be discussed in Chapter 9 (in
connection with the infant industry argument).

One might also like to take account of so-called 'non-
economic' arguments for fostering industrialization. Actually
the techniques of economics can be used to analyse, or at
least define more precisely, *all* arguments, and the use of the
term 'non-economic' sometimes means no more than that
economists have not got around to analysing an argument
properly, rather as miracles are phenomena that scientists
have not yet caught up with.

Adam Smith wrote that 'defence, however, is of much more
importance than opulence,' and so it has been traditional to

[25] Little et al., *Industry and Trade in Some Developing Countries*, Ch. 5.

allow defence as a legitimate argument for protection—
whether to maintain manufacturing industries which may be
needed in time of war, to maintain agriculture so as to ensure
food supplies in an emergency (a contributory explanation
of high agricultural protectionism in some European countries,
such as Switzerland), or to maintain a sturdy peasantry to
supply Prussian cannon-fodder. Yet in all these cases economic
analysis can be used to determine the minimum cost ways
of attaining desired defence objectives. The question is whether
protection is 'cost-effective' to achieve a given objective.
Maintaining emergency stockpiles may well be cheaper than
keeping uneconomic industries going in peace-time. In few
countries are the military authorities given the opportunity to
balance the costs of defence-motivated protection against
alternative ways of spending the sums involved.

Harry Johnson[26] has suggested that in some countries
industrial production 'appears as a collective consumption
good yielding a flow of satisfaction to the electorate independent of the satisfaction they derive directly from the consumption of industrial products.' While he suggests that the origin
of this preference may lie in 'the power of owners of and workers
in industrial facilities to achieve a redistribution of income
towards themselves by political means, or the belief that
industrial activity involves beneficial 'externalities' of various
kinds,' there is also the suggestion that, in some sense,
industrial production is an end in itself, perhaps having to do
with 'nationalist aspirations and rivalry with other countries.'
Industrial production, apart from the goods it actually
produces which can be consumed or traded, contains then an
element of a public good: additional satisfaction obtained by
one citizen does not reduce the satisfaction obtained by another,
and the private gain derived from production will fall short of
the collective gain.

Johnson uses this approach to explain various features of
tariff bargaining, and it could perhaps also be used to explain
the policies of some inexperienced governments of recently

[26] Harry G. Johnson, 'An Economic Theory of Protectionism, Tariff
Bargaining, and the Formation of Customs Unions', *Journal of Political
Economy*, 73, June 1965, pp. 256–83 (reprinted in his *Aspects of the Theory of
Tariffs*, 1971).

decolonized countries in the nineteen fifties. An economist who finds the government he is advising to be hopelessly muddled or irrational in its desire for a steel mill can comfort himself with the reflection that the mill might not produce much steel or employ many people, but 'since they want it so much it must be a collective consumption good.'

The weakness of this approach is, first, that there *are* acceptable and logical economic reasons for wishing to foster manufacturing industry in general, and the belief by many governments that these reasons apply is sufficient to explain the common 'preference for industrial production,' and secondly that any government would certainly deny that it wants industrial production for its own sake. It will have essentially economic reasons, such as the infant industry argument, even though they may be poorly based on facts or even be fallacious. It is sometimes analytically convenient to define a proximate target—such as a marginal preference for industrial production—without going into the reasons for the target. This approach, indeed, will be used at various points in this book. But this does not mean that the reasons must be utterly irrational or outside economics—as a view of industrial production as a collective consumption good surely is.

7

THE TERMS OF TRADE ARGUMENT FOR
TAXES ON TRADE

LET US now remove the small country assumption maintained in this book so far. If the elasticity of demand for exports is less than infinite, a restriction of export supply will raise the foreign prices of exports; and if the elasticity of import supply is less than infinite, restricting import demand will lower import prices. This is the basis of the terms of trade argument for taxes on trade. In a model with only two traded goods, an export tax is symmetrical with a tariff, so that either form of tax can restrict trade to the same extent and improve the terms of trade.

It is obvious that if the elasticity of demand for exports is less than unity, it will pay to restrict the quantity of exports since export income will be raised while fewer resources are used for export production. But the argument goes further than this since it will pay to restrict export supply or import demand even when the export demand elasticity is greater than unity. Nevertheless, this trade restriction cannot be unlimited, for if no trade remains, there can be no benefit from improved terms of trade.

There must be an optimum degree of trade restriction at which the marginal gain from improved terms of trade is just equal to the marginal loss from reduced use of the international division of labour. So we have the well-established concept of the 'optimum tariff' designed to exploit optimally a country's monopoly power. It has been a theme of this book so far that optimum tariffs or export taxes may be positive if certain constraints on the use of other policy devices are accepted, or if these other devices involve high costs. This has been true even when terms of trade effects have been assumed away. So the term 'optimum' must be treasured; and to distinguish the narrower concept of the optimum tariff or export tax designed to improve the terms of trade, the latter will be specifically called the *orthodox* optimum tariff or export tax here.

The theoretical literature on this subject is vast, and the lineage ancient. The basic idea that a tariff can turn the terms of trade in a country's favour, and that this may provide a case for a tariff, can be found in writings by Torrens, John Stuart Mill, and Sidgwick.[1] But the modern theory really originates with Bickerdike, who first made explicit the idea of an optimal degree of trade restriction, and that this can be attained by either a tariff or an export tax.[2] A large number of contributions in the nineteen thirties and forties—notably by Lerner, Kaldor, Scitovsky, Kahn, and Graaff[3]—developed and refined the theory and brought out both its significance and limitations. The best available expositions are by Scitovsky and Graaff.[4]

[1] John Stuart Mill, *Essays on Some Unsettled Questions of Political Economy* (1844), London 1948, pp. 24–36; Henry Sidgwick, *The Principles of Political Economy*, 2nd ed. London 1887. M. C. Kemp, in 'The Gain from International Trade and Investment: A Neo-Heckscher-Ohlin Approach', *American Economic Review*, 56, September 1966, p. 788, cites Torrens' *Essay on the Production of Wealth* (1824) and *The Budget. On Commerical and Colonial Policy* (1844).

[2] C. F. Bickerdike, 'The Theory of Incipient Taxes', *Economic Journal*, 16, December 1906, pp. 529–535, and 'Review of A. C. Pigou's *Protective and Preferential Import Duties*', *Economic Journal*, 17, March 1907, pp. 98–102. Also F. Y. Edgeworth, 'The Theory of International Values', *Economic Journal*, 4, March 1894, pp. 35–50.

[3] A. P. Lerner, 'The Diagrammatical Representation of Demand Conditions in International Trade', *Economica*, N.S., 1, August 1934, pp. 319–334; Abba P. Lerner, *The Economics of Control*, Macmillan, New York 1944, pp. 382–5; Nicholas Kaldor, 'A Note on Tariffs and the Terms of Trade', *Economica*, 7, November 1940, pp. 377–380; T. de Scitovsky, 'A Reconsideration of the Theory of Tariffs', *Review of Economic Studies*, 9, 1941–2, pp. 89–110, reprinted in H. S. Ellis and L. A. Metzler (eds.), *Readings in the Theory of International Trade*, Blakiston, Philadelphia 1949; R. F. Kahn, 'Tariffs and the Terms of Trade', *Review of Economic Studies*, 5, 1948–49, pp. 14–19; J. de V. Graaff, 'On Optimum Tariff Structures', *Review of Economic Studies*, 17, 1949–50, pp. 47–59.

[4] Scitovsky, 'A Reconsideration of the Theory of Tariffs', in Scitovsky, *Papers on Welfare and Growth*, Stanford University Press, 1964, (omits the lengthy discussion of community indifference curves which was in the original article); J. de V. Graaff, *Theoretical Welfare Economics*, Cambridge University Press, 1957, Ch. IX. Many textbooks expound the basic theory in terms of the usual two-good international trade geometry. See, for example, J. E. Meade, *A Geometry of International Trade*, Allen & Unwin, London 1952. For a mathematical treatment pursuing many intricacies, see M. C. Kemp, *The Pure Theory of Trade and Investment*, Prentice-Hall, 1969, pp. 296–313. See also the Appendix to this chapter.

I
THE OPTIMUM TRADE TAX IN THE TWO-SECTOR MODEL

(1) *Simple Theory*

The theory of the optimum tariff or export tax in the two-sector model is usually expounded in terms of offer or reciprocal demand curves, and the reader is referred to the many textbooks that follow this approach, and also to the Appendix to this chapter. Here a different method of exposition will be attempted, the aim being to explain in a simple way some

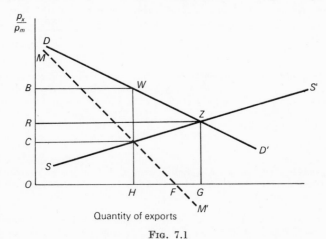

Quantity of exports

Fig. 7.1

finer points of the theory that are often found puzzling. With only two traded goods, there is a complete symmetry between a tariff and an export tax, and the analysis could be presented in terms of either. It is assumed that balance of payments equilibrium is maintained, perhaps by domestic price flexibility or exchange rate adjustment. It will be convenient to present the analysis in terms of the optimum export tax here.

In Figure 7.1 the quantity of a country's exports is shown along the horizontal axis and the price of exports in terms of imports (the terms of trade) along the vertical axis. Thus the import price is the numeraire. DD' is the foreign demand curve for exports. It shows the average revenue obtainable from exporting, expressed in terms of imports, and can be derived from the foreign reciprocal demand curve (or offer curve) of the usual diagram. The marginal revenue curve is MM'. The domestic supply curve of exports is SS'; it can be derived from the domestic offer curve in the usual diagram. Underlying it are the domestic demand and supply curves for exportables, all expressed in terms of the relative price of exports to imports. The underlying demand curve is based on the assumption of given tastes and distribution of income, and embodies both income and substitution effects. Free trade equilibrium is at Z, with exports of OG and a price of OR (implying imports of $RZGO$).

The domestic supply curve shows the marginal cost of supplying various quantities of exports, being equal at every point to the marginal cost in terms of importables of producing exportables and of consuming exportables. At the free trade point this marginal cost is equal to average revenue and not marginal revenue. Thus the marginal rates of transforming exportables into importables at home and abroad are not equal, and free trade is not optimal. So it is in the country's interests to behave like a monopolist and restrict the export supply until marginal revenue is equal to marginal cost. The optimum situation is where exports are OH and the price has risen to OB. This is attained by an export tax of BC/OB. The optimum point on the demand curve is W.

The elasticity of the demand curve at W is OB/BC (average revenue divided by the excess of average over marginal revenue). If t_x = the optimum export tax rate and η_x = elasticity of export demand, it follows that at W, $t_x = 1/\eta_x$. This is the formula for the optimum export tax, using imports as the numeraire.

Before going on, let us note some assumptions we have made, all but one of which will be removed later.

(1) The value of exports is always equal to the value of imports, so that the balance of trade is in equilibrium; one can assume that this is maintained by exchange rate adjustment.

(2) The foreign demand curve, DD', is given; hence foreigners do not alter their taxes on trade in response to the home country's export tax or tariff.

(3) There is perfect competition in the country; hence government action is required if the country is to behave like a monopolist.

(4) There are no domestic divergences; hence the supply curve SS' shows marginal social cost of exports.

(5) The revenue raised by the tax is spent by the government wholly on the importable product; hence it does not affect the domestic demand for exportables and thus the SS' curve.[5]

(6) A desired income distribution is maintained by a costless income distribution policy; hence we are concerned with Pareto-efficiency and not equity, and the domestic demand pattern (which depends, among other things, on income distribution) is given.

(7) We are interested only in the welfare of the country concerned, not that of foreign countries.

(2) *Domestic Conditions are Relevant*

Assume for the moment that the elasticity of the foreign demand curve, DD', over the relevant range from Z upwards is everywhere the same. This is possible, though it clearly is not so in the diagram. If it were so and if the elasticity were known, we would immediately know the optimum export tax from the formula given above. It has puzzled many students and used to puzzle the author why this optimum tax rate does not depend in this case also on domestic demand and supply conditions. Do the position and elasticity of the SS' curve not matter at all?

The answer is that the *rate* of tax depends only on the elasticity of the foreign demand curve. But the resulting point of equilibrium W, and thus the price, the extent of protection and the amount of trade depend also on the domestic supply curve. If SS' were steeper below Z, it would cut MM' at an

---

[5] This assumption is not essential to any of the following arguments and so will not be removed later; it happens to be convenient for the particular geometric exposition used. If the revenue were spent partly or wholly on exportables, or if it were refunded to the general public which spent it partly on exportables, the supply curve SS' would shift to the left, and this would reduce the optimum quantity of exports. When the offer curve technique is used it is easier to show the consequences of various ways of spending the revenue.

output greater than OH, but the rate of the optimum tax would be the same. What one can say is that if the foreign elasticity at the free trade point Z is known, then the optimum rate of tax (given the assumption of a constant elasticity along DD') is also known. But unless we know the elasticity of the domestic supply curve, we will not be able to foresee the effects on prices and outputs of such a tax.

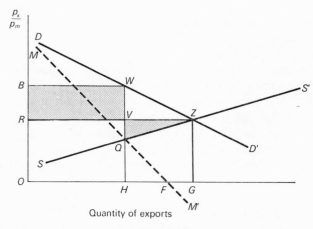

Quantity of exports

FIG. 7.2

Furthermore, it could be shown that the extent of the welfare gain from imposing the optimum tax depends on the elasticity of the domestic SS' curve; the higher the elasticity the greater the gain. If the elasticity were zero, there would be no reduction in export supply and hence no rise in the foreign price, and so no gain from the tax.

In Figure 7.2 the gain might be represented as follows. On the quantity of exports that remains after the tax, namely OH, the terms of trade have improved by RB, so that the total gain (shaded) is $RBWV$. But exports have been reduced by HG. The cost of these exports in terms of production of the importable forgone was $HQZG$, while the total receipts in the export market were $HVZG$, so that by forgoing these exports and producing more of the importable instead, a loss of QVZ (also shaded) is incurred. This latter cost is the cost of protection and must be set against the terms of trade gain.[6] The effect of the export tax is thus to yield a net gain of $RBWV$ minus QVZ. It can be readily verified that this net gain will be higher the flatter the supply curve through Z.

(3) Optimum Trade Tax Rate and Foreign Elasticity Determined Simultaneously

So far the optimum tax rate could be simply derived if one knew the elasticity of the foreign DD' curve, which was assumed to be the same all along the relevant part of the curve. But this simple approach is no longer adequate when the elasticity changes along the curve.

The normal and reasonable assumption is that the elasticity falls as the price falls. This, for example, is the property of a straight line curve. It could be shown that, leaving aside very peculiar cases, at very high prices the elasticity must be greater than one and at very low prices less than one. In general, a demand curve need not show a smooth continuous downward movement of the elasticity as the price falls, though this will be assumed here. In Figure 7.1 the elasticity is unity at F (where marginal revenue is zero), greater than unity to the left of F and less than unity to the right of F.

With this new assumption of a changing elasticity, knowing the elasticity at Z does not tell us what the optimum rate of export tax is. We know only that at the optimum point on the curve, wherever it is, the formula will be satisfied. Our formula seems to tell us that at F (where $\eta_x = 1$) the optimum tax is 100 per cent, so that domestic export producers would be getting a zero price. This is correct, but not so puzzling. As the tax rate is raised towards 100 per cent, the elasticity will rise and so the optimum tax rate will fall below 100 per cent. The formula also seems to say that the optimum tax rate is greater than 100 per cent when $\eta_x < 1$, but such a tax is not possible. One cannot tax away more than the whole of export income. The optimum tax is actually 100 per cent as long as η_x is unity or less.

---

[6] This assumes that domestic consumption of the exportable remains unchanged. If it changed owing to substitution effects, but income effects were negligible, the method would still be acceptable. With income effects we have the usual measurement-of-consumers' surplus problem.

Figure 7.3 is helpful for understanding these relationships. The horizontal axis shows the rate of export tax and the vertical axis the elasticity of the demand curve.

The curve QQ' shows how the elasticity rises as the tax is raised. Thus, as drawn, at the zero tax (free trade) point, equivalent to Z in Figure 7.1, the elasticity is less than unity, namely OQ. When the tax is ON the elasticity is unity (the point F in Figure 7.1). The curve QQ' need not slope continuously upward, though this is likely to be its general tendency:

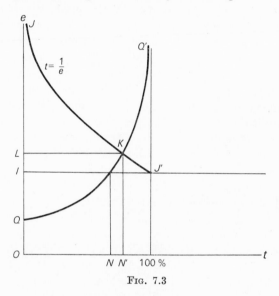

FIG. 7.3

reducing the volume of trade through raising the tax is likely to raise the elasticity.

The curve JJ' is the relationship derived from the optimum tax formula $t_x = 1/\eta_x$. It shows that the optimum tax is 100 per cent when the elasticity is unity and is zero when the elasticity is infinite. The curve should not be carried on towards the right of the 100 per cent line, as might be suggested by the formula; rather at J' it continues vertically downwards.

We can imagine starting in free trade with elasticity OQ and the optimum tax apparently 100 per cent. As the tax is raised, the elasticity rises; once the elasticity is greater than unity, the optimum tax declines from 100 per cent, and the two meet at K, with an optimum tax of ON' and an elasticity of OL. Thus ON' turns out to be the optimum tax. It is important to note that the curve QQ' depends not only on the DD' curve of Figure 7.1 but also on the domestic supply curve, SS'. For any given tax rate, the elasticity of DD' depends on where SS' intersects MM' (Figure 7.1), which depends on the shape of SS', and not just on the characteristics of DD'.

It can be said that, irrespective of the precise shape of SS'—and hence of QQ' in Figure 7.3—if the elasticity of export demand is unity or less, the tax should be raised; we know that it is below the optimum. When the elasticity is less than unity, this is obvious. It is possible then to raise export income, and hence to obtain more imports, by reducing the supply of exports; it must pay to restrict trade in that situation, though how much would depend also on domestic demand and supply conditions. Conversely—and this is a point of practical significance—*if* the tax is at the optimum, then the elasticity of export demand must be greater than unity.

A change in domestic demand conditions will now change the optimum rate of tax. If domestic demand shifts towards the country's exportable product, the export supply curve SS' in Figure 7.1 will shift to the left and hence the quantity exported for a given rate of tax will decline. This means that there is a movement upwards along the foreign DD' curve, so that the foreign elasticity is likely to rise (though, conceivably, it could fall). This effect can be represented in Figure 7.3. The QQ' curve will shift upwards and to the left, since for a given tax rate the elasticity is now higher. The rate of the optimum tax (given by the intersection of QQ' with JJ') will thus fall.

It is rather interesting that if one makes the reasonable assumption, which we have been making here, that the elasticity of foreign reciprocal demand rises as the quantity of exports is reduced, then a shift in the domestic demand pattern which would reduce the volume of trade and improve the terms of trade should lead to an adjustment in the trade tax so as to modify the effect on the volume and terms of trade. The volume of trade should still fall, and the terms of trade should still improve, but not by as much as they would have if the export tax had stayed unchanged. Similarly, a shift in the demand pattern towards importables, which naturally would increase trade and worsen the terms of trade, should be moderated by an appropriate rise in the export tax. The terms of trade should be allowed to deteriorate, but not by as much as in the absence of any change in the tax.

(4) *Optimum Tariff and Export Tax Related*[7]

In a two-good model trade can be restricted by either an export tax or a tariff. When fixed at comparable rates they will have the same effect on the terms of trade. Hence the whole of the discussion so far could have been carried out in terms of a tariff rather than an export tax, and indeed in the theoretical literature this usually has been done.

From the foreign offer curve could have been derived a supply curve of imports represented in a diagram where the vertical axis shows the price of imports in terms of exports. One would then derive from the domestic reciprocal demand curve a demand curve for imports. On the lines of the argument developed above the optimum rate of import tariff t_m would be equal to the inverse of the import supply elasticity ϵ_m.

Furthermore, there are two other formulae one might use. The export demand elasticity η_x used above has been derived from the same foreign offer curve as the import supply elasticity ϵ_m. From the relationship between the elasticity of the demand curve and the elasticity of the supply curve into which it can be converted in a two-good model, it follows that there is a unique relationship between η_x and ϵ_m, namely $\eta_x - \epsilon_m = 1$. Bearing in mind that $t_x = 1/\eta_x$ and $t_m = 1/\epsilon_m$, it follows that we can express the optimum export tax t_x in terms of ϵ_m or the optimum tariff t_m in terms of η_x. Hence $t_x = 1/(1+\epsilon_m)$ and $t_m = 1/(\eta_x - 1)$.

The latter formula is often used. It tells us, for example, that if the elasticity of export demand (expressed in terms of imports) is unity, the optimum tariff is infinite. This is the

---

[7] For further details and exposition see the Appendix to this chapter; and also H. G. Johnson, 'Alternative Optimum Tariff Formulae', in Johnson, *International Trade and Economic Growth*, Allen & Unwin, London 1958, pp. 56–61, where the various formulae are derived.

equivalent in our main formula $(t_x = 1/\eta_x)$ of the case where the optimum export tax is 100 per cent when the export demand elasticity is unity.

II
MULTI-COMMODITY MODEL

So far we have presented the essence of orthodox optimum tariff theory. Its formal development has always been in terms of the two-sector model, but it would be of little practical use if it could not be expanded to a multi-commodity model. This can be done easily if one makes two rather severe assumptions. (1) For each import and each export there is respectively a foreign supply curve or demand curve which is independent of all the other curves. (2) The elasticity of each of these curves over the relevant range is constant.

These assumptions mean that for each export there will be an optimum export tax rate as given by the formula, and for each import similarly an optimum tariff rate. These yield the orthodox optimum protective structure. The exchange rate maintains balance of payments equilibrium. The optimum rate for one product assumes that the appropriate optimum taxes are being imposed on all other products.[8]

The interesting feature about this solution is that one can apparently look at each traded product separately and determine its optimum rate of tax from the standard formulae. But this may be a misleading impression. It is true that the *rate* of optimum tax is uniquely determined by the relevant foreign elasticity and is independent of the positions and elasticities of the curves for other products. But the quantities traded and the domestic and foreign prices that will result

[8] This simple approach originated in A. P. Lerner, *The Economics of Control*, pp. 382–5, and is discussed critically by Graaff in 'On Optimal Tariff Structures', who stresses the assumption required that cross-elasticities of the foreign curves must vanish identically.

A model which is a bridge between the two-sector and the multi-good model is expounded in the Appendix to this chapter in connection with the Bickerdike-Graaff optimum tariff formula.

The optimum tariff-export tax structure in a multi-commodity model is also discussed in I. F. Pearce, *International Trade*, Macmillan, London, 1970, pp. 201–208.

from an optimum tax will depend on what is happening to other products because the *domestic* demand and supply curves are interrelated in a general equilibrium system. If the world demand for wheat falls, so that the quantity of wheat produced and exported declines, the supply curve of maize production may shift to the right, and this will influence the extent to which a given export tax on maize reduces maize exports below their level in the absence of such a tax.

If the elasticities are not constant along the foreign demand and supply curves (assumption (2) removed) the matter becomes much more complicated. It must be said immediately that the problem that results is extremely academic. Elasticities of export demand and import supply are rarely known with any precision, whether at any one point or at various points of a hypothetical curve. In practice the general idea has to suffice that trade taxes should, from the point of view of the orthodox optimum structure, be relatively higher where the elasticities are relatively lower, with some vague idea of the magnitudes of the elasticities.

The theoretical problem is now that for each product we cannot know the elasticity unless we know the full optimum. In principle one needs the solution to a complete general equilibrium optimizing exercise, out of which will come the set of optimum trade tax rates. All one can say is that the property of the optimum can be clearly described: for each good, at the optimum, the rate of trade tax will be the inverse of the elasticity of import supply or export demand at that point. But this does not help one to know the optimum structure of rates before it is imposed. Starting in free trade or with some existing non-optimal trade tax structure one could imagine reaching the optimum structure by successive approximations.

The matter becomes really complicated if one cannot assume that the foreign demand and supply curves are independent of each other (assumption (1) removed). Clearly, in the case of goods that are close substitutes, there will be interrelationships, and it may be best, at least conceptually, to group goods together, and treat them as one good, whenever the gaps in the chain of substitution are narrow. But many countries export products which, while they might compete closely for factors of production, are sold in quite different markets. They may be subject to common macro-economic influences on their demand curves, but it is certainly reasonable to think of the demand for one product being independent of the amounts put on the market for another.

Can one relate the orthodox optimum protective structure to the concept of effective protection? It may seem strange that this concept, which is so central to multi-commodity protection theory, has not been mentioned so far. It needs first to be noted that the precise theory of effective protection depends on the small country assumption, hence making the orthodox optimum trade taxes zero. But the general idea of varying degrees of protection for value added still survives, so that a question does remain. The answer is happily quite simple: from the present point of view one can ignore effective protection.

One must go back to first principles. For every import and export there is an optimum rate of *nominal* tariff or export tax which in fact produces an *adjusted* import supply curve or export demand curve facing domestic producers and consumers that indicates the *marginal* social cost of importing or the *marginal* social return to exporting. The unadjusted foreign curve indicates the *average* social costs and returns to which domestic producers and consumers then equate their marginal costs and returns. From a social point of view the unadjusted curves give thus the wrong signals to private producers and consumers. The adjustments, in the form of a set of optimum tariffs and taxes, put the signals right. The set of effective protective rates that will result from this adjustment cannot be simply described, other than that it will provide inducements to shifts in the production pattern that will be optimal.

It follows therefore that the input-output relationships between traded goods do not alter the general principles for determining the optimum protective structure. This is not surprising. The marginal divergences that the structure is designed to correct are *trade* divergences; foreign trade prices— which are nominal prices—are priced wrongly in the domestic market. Thus they must be put right. Only if the distortions were domestic *production* divergences would the adjustments have to be made directly in terms of effective rates.

III
PRIVATE MONOPOLY

So far we have assumed that all domestic industries and traders are competitive. The aim of the optimum tax or tariff has been to exploit the country's monopoly or monopsony power, something which it has been assumed private industry or traders do not do. Now let us allow for private monopoly, considering first an export industry.

If there is a sole domestic exporter of a product, he will bear in mind the effect extra export sales have on his price and will maximise his profits by equating marginal cost not with average but with marginal revenue. An export tax is not needed to induce him to restrict sales so as to exploit his monopoly power. This is relevant for countries that export highly differentiated manufactures, especially capital-goods. There is no need for a country to induce its aircraft manufacturer to export less so as to exploit any monopoly power he may have. Countries that have agricultural marketing boards with monopolistic powers that aim to maximise profits for their producers or the government also do not need export taxes in addition, the profits of the boards being, in fact, export taxes.

The export industry may be 'somewhat' monopolized; it may be an oligopoly, or there may be monopolistic competition so that any one firm takes *some* account of the price fall resulting from extra sales, but not complete account since it suffers only part of the adverse effect on the prices of previous sales. Hence, unless the firms combine to behave like a monopoly, the optimum export tax is positive, but it is less than that indicated by our formula, and the fewer the firms the lower it is. For each export one might have an optimum tax based on the estimated export demand elasticity modified by some kind of 'monopoly factor'.

The same sort of argument would apply if there were, for example, two exports X_a and X_b, each of which is exported by a monopoly but where the foreign demands are interrelated. Extra exports of X_a would tend to reduce the price of X_b, and vice versa. The producer of X_a would take into account the price fall of his own product when he exports more of it, but not the price fall of X_b. One could think here of a combined demand curve for the two products. The optimum export tax will again be positive but less than indicated by the formula.

It is also possible that the two exports are complements, not substitutes, in foreign demand. This suggests the unexpected but plausible possibility that sometimes an export subsidy, not a tax, may be optimum. If there are complements, extra supplies of one export will shift the other's demand curve to the right. The producer of X_a would take into account the price fall of his own product resulting from extra sales but not the price rise in X_b. He would be restricting his sales below the social optimum unless encouraged to sell more by a subsidy.

While monopsony in importing is likely to be less common than monopoly in exporting, there can be monopsony or oligopolistic buying in the case of intermediate goods where the purchasers are monopolies or oligopolistic producers. And in some countries there are monopoly trading firms, sometimes state-owned, even though final buyers are competitive. So the same argument applies on the import side. The aim of the optimum tariff is to restrict demand so that the foreign import price will fall. If private buying is already monopolized there is no need for this.

IV
FOREIGN REACTION AND RETALIATION

The orthodox optimum tariff argument is concerned with maximising *national* income; world income as a whole (in the potential sense) is reduced, but at the same time is redistributed towards the trade-restricting country. The orthodox optimum tariff is thus from a world point of view a second-best way of income redistribution. Maintaining free trade and then transferring income direct would be first-best. Given the difficulty of doing this, from the national point of view, the orthodox optimum trade tax is first-best. It is a way in which—given the relevant assumptions—a country can unilaterally bring about the desired redistribution.

But can the assumptions be given? The simple model assumes that foreign tariffs are given and that foreign governments will happily stand by and allow income to be redistributed against them. Still staying within the framework of the general type of model, one could then proceed in one of three ways.

Firstly, one might assume that foreigners apply optimum tariff policy in any case. They take as given our country's tariffs and fix their optimum tariffs in the light of these. Thus we impose what appears in the first instance to be an optimum tariff; this may change the elasticities of the curves facing the foreigners, so they alter their tariffs; this in turn may alter the elasticities facing us, and so requires us to change our tariffs further, and so on.

It is perfectly possible that the elasticity facing the foreigner remains unchanged as a result of our tariff, so that the foreign tariff would not in fact alter when we impose an optimum tariff. It is also possible that the elasticity facing the foreigner rises, in which case the foreign tariff will actually fall. There is, in this process, no element of foreign 'retaliation' but only one of foreign *reaction*. Taking the foreign reaction into account, the gains from an orthodox optimum tariff may be greater or less than in the absence of the foreign reaction.

Secondly, one might assume that if our country does not impose a tariff the foreign country will not do so either, but that the adoption of orthodox optimum tariff policy by us will provoke foreigners to embark on the same policy. Thus we would start with free trade, and once we abandon it would provoke the reaction process described above. There is then an element of 'retaliation', the foreign country retaliating by initiating the reaction process.

Johnson[9] has worked out a two-country model along these lines in great detail. It is obvious that the gain from an optimum tariff to a country will be less than when there is no retaliation from foreigners causing them to impose optimum tariffs, and on balance both countries may well lose compared with no tariffs at all. Johnson shows that the reaction process might converge to a policy equilibrium, or alternatively there might be a tariff cycle. In equilibrium some trade must remain. While it would seem probable that both countries would

[9] H. G. Johnson, 'Optimum Tariffs and Retaliation', *Review of Economic Studies*, 21, 1953–4, pp. 142–153 (reprinted in his *International Trade and Economic Growth*).

finally lose compared with mutual free trade, he has discovered special cases where one country would gain in spite of the other country having imposed a tariff for the first time.

If one really thinks in terms of two countries only, both these approaches seem quite inadequate. They assume myopic behaviour by the governments of the two countries and disregard the possibility of agreement between them for mutual gain. But Scitovsky,[10] who pioneered this analysis, has suggested that the model should really refer to a many-country world, in which case each country would not expect to get much reaction to its tariffs from foreign countries; it would simply take other countries' tariffs as given and optimize subject to these. Furthermore, it would not expect much benefit from negotiating mutual tariff reductions with just one other country, so that only multilateral negotiations—which are always much more difficult—would have much purpose.

This interpretation makes the approach seem more sensible, though there are still difficulties. Consider first our simple reaction model with no element of retaliation. If a country is so small that any change in its tariff has no effect on other countries' tariffs then it is a *small country* and the orthodox optimum tariff is zero. If it is large enough for the optimum tariff to be positive then the possibility of foreign reactions which affect the level of its own optimum tariff cannot be ruled out. But in that case this reaction should be expected in the first place.

If a tariff increase by our country is likely to provoke increases in the tariffs of some of our trading partners as a result of the reaction process, then our true optimum tariff rate will be less than the orthodox optimum rate. Since it may be optimum for foreigners to reduce their tariffs when our country raises its tariff, it is also possible for the true optimum rate to be above the orthodox optimum rate. But these complications may be trivial. They arise only because elasticities of offer curves are allowed to change as a result of tariff-induced shifts in the curves or along the curves. But elasticities are never known precisely in any case. A reasonable assumption for a country may be that the elasticities facing foreigners remain unchanged when it alters its tariffs, so

[10] T. de Scitovsky, 'A Reconsideration of the Theory of Tariffs'.

that—in the absence of retaliation or bargaining responses—foreign tariffs will not alter.

The difficulty is greater in the case of the Johnson model where foreigners are assumed to retaliate to our country's optimum tariff by embarking on their optimum tariff policy for the first time. The main assumption is that foreigners choose to react with an orthodox optimum tariff rather than a bargaining tariff; they do not expect to be able to alter our country's basic policy. This may correctly represent some actual situations. The less convincing assumption is that our country is assumed to ignore, or fail to predict, possible foreign retaliation. If such retaliation can be foreseen, it may still be optimal for our country to impose a tariff, but not necessarily at the orthodox optimum level.

The third possibility is to accept that countries are in oligopolistic relationships to each other. Each country will bear in mind the likely reactions of other countries to its policies, and in addition will aim to alter other governments' policies. Relevant models need then to come from game theory. It is in these terms that retaliation and bargaining tariffs have to be explained. One might imagine country B to start with an orthodox optimum tariff. Country A then imposes, or threatens to impose, a bargaining tariff which is high enough for B to agree to mutual tariff disarmament. The bargaining tariff may well be far above A's orthodox optimum tariff, and so may inflict a short-term loss on A as well as on B.

Countries rarely impose substantial tariffs all at once designed explicitly to improve their terms of trade and tax the foreigner. Rather, they use tariffs or import restrictions in preference to devaluation or export subsidies to encourage import-substitution in the process of growth so as to avoid the terms of trade deterioration *that might otherwise take place*. In addition, it has been common for countries to cope with balance of payments difficulties by using import restrictions rather than devaluation in order to avoid adverse terms of trade effects. Again, the aim was not explicitly to improve the terms of trade but to avoid a deterioration. The same result is also achieved when a country deals with a balance of payments problem by devaluing, and then supplements this with export taxes to avoid a terms of trade deterioration.

This sort of behaviour makes retaliation much less likely. On the other hand, it appears to make the simple tariff-reaction model relevant, at least as suggesting one element in long-term tariff adjustment. Since foreigners' reactions to changes in elasticities facing them, provoked by our country's tariff changes, will be delayed until the foreigners want to make balance of payments adjustments themselves, it seems conceivable that our country would ignore such prospective reactions.

V
GROWTH AND THE TERMS OF TRADE

It can be shown that—given fairly plausible assumptions—if there is a long-term tendency for a country's terms of

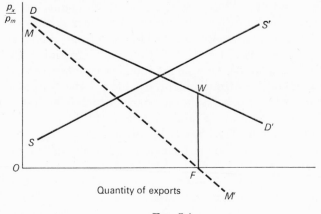

Fig. 7.4

trade to decline, it should steadily increase its degree of protection. The orthodox optimum trade tax rate will rise sufficiently to modify the terms of trade deterioration that would otherwise take place, but not sufficiently to prevent it altogether. To explain this, we shall confine ourselves to the two-sector model. We assume no foreign reactions or retaliation.

Figure 7.4 shows (like Figure 7.1, p. 160) the home country's general equilibrium export supply curve SS' and the general equilibrium export demand curve, DD', that it faces. Both curves may shift over time as a result of growth. Suppose that the home country's growth is 'unbiased'; this means that at constant prices its domestic demand for and supply of exportables will expand at the same rate as total income and output expand, and hence the difference between demand and supply, namely exports available at that price, will also expand at that rate.[11] Thus the rate at which the SS' curve shifts to the right is given by the country's general rate of growth. If growth is 'biased' the SS' curve will shift to the right at a higher or lower rate, and it is even possible that it shifts to the left. Similarly the foreign export demand curve, DD', may shift, perhaps also moving to the right.

Given free trade, the net result may be that the terms of trade change over time, the direction of movement depending on relative shifts. If growth is unbiased both at home and abroad, the terms of trade will move against the faster-growing country. To simplify for the moment, we shall now assume that there is no growth or change abroad, so that the DD' curve stays in place, but there is growth in the home country causing the SS' curve to shift to the right over time. In free trade the intersection of the moving SS' curve with the static DD' curve will then trace out the terms of trade and export quantities over time.

Let us now introduce the orthodox optimum export tax. If it is imposed at all times, the quantity of exports will be traced out by the intersection of the SS' curve with the MM' curve (the marginal revenue curve), the terms of trade thus also falling over time, though being at every point in time better than if free trade had been maintained. Thus one can clearly describe the path of the trade volume and of the terms of trade with an optimum trade tax.

If the elasticity of the DD' curve remained unchanged over the relevant range, the *rate* of optimum tax would stay constant even though growth is worsening the terms of trade.

---

[11] The standard Hicks-Johnson analysis of the effects of growth on the terms of trade using the concepts of 'trade' and 'anti-trade' biases is being used here. See H. G. Johnson, *Money, Trade and Economic Growth*, Allen & Unwin, London 1962, Ch. 3.

7

On the other hand, if we make our usual assumption that the foreign elasticity falls as the price falls, the rate of optimum tax will rise in the process of growth. It will eventually have to rise sufficiently to prevent the elasticity of demand becoming unity or less.

In free trade, as the supply curve shifts to the right, the unity point (W in Figure 7.4) would eventually be passed. But an optimum tax would prevent this. The quantity of exports could never get beyond OF, a level which it could reach only if the marginal cost of export supply became zero, the supply curve coinciding with the horizontal axis.

We could also let the foreign demand curve shift. Suppose that it moves to the right iso-elastically, and that its elasticity falls as the price falls. Then the rate of optimum tax will rise if the SS' curve moves more to the right, so that the terms of trade deteriorate, while the rate of optimum tax will fall if the DD' curve moves more to the right.

This is the most general statement of our result. If in free trade the terms of trade would deteriorate over time then the rate of tax should be increased over time, while if they would improve, the rate of tax should be reduced. But as long as the foreign elasticity exceeds unity, the whole of the deterioration in the first case should not be prevented. The argument hinges completely on the assumption that the elasticity of any given foreign offer curve (from which the DD' curve is derived) falls as the volume of trade increases.

Given free trade, a country's growth may actually lower its real income. This has been called *immizerizing growth*. The loss from worse terms of trade could more than offset the gain from expanded production possibilities. Of course immizerizing growth could also be brought about in other ways—such as increasing external diseconomies (congestion and pollution) in the process of growth—but these do not concern us here. If a country insisted on maintaining free trade, it would then benefit from restraining its own growth. It has been shown by Bhagwati that necessary, though not sufficient, conditions for such immizerizing growth are that the foreign elasticity of demand for exports is less than unity, so that in Figure 7.4 the supply curve movement has brought equilibrium to the right of W, or alternatively that growth actually reduces the domestic production of importables at constant prices.[12]

The relevant point to note here is that once the optimum export tax or tariff is consistently applied, the possibility of immizerizing growth disappears. A tax continually modified to be optimal will ensure that at every point in time the country makes the best of its opportunities: it maximizes its welfare, given the two constraints of domestic production possibilities, which have expanded owing to growth, and transformation possibilities through foreign trade. Since growth expands the domestic production possibilities, it cannot—*given the imposition of an optimum tax at every point in time*—make the country worse off.

VI
BALANCE OF PAYMENTS POLICY AND OPTIMUM TARIFFS

There is an intimate connection in practice between countries' balance of payments policies and their orthodox optimum trade tax policies. Countries rarely impose *ab initio* tariffs or export taxes to improve the terms of trade, if only to avoid foreign retaliation or the adverse domestic income effects for exporting or import-consuming interests. But when faced with a balance of payments problem, they are likely to take possible terms of trade effects into account in choosing their policies, and may end up imposing tariffs or export taxes which might be regarded as something like orthodox optimum trade taxes.

Suppose, to simplify, that a country starts with a balance of payments deficit and no trade taxes or subsidies. Assume away all complications, such as foreign reactions or retaliation,

---

[12] Jagdish Bhagwati,'Immizerizing Growth: A Geometrical Note', *Review of Economic Studies*, 25, June 1958, pp. 201–5. The main ideas were anticipated by Harry Johnson; see *International Trade and Economic Growth*, p. 69 and pp. 142–3 (the latter originally published in 1953). Also, see M. C. Kemp, *The Pure Theory of Trade and Investment*, pp. 107–9. Kemp has noted that the idea has been discussed by Mill, Edgeworth and Bastable. Note especially F. Y. Edgeworth, 'The Theory of International Values', *Economic Journal*, 4, March 1894, p. 40, where he draws out the implications of Mill's remarks on the effects of technical change in an export industry on the terms of trade: 'the exporting country is damnified by the improvement'.

domestic monopoly, domestic divergences of all kinds, concern for the domestic income distribution, and effects of devaluation on capital movements. First-best policy then consists of a composite package of (a) the orthodox optimum set of trade taxes, and (b) devaluation necessary to maintain balance of payments equilibrium.

An alternative to devaluation, having much the same effect on visible trade, is a uniform tariff and export subsidy, both at the same rate. In that case the optimum package will consist of *two* sets of tariffs (which can be amalgamated in practice), and both export subsidies and export taxes. On balance some exports would be subsidized and some taxed. Broadly, the export subsidies and taxes might be thought to cancel out for most exports, or for a category of 'traditional' exports where the country has a significant share in the world market, so that the optimum package would in fact consist wholly of tariffs. The way the matter is usually seen is that (a) tariffs were imposed to improve the balance of payments and (b) they were not accompanied by export subsidies, or were preferred to devaluation, because the latter would worsen the terms of trade.

If the elasticity of demand for exports is less than unity, so that fewer exports would lead to more foreign exchange earnings, there appears to be both a balance of payments and a terms of trade argument for an export tax. But this is a confusing way of looking at such a case. Once the orthodox optimum export tax is imposed, the elasticity will cease to be less than unity. If the balance of payments is still in deficit and devaluation is not possible, it will then be appropriate to *reduce* the export tax somewhat below its orthodox optimum level.

It might also be noted here that a devaluation should always succeed in improving the balance of payments if the orthodox optimum set of trade taxes is being applied at the same time, at least provided (a) the internal cost-level is stable and (b) there is no kink in the foreign demand curve.

The familiar elasticity formula for a devaluation to improve the balance of payments expresses the conditions when the devaluation-induced fall in the value of imports in foreign currency terms is greater than any fall in the value of exports, the latter also in foreign currency terms. But a devaluation can cause the value of exports to fall only if the foreign demand elasticity for decreases in price is less than unity. And if the orthodox optimum export tax rate is imposed, the elasticity for increases in price must be greater than unity. It is of course possible that a reduction in export supply would have little effect in raising the price while an equivalent increase in supply would lead to a large price drop. But leaving such a 'kinky' situation aside, and hence assuming that the elasticity is approximately the same for movements in both directions, the imposition of the orthodox optimum tax ensures an export demand elasticity greater than unity for the purposes of the devaluation formula.

In the nineteen-fifties and early nineteen-sixties, export pessimism in India and Pakistan, and also many other less-developed countries, was rather extreme, and it was widely believed that the elasticity of demand for exports was unity or even less, so that extra exports could not improve the balance of payments and might even worsen it.[13] This made it inevitable, and even logical, that balance of payments difficulties would be dealt with purely by restricting imports. While the extreme export pessimism was almost certainly ill-founded, the import-substitution bias of those years could be interpreted as an attempt to approximate an orthodox optimum export tax policy. Of course, if the export demand elasticity was really believed to be unity or less, the degree of restriction was actually below the optimum.

VII
PRACTICAL SIGNIFICANCE OF THE TERMS OF TRADE ARGUMENT

(1) *Elasticity Estimates for Advanced Countries*

To form a clear view about the significance of the terms of trade argument for trade taxes, one needs to have some idea of the magnitudes of the relevant elasticities. Quite a few

---

[13] Little et al., *Industry and Trade in Some Developing Countries*, pp. 231–4, and references cited there. The belief in a 'foreign exchange bottleneck' gave rise subsequently to the well-known 'two-gap' models for determining foreign aid requirements.

calculations of demand elasticities in foreign trade have been made, especially for advanced countries.[14] The purpose of the calculations has usually been to see how effective exchange rate adjustments are likely to be, and especially, whether it is possible that a devaluation will actually worsen the balance of payments. In the immediate post-war period, 'elasticity pessimism' was induced by calculations which were later shown to be biased downwards. Interest in such calculations has waned since the simple elasticities approach to balance of payments theory has become modified by the absorption approach.

There are great statistical problems, and it is apparent that not too much reliance can be placed on any of the figures that have been calculated; curves shift, circumstances change, and other things are, regrettably, never equal.[15] But the general conclusion that has emerged (insofar as the figures mean anything) is that elasticities appear generally to be high enough to ensure that a devaluation would improve the balance of payments, at least if appropriate policies of internal balance are being followed. In particular, export demand elasticities have usually been found to be well above unity, which is a sufficient condition for a favourable balance of payments effect.

The figures are not usually looked at from an orthodox optimum tariff point of view. Calculated elasticities that are high from a balance of payments point of view then appear quite low and seem to suggest quite high rates of tariff or

[14] See H. S. Cheng, 'Statistical Estimates of Elasticities and Propensities in International Trade: A Survey of Published Studies', *International Monetary Fund Staff Papers*, 7, April 1959, pp. 107–158, for a survey covering studies 1937–1957. Later studies are listed in Edward E. Leamer and Robert M. Stern, *Quantitative International Economics*, Allyn and Bacon, Boston, 1970, pp. 51–55, 74–75. To this list should be added H. S. Houthakker and S. P. Magee, 'Income and Price Elasticities in World Trade', *Review of Economics and Statistics*, 5, May 1969, pp. 111–125, who seem to have obtained some extraordinarily low export demand elasticities.

[15] See G. Orcutt, 'Measurement of Price Elasticities in International Trade', *Review of Economics and Statistics*, 32, May 1950, pp. 117–132. Also, the discussion and references cited in S. J. Prais, 'Econometric Research in International Trade: A Review', *Kyklos*, 15, 1962, pp. 561–577; and Leamer and Stern, *Quantitative International Economics*, Chs. 2 and 3, and many references in Cheng, 'Statistical Estimates of Elasticities and Propensities in International Trade'.

export tax. For example, in 1957 Harberger, in a review of various calculations designed to discover whether the international price mechanism worked, concluded that 'I would hazard the rule-of-thumb judgment that in the relatively short-run the elasticity of import demand for a typical country lies in or above the range -0.5 to -1.0, while its elasticity of demand for exports is probably near or above -2.'[16] This quotation underlines the fact that many of the calculations refer to the short-run, while longer-run elasticities would be more relevant for our purpose. But the main point is that an export demand elasticity of 2 means that the optimum export tax rate is 50 per cent! Harberger in fact doubted that countries had as much monopoly power as this figure implied, and felt 'quite confident that long-run elasticities of export demand are substantially greater than 2.'

An analysis by Kreinin[17] of the consequences for the United States of the reductions in her tariffs negotiated in 1956 concluded that these led to substantial increases in prices charged by foreign suppliers, perhaps close to half of the benefit from tariff concessions, with the other half yielding reduced prices to U.S. consumers. His analysis ignored the benefits to the U.S. from other countries' tariff reductions.

In 1965 Floyd published some rough estimates of U.S. foreign trade elasticities for the purpose of analysing the problem of the over-valuation of the dollar.[18] He obtained an overall import supply elasticity of 6·1 and an export demand elasticity ranging from -9.9 to -5.1. Using the higher limit of the latter, these figures would mean that the optimum tariff is about 17 per cent and the optimum export tax 10 per cent, and if only a tariff were used, the optimum rate would be about 30 per cent.[19]

Basevi[20] made use of these figures, and also of an estimate

[16] Arnold C. Harberger, 'Some Evidence on the International Price Mechanism', *Journal of Political Economy*, 65, December 1957, p. 521.

[17] M. E. Kreinin, 'Effect of Tariff Changes on the Prices and Volume of Imports', *American Economic Review*, 51, June 1961, pp. 310–324.

[18] J. E. Floyd, 'The Overvaluation of the Dollar', *American Economic Review*, 55, March 1965, pp. 95–107.

[19] The 30 per cent rate is obtained by using the Bickerdike-Graaff formula given in the Appendix.

[20] Giorgio Basevi, 'The Restrictive Effect of the U.S. Tariff and Its Welfare Value', *American Economic Review*, 58, September 1968, pp. 840–852.

of an average U.S. tariff of 15 per cent, so it is not surprising that he concluded that there was a net gain to the U.S. from her tariffs. When Walker[21] subsequently reviewed Basevi's results, he argued first that, on the basis of more recent calculations, the export demand elasticity is likely to be more like $-2\cdot0$ (which implies a 50 per cent export tax or a 133 per cent average tariff, if only tariffs are used), and secondly, that the gains to the United States from other countries removing their tariffs simultaneously need to be taken into account. In this case there would be a net gain to the United States from reciprocal tariff elimination, most of which would result from a net terms of trade improvement.

One should really qualify the apparently high optimum trade tax figures that appear to have emerged for advanced countries, notably the U.S., in two ways: first, one should remember that they are usually based on short-run elasticities since in the long-run the elasticities may be much higher; and secondly one must take into account foreign retaliation or reciprocal concessions.

(2) *Implications for Less-Developed Countries*

Most less-developed countries, like the smaller developed countries, are unlikely to be able to affect their import prices very much, and in the longer-run possibly not at all. But they are often significant world suppliers of particular export products. Hence, for these countries, the orthodox trade tax argument is an argument for export taxes, not tariffs. Since the export demand elasticities are likely to differ very much between different exports, there will be an optimum structure of export taxes with greatly differing rates.

While the use of tariffs would have the net effect of restricting export supply and so improving the terms of trade, no system of tariffs could yield exactly the same results as the optimum export tax structure. Hence tariffs would be second-best devices. This is an important conclusion. For

[21] Franklin V. Walker, 'The Restrictive Effect of the U.S. Tariff: Comment', *American Economic Review*, 59, December 1969, pp. 963–966. Walker based his estimates on calculations in H. Junz and R. Rhomberg, 'Prices and Export Performance of Industrial Countries, 1953–63', *International Monetary Fund Staff Papers*, 12, July 1965, pp. 224–71.

most of the countries that make extensive use of restrictions on imports, whether through tariffs or quantitative restrictions, even the terms of trade argument for tariffs turns out not to be a first-best argument.

The usual figures of demand elasticities for exports of primary commodities by less-developed countries concern the short-run, and are quite low. They may give a correct indication that in the short-run the less-developed countries would gain from restricting their export supplies beyond existing levels, or at least at the levels attained by existing trade restriction regimes. But in the longer-run, alternative supplies will come forth, whether of identical products from other less-developed countries, or of synthetic substitutes, and research will be initiated in response to high prices to develop new sources of supply and new substitutes. In assessing this matter, one may have to balance short-run gains against long-run losses.[22]

The demand elasticities facing less-developed countries as a whole will be much lower than those facing any individual less-developed country. It is one purpose of international commodity agreements to exploit this fact and to allow all the supplying countries to improve their terms of trade relative to the buying countries. The familiar difficulty is one of enforcement and ensuring complete membership of the exporting cartel. Only if the principal buying nations are also part of the scheme, and agree to enforce it by refusing to purchase increased quantities from outsiders, does such an arrangement become practicable. But then the buying nations would be collaborating in an arrangement to turn the terms of trade against themselves.

In general, less-developed countries do not have to be concerned with retaliation and foreign reactions of the type discussed earlier. Perhaps an exception must be made in the case of bilateral and group preferential arrangements, such as Commonwealth preference and the various arrangements the E.E.C. has made with less-developed 'associate countries'. Nevertheless, in general, their tariffs and export restrictions

---

[22] A comprehensive general discussion of the less-developed countries' scope for increasing their exports, including case studies of some success stories, is in *Industry and Trade in Some Developing Countries*, Ch. 7.

are unlikely to cause tariffs to be raised against them, and similarly tariff reduction is unlikely to lead to large reciprocal benefits. But this does not mean at all that one can generally assume the tariffs and import restrictions facing less-developed countries to be constant when considering their own optimum trade restriction policies. Another consideration needs to be introduced into our analysis.

(3) *Import Restrictions in Developed Countries and Voluntary Export Restraint*

The export pessimism which has been so prevalent in less-developed countries and has for so many years dominated their economic policies is explained only partly by a belief that the ordinary export demand elasticities for their primary commodities are low. A more important factor, especially in recent years, has been the sadly justified belief that increased exports, especially of labour-intensive manufactured products, provoke import restrictions in the developed countries.[23] These restrictions have applied particularly to textiles. The motive for them has been quite clear: it has nothing to do with the terms of trade but has been sectional income maintenance and employment.

Such restrictions, and the expectation of them, can be incorporated in our formal analysis. The orthodox export tax argument turns out to be very relevant. If one takes these restrictions, or at least a certain behaviour pattern in response to increased exports of labour-intensive goods, as given, then one can incorporate them in the export demand curves facing less-developed countries, yielding *adjusted* export demand curves.

If the restrictions were quite rigid then the adjusted demand curves would at some point turn down vertically. In practice, as a country's exports increase, it may face more and more restrictions from more and more countries, so that it would be forced to unload extra supplies on to a declining segment of the total potential market. Hence the elasticity of

---

[23] *Industry and Trade in Some Developing Countries*, Ch. 8. It should be noted that, in spite of restrictions, exports of manufactures from less-developed countries to developed countries have greatly increased since the early nineteen sixties.

adjusted demand would decrease, though not necessarily to zero. Naturally a large exporter is likely to encounter more restrictions than a small one. The longer-run elasticity may be higher than the short-run one if some importing countries impose only temporary restrictions.

The optimum policy from the point of view of the exporting country is then quite clear and has been appreciated by Japan and several less-developed countries. If the potentiality of foreign restriction is given, it will pay to engage in voluntary export restriction, the equivalent of an export tax. The issue is clearly one of international income distribution: the choice may be between import licences issued to foreign importing interests or export licences issued to domestic exporting interests; or alternatively it may be between tariffs imposed by the importing countries, yielding revenues to their govern-ments and forcing down the prices received by exporters, or export taxes levied by exporting countries, yielding revenues to their governments, and raising the prices paid by importers.

The logic is the usual optimum export tax logic. It has to be stressed that the export restrictions are optimum only if the foreign restrictions are given. One can hardly calculate in any precise way the elasticities of the adjusted export demand curves—unless there is evidence that they would be zero—but one can make a general assessment of the situation in the light of the political and economic circumstances in the importing countries, and bear these in mind when framing export policies. The pattern of export restrictions for this purpose was long ago set by Japan, and is now followed by a number of other countries, notably Hong Kong.

VIII
SOME REFINEMENTS AND COMPLICATIONS

It was pointed out earlier that the theory of the optimum tariff has been developed to a high degree of refinement, and numerous permutations and possibilities have been explored rigorously in the literature of international trade theory. There are paradoxes galore, and many economists seem to have devoted their energies to pointing out that nothing simple can be said. Some (but not all) of the cases have practical relevance, and it is worth quickly reviewing some of the issues and contributions without in any way attempting to reproduce the rigorous proofs or arguments of the original papers. The theory will also be related briefly to the analysis of some of our earlier chapters.

(1) *Income Distribution and the Orthodox Optimum Trade Tax*

Some modern writers on the theory of the optimum tariff have stressed that for every income distribution there is a different optimum tariff (or export tax). In Section I above we assumed a given income distribution maintained by non-distorting means. The argument, often put rather obscurely, was that if this independently-determined income distribution were changed, the pattern of domestic demand would change, and this would bring about a change in the optimum tax rate.[24]

For example, suppose income distribution were shifted towards people who, for given relative prices, would purchase more exportables relative to importables than the community in general. So the community's desire to export at a given tax rate would decline; the effect would be the same as if community tastes had shifted towards exportables, a case we have already considered. In Figure 7.5 (identical with Figure 7.3) the QQ' curve would then shift upwards, and the optimum tax would decline.

A more interesting consideration is that internal income redistribution may not be costless, and possibly one cannot assume that it will take place at all. This consideration is generally ignored in the literature, which usually confines itself to Pareto-efficiency. One must then take into account the income distribution effects of the trade taxes.

Consider the two-sector model again and refer to Figure 7.5. Every export tax yields a different income distribution. Suppose tax ON' yields an income distribution which in turn yields the curve QQ'. A lower tax, OH, would shift incomes

---

[24] See Graaff, 'On Optimum Tariff Structures', who concluded 'Thus we have an infinity of optimum tariff structures, each corresponding to a different initial distribution of wealth, but all determined by the same rule' (pp. 56–7). The issue also underlies the complications about community indifference curves crossing in Scitovsky, 'A Reconsideration of the Theory of Tariffs'.

towards factors intensive in exportables. Assume, for exposition, that these are the people who have a relatively higher demand for exportables (although the opposite is just as likely). The curve that goes with the new income distribution is then above QQ', namely GG'. Hence the points K and F are both feasible points, and one can obtain a *feasible curve* RR' which shows how the elasticity varies with the export tax rate

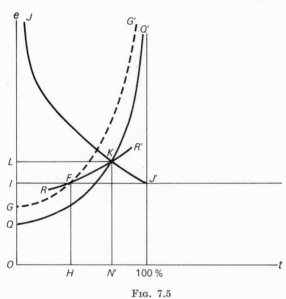

Fig. 7.5

assuming that the income distribution yielded by the tax cannot be changed.

If GG' represents the income distribution that is optimal for a given total national income, then the optimum tax rate from a distributional point of view is OH. On the other hand, the Pareto-efficiency conditions will be fulfilled only by a tax rate of ON'. Thus it is likely that the optimum tax, taking both equity and efficiency into account, will be between OH and ON'.

(2) *Tax Revenue and the Orthodox Optimum Trade Tax*

The theory of optimal taxation for revenue and of the maximum revenue tax could be integrated with orthodox optimum trade tax theory. In the absence of collection costs, the revenue raised by the orthodox optimum trade tax structure is genuinely costless: it does not create marginal divergences; rather it eliminates them. This is thus a first-best way of raising revenue. Bearing in mind that collection costs on trade taxes are likely to be less than on other sorts of taxes, this remains true when collection costs are introduced. But of course, from a fiscal point of view, the level of trade taxes may

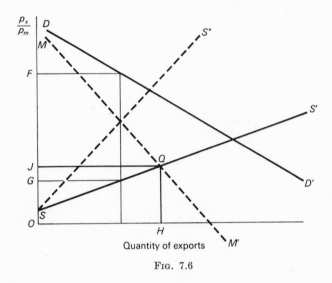

Fig. 7.6

need to be higher than indicated by the orthodox optimum rates.

If the need for revenue were very great, the maximum revenue rates would have to be applied, and (in a simple model at least) these are always greater than the orthodox optimum rates. The relationship between the maximum revenue and the orthodox optimum rate is shown in Figure 7.6. The orthodox optimum rate is obtained by equating marginal revenue with the SS' curve, which shows marginal cost from a social point of view. The maximum revenue rate is obtained by equating marginal revenue with the *marginal* curve derived from SS', namely $SS*$, the maximum revenue rate being FG/OF.

From a national point of view, SS' is a marginal cost curve, and the surplus above SS' (the area SJQ at output OH) is not a cost. By contrast, from the point of view of the tax-raising authority, the supply curve SS' is an average cost curve, and the surplus above it for any given price is just a cost.[25]

(3) Domestic Divergences and the Orthodox Optimum Trade Tax

The analysis of the orthodox optimum tariff—which is concerned with *trade* divergences—needs to be combined with that of *domestic* divergences. For example, external economies may be yielded by domestic import-competing production. The first-best solution is then to combine a production subsidy with the orthodox optimum trade tax rate determined in the usual way. The rate of tariff (or export tax) required will vary with the extent of the domestic divergence: surprisingly, it may actually be necessary to *reduce* the tariff rate when the need for protection increases. A high rate of external economies means a high subsidy rate, hence high import-replacement, and hence a reduced volume of trade. But the lower the volume of trade, the higher the foreign elasticity (making our usual assumption that this elasticity rises as the volume of exports falls), and hence the lower the rate of optimum trade tax.[26]

(4) The World Optimum

It has been stressed here that orthodox optimum trade tax theory is concerned with the *national* and not the *world* optimum. It might seem at first sight that it has no role if one is concerned with world welfare. But this is not so. One could attach distributional or welfare weights to the national incomes of different countries and then build up criteria for maximizing world welfare by taking these distributional weights into account. This is the method Meade[27] has used.

Suppose that we have two country groups—the UNCTAD

[25] See H. G. Johnson, 'Optimum Welfare and Maximum Revenue Tariffs', *Review of Economic Studies*, 19, 1950–51, pp. 28–35. This article uses the offer curve technique, and proves algebraically that the maximum revenue tariff will always be higher than the optimum tariff.

[26] W. M. Corden, 'Tariffs, Subsidies and the Terms of Trade', *Economica*, 24, August 1957, pp. 235–242.

[27] *Trade and Welfare: Mathematical Supplement*, Ch. XVI.

group and the OECD group—with a higher welfare weight attaching to the former. A tariff or export tax imposed by the UNCTAD group may then increase world welfare. If one attaches precise weights, there will be an optimum trade tax structure which trades off the adverse efficiency effects against the favourable equity effects. The problem is a second-best one, since a direct income redistribution from the OECD group to the UNCTAD group, if financed and disbursed in a minimum cost way, would minimise the distortions. In fact we have the familiar case discussed in Chapter 5 of trade restrictions being used to alter income distribution and inflicting by-product distortion costs. This time it is world rather than internal income distribution with which we are concerned.

Meade has derived a general formula for the optimum rate of trade tax in a two-country two-product model with explicit welfare weights for the two countries. The orthodox optimum tariff rate falls out as the special case where the distributional weight attached to the foreign country (OECD here) is zero.

(5) *A Tariff May Worsen the Terms of Trade*

The essence of the orthodox optimum tariff concept is that a tariff or an export tax will improve the terms of trade. In the small country case, of course, it will not do so. It is more surprising that many circumstances can be uncovered where a tariff would actually worsen the terms of trade. In all cases this can happen only if the trade tax structure or the method of raising and spending revenue are not optimal. Some of these cases seem of minimal importance, so we can be brief here. The first four cases are concerned with a shift in the domestic demand pattern towards importables induced by a tariff, and some of the other cases require a model with at least three goods.

(1) The government may spend the tariff revenue mainly on imports, so that the rise in government imports more than offsets the reduction in private imports resulting from the tariff, and on balance imports may increase and hence the terms of trade may worsen.[28] This is possible only if the

[28] A. P. Lerner, 'The Symmetry between Import and Export Taxes', *Economica*, N.S. 3, August 1936, pp. 306–13.

price elasticity of the private demand for imports is less than unity.

(2) If the marginal propensity to consume importables by that factor which is intensive in the importable industry is sufficiently higher than the marginal propensity to consume importables by the other factors, then, owing to the income distribution shift brought about by the tariff, the demand for imports may rise even though income, substitution and production effects have induced an opposite tendency.[29]

(3) Tariff revenue may be redistributed mainly to that section of the community which has a relatively high marginal propensity to consume importables, so that, again, on balance the demand for imports may go up.[30]

(4) The importable may be an inferior good, so that the adverse income effect of the tariff is to increase demand for the importable on that account; in spite of the substitution and production effects pulling the other way, demand for imports may then rise.[31]

(5) If the importable is labour-intensive and there is a backward-bending supply curve for labour, a tariff may reduce the output of the importable, and on balance imports might go up.[32]

(6) In a three-commodity model, with two exportables and one importable, a tariff may increase output of one of the exportables, and this may lead to increased exports of that product and so worsening of the terms of trade on that account. Output of the other exportable must go down, but it may face a high foreign demand elasticity. Hence the pattern of exports

[29] J. Bhagwati and H. G. Johnson, 'A Generalized Theory of the Effects of Tariffs on the Terms of Trade', *Oxford Economic Papers*, 13, October 1961, pp. 225–53.

[30] Bhagwati and Johnson, loc. cit.

[31] This was conjectured by Marshall (and also Graaff), and Kemp has shown that it can be true provided the government spends some part of the tariff proceeds. See M. C. Kemp, 'Note on a Marshallian Conjecture', *Quarterly Journal of Economics*, 80, August 1966, pp. 481–4. Kemp has also shown, with great sophistication, that if the *exportable* is an inferior good a tariff may worsen the terms of trade; but in that case there must, apparently, be market instability. See Kemp, *The Pure Theory of International Trade and Investment*, pp. 307–9.

[32] M. C. Kemp and R. W. Jones, 'Variable Labour Supply and the Theory of International Trade', *Journal of Political Economy*, 70, February 1962, pp. 30–36; also Bhagwati and Johnson, 'A Generalized Theory of the Effects of Tariffs on the Terms of Trade'.

has been shifted towards that product which faces the relatively low foreign demand elasticity.[33]

(7) In a model with at least two importables, a tariff on one only may increase imports of the other if they are substitutes in production or consumption. If the import pattern shifts towards those imports that face relatively low foreign supply elasticities, the terms of trade may worsen.

(8) A tariff will normally lead to reduced imports of the protected good but to increased imports of intermediate inputs used by it. The adverse terms of trade effect of the extra intermediate imports may more than outweigh the favourable effect of the reduced final good imports.

(9) A tariff on the intermediate good will lead to reduced imports of it but to extra imports of the final good (owing to reduced effective protection for the latter). Again, there will be opposing terms of trade effects.

(10) If there are strongly increasing returns in the foreign country, increased imports by the home country will lower the relative price of its imports and so improve the terms of trade; a tariff would worsen the terms of trade. If there is competition at home, it will pay to impose an import subsidy; the appropriate import subsidy will equate the marginal cost of imports (which must be below the average cost) to the home country's demand price.[34]

(6) Second-Best Problems

Many writers have considered second-best problems where some constraints are imposed on the use of tariffs or export taxes but where there is an opportunity to improve the terms of trade.

Meade[35] has developed a precise mathematical model with two exports and two imports showing what the optimum trade tax is on one of the goods when the taxes on the other

[33] F. H. Gruen and W. M. Corden, 'A Tariff that Worsens the Terms of Trade', in I. A. McDougall and R. H. Snape (eds.), *Studies in International Economics*, North-Holland, Amsterdam 1970; also *The Theory of Protection*, pp. 95–97.

[34] Murray C. Kemp, 'Notes on the Theory of Optimal Tariffs', *Economic Record*, 43, September 1967, pp. 395–404; also Kemp, *The Pure Theory of International Trade and Investment*, pp. 296–313. Note also the possible case for an export subsidy mentioned on p. 172 above.

[35] *Trade and Welfare: Mathematical Supplement*, Ch. XVII.

three are constrained at non-optimum levels. The present author[36] has analysed the case where there are potential terms of trade effects on the export but not the import side, but only tariffs can be used. The second-best tariff structure will be non-uniform if the export demand elasticities differ. Scott[37] has shown that a country that exports raw cotton and imports cotton textiles may be able to raise the world price of raw cotton by an import subsidy on cotton textiles. If an export tax is ruled out, some import subsidy will then be optimal.

Friedlaender and Vandendorpe[38] have explored a case possibly relevant to E.E.C. countries that make use of value-added taxes at varying rates but are not free to vary taxes on trade. If a country cannot use trade taxes to improve its terms of trade, it can do so by taxing consumption of importables at a higher rate than consumption of exportables, or taxing production of exportables at a higher rate than production of importables. There will then be a second-best optimum set of such taxes.

APPENDIX
THE OPTIMUM EXPORT TAX
AND OPTIMUM TARIFF

This Appendix tries to sort out a little more precisely various ideas in this chapter, especially the optimum tariff concept and formulae in the two-traded-goods model.

(1) *Elementaries: Optimum Export Tax*

In Figure 7.7 the horizontal axis shows the country's export quantity (X) and the vertical axis the quantity of imports (M). The foreign offer curve is OF.

Pareto-optimality requires that the marginal rate of transformation in foreign trade (slope of the offer curve) is equal to the marginal rate of transformation in domestic production and of substitution in domestic consumption. With the usual assumptions, notably perfect competition and absence of external effects, the latter will equal the

---

[36] *The Theory of Protection*, pp. 194–5.

[37] M. Fg. Scott, 'Comparative Advantage and the Use of Home-Produced versus Imported Materials', in W. A. Eltis, M. Fg. Scott, and J. N. Wolfe (eds.), *Induction, Growth and Trade: Essays in Honour of Sir Roy Harrod*, Clarendon Press, Oxford 1970.

[38] A. Friedlaender and A. Vandendorpe, 'Excise Taxes and the Gains from Trade', *Journal of Political Economy*, 76, October 1968, pp. 1058–68.

domestic price ratio. Hence the condition of the optimum is that the domestic price ratio is equal to the slope of the offer curve.

The orthodox optimum export tax will achieve this result. In Figure 7.7 let J be the optimum point attained by this tax. The domestic price ratio is given by the slope of RJ, which is equal to the *marginal* rate of transformation in foreign trade at that point, while the slope of OJ gives the terms of trade—the *average* rate of transformation in foreign trade. A quantity of exports OG sells for OS in

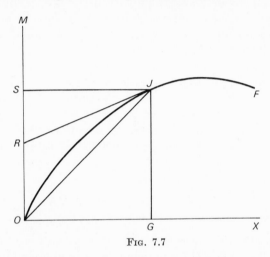

Fig. 7.7

the foreign market, but the amount received by the exporter on the basis of the domestic price ratio is only RS. Hence the export tax is OR, or OR/OS when expressed in the usual way as a proportion of the world price.

The elasticity of demand for exports in terms of imports at J is OS/OR. This is arrived at as follows. Average revenue is OS/OG, and marginal revenue is RS/OG. Since $\eta_x = A/(A-M)$ where η_x is the elasticity of demand, A is average revenue and M is marginal revenue,[39] it follows that $\eta_x = OS/OR$. We saw above that the optimum export tax, t_x, is OR/OS. Hence $t_x = 1/\eta_x$.

[39] This is well-known from elementary monopoly theory. Where P refers to price and Q to quantity,

$$\eta_x = -(dQ/Q) \cdot (P/dP) \qquad (1)$$

$$A = P \qquad (2)$$

$$M = (P\,dQ + Q\,dP)/dQ \qquad (3)$$

and from (1), (2), and (3),

$$\eta_x = A/(A-M)$$

(2) *Elementaries: Optimum Tariff*

Now look at the same thing from the point of view of M rather than X. The rate of tariff on M (namely t_m) which is symmetrical with the optimum export tax can be shown to be OR/RS. The domestic price of imports at J is OG/RS, and the world price is OG/OS. The tariff rate is the excess of the domestic over the world price as a proportion of the latter, that is $(OG/RS-OG/OS)/(OG/OS)$, which comes to OR/RS. Comparing this with t_x (which was shown to be OR/OS), we find then that $t_m = t_x/(1-t_x)$.

The elasticity of supply of imports in terms of exports at the point J can be shown to be RS/OR. Average revenue is OG/OS and marginal revenue is OG/RS. Since $\epsilon_m = A/(M-A)$, where ϵ_m is the elasticity of supply, A is average revenue and M is marginal revenue,[40] it follows that $\epsilon_m = RS/OR$. Since the optimum tariff t_m is OR/RS, $t_m = 1/\epsilon_m$.

All this is elementary and has been often expounded.[41]

(3) *The Bickerdike-Graaff Formula*

In the two-good model the export demand elasticity and the import supply elasticity are derived from the same foreign offer curve and express the same price-quantity response. Furthermore, a tariff is completely symmetrical with an export tax, so that one of them will do the job completely. By contrast, in the multi-good model (beginning of Section II of this chapter) each export is assumed to have its own foreign demand elasticity and each import its own foreign supply elasticity, and there is an optimum structure consisting of both tariffs and export taxes.

Consider now a model which is a transition between these two models.[42]

---

[40] Again, from elementary monopoly theory, where P refers to price and Q to quantity:

$$\epsilon_m = (dQ/Q) \cdot (P/dP) \tag{1}$$

$$A = P$$

$$M = (P\,dQ + Q\,dP)/dQ \tag{2}$$

and from (1), (2), and (3),

$$\epsilon_m = A/(M-A)$$

[41] See especially H. G. Johnson, 'Alternative Optimum Tariff Formulae', in Johnson, *International Trade and Economic Growth*, Allen and Unwin, London, 1958.

[42] This owes a great deal to Johnson, 'Alternative Optimum Tariff Formulae' though it goes beyond his discussion. See also Graaff, 'On Optimum Tariff Structures'.

There are only two traded goods, X and M. The country faces a foreign export demand curve and a foreign import supply curve, as shown in Figures 7.8a and 7.8b. Exports and imports are functions of money prices p_x and p_m respectively. Hence there must be a *numeraire* with a constant money price, which implies a non-traded good. Furthermore, cross-elasticities must be zero, the two curves being quite independent. Hence there are elements of partial equilibrium in this model.

The value of exports is always equal to the value of imports (the two shaded areas are equal). For every point on the DD' curve

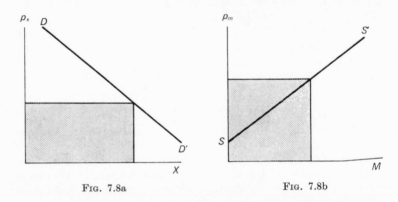

FIG. 7.8a   FIG. 7.8b

(Figure 7.8a) there must be a corresponding point on the SS' curve (Figure 7.8b) to ensure this. In this way the usual foreign offer curve can be derived from these two curves showing the quantity of M the world is willing to supply for any given quantity of X.

Since there are only two traded goods, a tariff and an export tax are symmetrical. Thus we should be able to derive the optimum tariff formula which will have as ingredients both the export demand elasticity z_x and the import supply elasticity σ_m. We now give Johnson's proof[43] to arrive at a classic formula—the 'Bickerdike-Edgeworth-Kahn-Little-Graaff result' as Johnson calls it. But we shall shortly go beyond Johnson to show what it *really* means.

The condition of balance of trade equilibrium is

$$X p_x = M p_m \tag{1}$$

Differentiating (1)

$$X \, \mathrm{d}p_x + p_x \, \mathrm{d}X = M \, \mathrm{d}p_m + p_m \, \mathrm{d}M \tag{2}$$

[43] 'Alternative Optimum Tariff Formulae', pp. 59–60.

Rearranging (2)

$$p_x \, \mathrm{d}X\left(1+\frac{X \, \mathrm{d}p_x}{p_x \, \mathrm{d}X}\right) = p_m \, \mathrm{d}M\left(1+\frac{M \, \mathrm{d}p_m}{p_m \, \mathrm{d}M}\right)$$ (3)

Denoting the export demand elasticity by

$$z_x = -\frac{p_x}{X}\frac{\mathrm{d}X}{\mathrm{d}p_x}$$ (4)

and the import supply elasticity by

$$\sigma_m = \frac{p_m}{M}\frac{\mathrm{d}M}{\mathrm{d}p_m}$$ (5)

From (3), (4), and (5)

$$\frac{p_x}{p_m}\frac{\mathrm{d}X}{\mathrm{d}M} = \frac{1+\dfrac{1}{\sigma_m}}{1-\dfrac{1}{z_x}}$$ (6)

From (1) and (6)

$$\frac{M}{X}\frac{\mathrm{d}X}{\mathrm{d}M} = \frac{1+\dfrac{1}{\sigma_m}}{1-\dfrac{1}{z_x}} = \alpha$$ (7)

Referring back to Figure 7.7, α is OS/RS. We have already shown that the optimum tariff t_m is OR/RS; hence:

$$t_m = \alpha-1$$ (8)

From (7) and (8)

$$t_m = \frac{\dfrac{1}{\sigma_m}+\dfrac{1}{z_x}}{1-\dfrac{1}{z_x}}$$ (9)

This is the Bickerdike-Graaff result.

(4) *Optimum Export Tax and Tariff Combined*

In order to appreciate what this result really means, let us approach the matter in another way. It can be shown that one could get the same result by combining an optimum tariff t'_m with an optimum export tax t'_x, the first taking into account only the elasticity on the import side and the latter only the elasticity on the export side. Thus, at the optimum

$$t'_m = 1/\sigma_m \tag{10}$$

$$t'_x = 1/z_x \tag{11}$$

The balance of trade must of course be kept in equilibrium through factor-price or exchange rate adjustment.

If one wants to achieve the same result with a tariff alone (as implied in the Bickerdike-Graaff approach), one must convert the export tax t'_x into a tariff rate t''_m with which it is symmetrical. The relationship between tariff and export tax rate was derived earlier from Figure 7.8. Hence

$$t''_m = \frac{t'_x}{1 - t'_x} \tag{12}$$

If t'_m were zero (i.e., $\sigma_m = \infty$), the optimum tariff t_m would be equal to t''_m. But when there are terms of trade effects on both the import and the export side, the optimum tariff t_m will be made up from two components, and we have:

$$t_m = t'_m + t''_m + t'_m t''_m \tag{13}$$

(One supposes t'_m first to be imposed and the resulting domestic import price to be increased by a proportion t''_m, the latter increase being the effect that is symmetrical with the export tax.) From (10), (11), (12), and (13)

$$t_m = \frac{\dfrac{1}{\sigma_m} + \dfrac{1}{z_x}}{1 - \dfrac{1}{z_x}} \tag{9}$$

So we have arrived at the Bickerdike-Graaff result. In other words, the classic Bickerdike-Graaff optimum tariff rate t_m gives us the same result as an optimum *structure* of a tariff t'_m and an export tax t'_x where each of the latter is the inverse of the relevant foreign (supply or demand) elasticity.

8

MONOPOLY, MARKET STRUCTURE, AND ECONOMIES OF SCALE

DO TRADE restrictions affect the degree to which private monopoly power in an import-competing industry is exploited, and, if so, how should this affect policy recommendations? More generally, how do trade restrictions affect market structure and the degree of competition? These questions have been avoided in international trade theory by the common assumption of perfect competition.[1] Once one allows for the possibility that a product has only a single domestic producer, or only a limited number of them, one can also introduce internal economies of scale, another common phenomenon that trade theory has tended to avoid. We also consider here the effects of tariffs on efficiency within firms (so-called *X-efficiency*) and finally the whole question of dumping.

I
THE EFFECTS OF PROTECTION ON MONOPOLY PROFITS

A tariff can affect monopoly profits earned by an import-competing industry. This subject will mainly be examined here in a partial equilibrium way since there seems no other way of doing it.

(1) *Protected Monopoly and Economies of Scale*

One can imagine an actual or potential domestic industry which consists of only one firm and which faces a given world

---

[1] Surveys of the relatively sparse literature relating monopoly or monopolistic competition to international trade theory are in R. E. Caves, *Trade and Economic Structure*, Harvard University Press, Cambridge, 1960, pp. 174–189 and in H. G. Johnson, 'International Trade Theory and Monopolistic Competition Theory' in R. E. Kuenne (ed.), *Monopolistic Competition Theory: Studies in Impact*, Wiley, New York, 1967. See also R. E. Caves and R. W. Jones, *World Trade and Payments: An Introduction*, Little Brown & Co., Boston, 1973, Ch. 11.

price of imports. It follows immediately that if it is sharing the
market with imports, it cannot do anything about the price it
faces.

The interesting and common case to explore is one where this
firm has a downward-sloping average cost curve over the
relevant range of output but is dependent for its existence on a
tariff. This is represented in Figure 8.1. The average cost curve

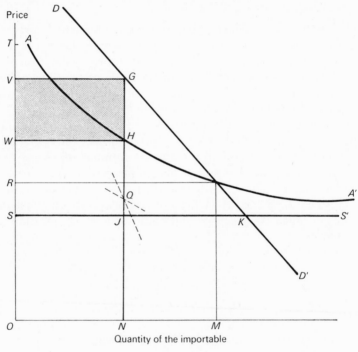

FIG. 8.1

is AA'; it includes normal profits and is assumed to trace out
both private and social costs of production. The import supply
curve is SS' and the domestic demand curve DD'. A tariff of
SR, or a little above it, would get domestic production
established and put an end to imports. The average cost of
production would be OR, at least provided the tariff is
sufficiently above SR to ensure that the whole domestic
market does go to the domestic producer.

If the tariff were raised above this necessary level, output would decline and monopoly profits would be earned. If there were no import competition monopoly profits would be maximized at price OV and output ON, the marginal revenue curve crossing the marginal cost curve at Q. Hence, if the tariff were high enough so that the domestic price could rise to OV, monopoly profits (shaded in Figure 8.1) would be maximized; further increases in the tariff would have no effect. Thus a tariff below SR would have no effect other than bringing in customs revenue. A tariff of SR, or a little above it, would get the firm established, a further increase up to SV would invite monopoly profits and cause output to go down again, and any increase beyond that would again have no effect.

If the country desires to have domestic production of the product but wants to minimize the costs to the consumer (and to the country) and also to avoid monopoly profits, it should fix the tariff at a level of SR, or marginally above it. This will be called the *made-to-measure tariff* level. Any tariff above this level will cause the average cost to be higher than it need be, owing to the scale of output being too low, and will cause the price charged to the consumer to be even higher, owing to monopoly profits being added on to average costs. The general point is that tariff policy can fulfil the role of controlling the prices charged by domestic protected monopolists.

One should really introduce some reason why the domestic firm is to be protected at all, even if it is only a second- or third-best argument for tariffs. There might be externalities associated with domestic production, or the social cost of labour may be less than the private cost. The net result, let us say, is that the social average cost curve is below the private one, and is less than the cost of imports. This is represented in Figure 8.2 (the social average cost curve is QQ').

Assuming that the only device available is a tariff, a tariff of SR should then be imposed, provided its by-product consumption-distortion is not too high. The tariff will bring about a gain ($R'SK'J'$, which is shaded) on the production side and a consumption-distortion cost of $K'H'K$. The industry should be given the tariff if the gain exceeds the cost. Because of economies of scale, we are in an all-or-nothing world here, so gain and cost cannot be traded-off against each other at the margin. At any rate, the main point is that the tariff should not exceed SR, for that would reduce the gain and raise the cost (unless income redistribution towards monopoly profits were desired!).

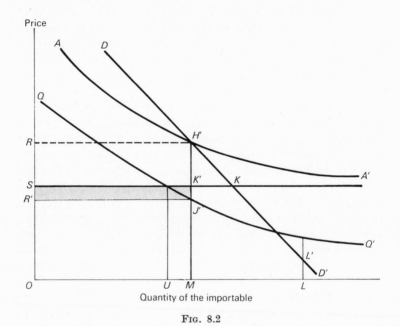

Fig. 8.2

The possibility of exporting should also be allowed for, though it does not really affect our main argument. If the private average cost curve fell below the import price, it might also fall below the f.o.b. export price, in which case there would be no need for a tariff to initiate domestic production. It could be that the social average cost curve, QQ' falls below the export price, even though the private curve AA' does not. In that case first-best policy would call for a production subsidy which would get the domestic firm going, allow it to capture the domestic market and to export as well. A combination of tariff and export subsidy would be second-best, and a tariff alone third-best.[2]

(2) *Upward-Sloping Import Supply Curve*

The preceding analysis can be varied or elaborated in a number of ways. One could allow for the case in which the sole domestic producer is able to exploit his monopoly power even when he is sharing the market with imports. We must then assume that his average and marginal cost curves are upward-sloping, and that the foreign import supply curve is also upward-sloping. The domestic producer will then face a downward-sloping demand curve which is derived by subtracting foreign supply at each price from total domestic demand; and if he maximizes profits, he will produce to the point where his marginal revenue equals marginal cost. Pareto-efficiency could be attained by combining the orthodox optimum tariff with a subsidy on output; the two instruments of tariff and output subsidy could then attain the two targets of optimal exploitation of national monopsony power and optimal domestic production. But this ignores income distribution effects: both the output subsidy and the tariff would increase monopoly profits.

The practical question is whether a tariff alone could fulfil a useful role. It will clearly increase monopoly profits. Furthermore, from a Pareto-efficiency point of view, there are two opposing effects at work. When one compares the monopolistic equilibrium with the social optimum, one finds that the price to consumers is too high, and hence the amount of domestic consumption is too low, while domestic production

[2] When the average social cost curve falls below the import price but not below the export price, first-best policy (from a Pareto-efficiency point of view) requires a production subsidy which, in the absence of various costs of revenue-raising, would bring output to the point where marginal social cost equals price, as given by the domestic demand curve. This follows from the well-known principle of marginal-cost pricing. Allowing for distortion and collection costs of revenue-raising, the subsidy would need to be less than that, and income distribution effects would also need to be taken into account. Even if the private and social cost curves coincided, first-best Pareto-efficiency policy would thus require a subsidy. The roles of output-related subsidy, lump-sum subsidy, import subsidy (negative tariff) and a tariff in achieving or getting closer to Pareto-efficiency in this type of model are fully explored in W. M. Corden, 'Monopoly, Tariffs and Subsidies', *Economica*, 34, February 1967, pp. 50–58.

is too low.[3] A tariff will raise the price to domestic consumers further, which is undesirable, but will also increase domestic output at the expense of imports, which is desirable. Taking into account its effect in increasing monopoly profits, a tariff does not appear an appropriate device in this situation even in the absence of any other devices. Of course this conclusion might need to be altered if there were some further argument for protection, perhaps based on a domestic production divergence.

(3) *Quality Variation*

Even though a tariff leads to a rise in the domestic price of imports, the price of the domestic product may not rise to the same extent, if at all, but rather its quality may deteriorate relative to the imported product. One may then get the misleading impression that some part of the tariff is redundant and perhaps that there is little or no distortion-cost of protection. But this would be quite wrong since like would not be compared with like. This is a trap to beware of in empirical studies of the effects of tariffs which may suggest that there is much 'water in the tariff' (redundancy). The tariff may actually be just sufficient to persuade domestic consumers to buy the inferior and cheaper domestic product, and so may be fully needed.

The basic argument remains that there is some made-to-measure tariff, and that any increase above it will yield monopolistic restriction of output. But the output and consumption reductions resulting from a tariff above this

---

[3] One can draw a diagram to verify this conclusion. The private monopolistic equilibrium is obtained as follows. Subtract the import supply curve horizontally from the domestic demand curve to obtain a net demand curve NN', which is the average revenue curve facing the monopolist. Draw the marginal revenue curve MM' derived from this. The monopolistic equilibrium output is given by the intersection of MM' with the monopolist's marginal cost curve CC'. The social optimum is obtained as follows. The import supply curve is the average cost of imports to the country. Derive the marginal curve GG' from it. Subtract GG' horizontally from the domestic demand curve to obtain the marginal social valuation curve VV'. This must be to the right of NN' and MM'. The social optimum is at the intersection of VV' with CC'. Actually, if the producer's marginal cost curve is very steeply upward-sloping the price to consumers in the social optimum could exceed that of the private optimum. This general case is analysed in *Trade and Welfare*, pp. 234–237.

level may be brought about by deterioration in quality rather than increase in price. Both can increase monopoly profits.

(4) *Rise in Wage Rate*

There may be social gains from the establishment of the industry owing to its wage payments. This point is important when one adapts the analysis here for cost-benefit analysis, and it yields us a possible argument for protection. But the discussion is particularly limited at this point by its partial equilibrium nature.

The supply curve facing the domestic producer may be upward-sloping, so that the wage-rate is raised by the establishment of the industry. The wage the industry pays may be equal to the *marginal* social opportunity cost of labour, but it will then exceed the *average* social opportunity cost. This means that the private average cost curve will be above the social average cost curve.

The essential point is that, owing to economies of scale, the establishment of the industry involves an all-or-nothing, rather than a marginal, decision, so that intra-marginal effects must be taken into account. The appropriate analysis is that described by Figure 8.2; average social cost will be below average private cost, and the divergence between private and social cost will increase with scale of output.

(5) *Import Quotas*

Protection may be by quota rather than by tariff. If the industry is to be protected and the maximum benefits of economies of scale are to be attained, imports will have to be prohibited entirely. The protected producer will then practise monopolistic output restriction and earn monopoly profits. Import prohibition makes it impossible to practise a made-to-measure policy unless the price of the product is controlled directly. Thus there is an advantage in using a tariff rather than quantitative restriction, even when both lead to the cessation of imports.

Alternatively, some imports may be allowed under quota. The demand curve facing the domestic producer will then be a curve parallel to the domestic demand curve DD' (Figure 8.1), the amount of permitted imports having been subtracted from DD'. The intersection of the marginal revenue curve derived from this new demand curve with the producer's marginal cost curve will determine his output. The larger the quota the lower domestic output and the lower monopoly profits. It is thus possible to squeeze profits by increasing the size of the quota, though it is not certain whether this would reduce the price to the consumer: while the demand elasticity facing the producer will rise (causing the price to the consumer to fall), if he has a downward-sloping marginal cost curve his marginal costs will also rise (causing the price to the consumer to rise). In any case, this approach will not allow the maximum benefits to be derived from economies-of-scale, so that—if there is to be domestic production and monopoly profits are to be avoided—it appears preferable to protect with a made-to-measure tariff.

The same analysis applies if the marginal cost curve is upward-sloping. In the absence of a quota or tariff, with a given import price, the sole domestic producer cannot exploit his monopoly power. As soon as a quota is established, he has an opportunity to exercise monopoly power and make monopoly profits because the demand curve facing him will cease to be horizontal.[4] This is true even though the quota may permit quite a lot of imports to enter.

In this case (with the import supply curve horizontal) the quota differs from a tariff in the following way: a tariff only creates a monopoly situation once it has excluded all imports; by contrast a quota does so even when imports are permitted. This would seem to be a strong argument for preferring tariffs to quotas if protection is desired.

All this has been concerned with the effects of quotas on the ability of an import-competing producer to exercise monopoly power. In addition, quotas can create traders' or exporters' monopolies.[5] Furthermore, quotas on imported inputs available to existing users of these inputs can create or strengthen the monopoly power of these input-using firms by making it difficult for new entrants to obtain imported inputs.

[4] This case is analysed in *The Theory of Protection*, pp. 204–6.
[5] See *The Theory of Protection*, pp. 202–8.

(6) *Product Differentiation*

It has been assumed here (as elsewhere in this book) that the import is a perfect substitute for the domestically-produced product. In fact, this is just a limiting case which highlights the main considerations but does not realistically describe markets for manufactured goods. We shall come back to this again shortly. Here let us vary the model slightly to allow for a case of close but not perfect substitution.

We have the same average cost curve for the domestic producer as before. We do *not* draw in any import supply curve. Rather, the producer faces a domestic demand curve for his product which takes into account the given tariff-inclusive prices of competing products. We can imagine this curve to be downward-sloping but quite flat. We assume that, with free trade in competing products, the demand curve is completely below his average cost (AA') curve, so that there would be no domestic production.

Tariffs on competing products will raise the demand curve. There will be some set of tariffs (possibly more than one set) which will just make his demand curve tangential to the AA' curve. This is the made-to-measure set. Domestic production should commence and will yield only normal profits. If the tariffs are raised further, so that the demand curve rises further, he will be able to make monopoly profits. To maximize these he *may* need to reduce output below the made-to-measure level. The flatter the demand curve, the more likely is such an output reduction. But it is no longer certain that the tariff increase will require him to reduce output. It is only a possibility, highlighted by the simple perfect-substitution model.

(7) *Monopoly Elsewhere*

The approach so far has been partial equilibrium, and monopolistic restriction elsewhere in the economy has been ignored. Such restriction elsewhere in the economy will have three effects from the point of view of our analysis.

Firstly, it will have shifted income distribution towards profits in these other monopolized industries. If capital and entrepreneurship are generally mobile, this might strengthen

8

the income distribution argument against monopoly in the industry under consideration, since each extra bit of monopoly is then likely to raise the general level of profits. On the other hand, if there is considerable immobility the income distribution argument could, conceivably, go the other way: profit receivers of one industry lose when prices of other products rise, and there may be some case for restoring their relative position.

Secondly, there are Pareto-efficiency considerations on the output side. If output is restricted below the social optimum in other industries, this will artificially lower costs to industries competing for similar resources; so the private cost curve of 'our' industry is, for this reason, likely to be lower than the social cost curve.

Thirdly, if the domestic prices of other products are raised by monopolistic restriction, this will artificially shift the demand curve facing 'our' industry to the right; the private demand curve will thus be to the right of the curve showing marginal social valuation of the product.

(8) *Dissipation of Monopoly Profits*

Finally, the monopoly profits potentially available as a result of a tariff or a quota may be dissipated by the protected industry through fragmentation of production, through managerial inefficiency, or through resources used up in 'rent-seeking.' We turn to the first in the next section, to the second in Section V, and to the third in Section VI.

II
FRAGMENTATION OF PRODUCTION

It has been noted in many less-developed countries and also in Canada that the provision of protection appears to bring into being more domestic producers than seem called for by the size of the market. There are several firms, all producing similar products at scales of output below the level of long-run minimum average costs. This apparent phenomenon of fragmentation of production calls for some analysis.[6] It appears to be particularly relevant for the study of the motor-car industry in many less-developed countries.

In Figure 8.3, SS' is the import supply curve and DD' the domestic demand curve, as usual. AA' is the average long-run cost curve of a representative domestic producer. In the

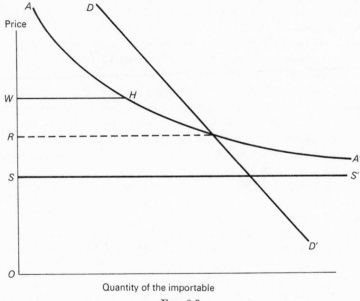

Quantity of the importable

Fig. 8.3

absence of protection there is no domestic production. A tariff of SR, or marginally above it, then brings into operation one producer.

Next, the tariff is increased further, say to SW. In our

[6] There is a large literature on this subject in Canada. The main argument is in H. E. English, *Industrial Structure and Canada's International Competitive Position*, Canadian Trade Committee, Montreal, 1964. See also H. C. Eastman and S. Stykolt, *The Tariff and Competition in Canada*, Toronto, 1967, and the earlier article by them, 'A Model for the Study of Protected Oligopolies', *Economic Journal*, June 1960. The most thorough discussions are in R. J. Wonnacott and P. Wonnacott, *Free Trade Between the United States and Canada: The Potential Economic Effects*, Harvard University Press, Cambridge, 1967, pp. 176–182 and pp. 234–245; and in B. W. Wilkinson, *Canada's International Trade: An Analysis of Recent Trends and Patterns*, Canadian Trade Committee, Montreal, 1969, pp. 110–113.

earlier story this led to monopoly profits. Now we can imagine
that further producers enter the industry. The costs of each
will be equal to or close to OW. Perhaps one or two firms make
some monopoly profits, but in general the potential monopoly
profits which would have been earned if the industry had been
confined to a single producer will be dissipated in increasing
costs owing to reduced scale of output. This implies that the
existing producers cannot impose entry barriers to exclude
new suppliers. The new producers may be subsidiaries of
large corporations engaged also in other activities.

Perhaps a more realistic way of getting to the same result is
as follows. Initially the domestic demand curve crosses AA'
at H, and a tariff of SW makes it just possible for one producer
to survive. Over time the market expands and the demand curve
shifts to DD'. Instead of the existing producer expanding
output, hence realizing the benefits of scale economies and
earning monopoly profits, new producers enter the industry,
and keep all their costs up.

Another possibility is that originally economies-of-scale
were exhausted at a low scale of output, so that it was optimal
at a given tariff for a number of firms to supply the market.
Then a technical change shifted the point of minimum average
costs to the right, so that the maximization of joint profits
required all but one firm to leave, but instead they simply
continue to divide up the market.

If the tariff were reduced from SW to SR this situation
could not last. Some producers would be taken over or forced
out, or existing producers would amalgamate, until only one
producer was left. Thus one can say that the maintenance
of a constant tariff when demand is rising or costs falling is a
cause of fragmentation of production.

Given the tariff, and assuming that the domestic price is set
by the tariff, the loss to consumers is the same when producers
earn monopoly profits as when the latter's production costs are
higher. But production fragmentation does lead to a real
resource loss, which is a net loss in real national income borne
by the producing firms. If the firms are foreign-owned then
much of the loss will be a foreign and not a national loss.
But reduced foreign profits will mean reduced tax payments,
so there will still be some national loss. Furthermore, some of

the potential monopoly profits might have been shared with employees in higher wages.

The real resource loss resulting from production fragmentation is caused by a special kind of inefficiency, namely inefficiency in the organization of the industry. The number of firms is Pareto-inefficient. Since there are potential mutual gains from reducing the number of firms, the question naturally arises how such a dissipation of potential monopoly profits can arise or last. Perhaps there are other considerations to take into account, so that the analysis here is over-simplified.

Firstly, observers may sometimes see production fragmentation where it does not really exist. A number of firms in an industry may be producing apparently very similar products subject to economies-of-scale in production. But are the products really the same? If they are somewhat differentiated then the question is whether the demand from consumers is really for a single homogeneous product that could be produced in one long production run. Perhaps they want a differentiated set. In the latter case we cannot definitely say that, given the tariff, there is 'undue' production fragmentation. Furthermore, one should be sure that economies-of-scale in production are not offset by diseconomies of scale and of geographic concentration in distribution, or possibly in management.

Secondly, such fragmentation may be non-optimal from the point of view of the firms concerned only if one takes a short-run view. They may expect the market to grow so that eventually there will be room for several firms, all producing at a point of minimum average costs, and all making good profits. But it is necessary to enter the market early to build up production and marketing experience, a distribution network, brand goodwill, and so on. There is a learning-by-doing effect, the profits forgone being a form of investment.

Thirdly, fragmentation may be imposed by governments. Some governments have deliberately invited a number of foreign firms to set up domestic production of similar products, promising protection of the local market as an inducement— sometimes with complete import prohibition. Their aim has been to encourage competition and avoid exploitation of domestic consumers by a single domestic firm. But the best safeguard for the consumer is the potentiality of import competition. Governments have mistaken the number of domestic *producers*, rather than the total number of suppliers of the domestic market (which includes suppliers of imports), as an indication of market concentration.

Inevitably some of the firms will become relatively stronger, but these have hesitated to drive out their weaker brethren with price or quality competition because of the fear of being branded monopolists, and possibly provoking government intervention. They may also fear that their protection against imports may be reduced once it becomes evident that they are able to make excessive profits under the shelter of existing protection.

Fourthly, the domestic producers may be mainly subsidiaries of American corporations, as in many industries in Canada. They cannot take each other over. They cannot easily force each other out of business, or prevent each other from entering the industry, using the usual devices of temporary price-cutting, and so on, because the large corporations have the resources to take losses for long periods, and they know how damaging such prolonged price-wars can be. Finally, the U.S. anti-trust laws discourage co-operation and market-sharing agreements which would permit one corporation to supply the market with one product line while another corporation is allocated another product line.

Fifthly, the production fragmentation situation may not be stable. The problem may be one of lagged adjustment and disequilibrium. The equilibrium may indeed be to have a single producer.

A question remains why some oligopolistic situations are stable, when each firm is operating at a scale less than minimum average costs, when the firms are not foreign-owned or controlled, and when the situation is not imposed or influenced by governments. This, of course, is the concern of a large literature on market structure. The following approach may be helpful.

Combined profits could indeed be increased by reducing the number of firms. But this opportunity to increase profits is forgone because entrepreneurs or managers enjoy the luxury of independence, of separate hierarchies, and so on. Profits are traded-off against this benefit. The preference for independence yields inefficiency in industry organization. Of course, it may be inaccurate to describe it as 'inefficiency' in this case. The managers may be very efficient in the trading-off process and may be maximizing their welfare. There is some similarity here with *X-inefficiency*, which concerns inefficiency within firms and will be discussed in Section V below. Thus the potential surplus yielded by a tariff above SR (Figure 8.3) will be taken out partly in profits and partly, or even wholly, in industry 'inefficiency.'

Yet this is not an entirely satisfactory answer. We may start with firms A and B, both of which wish to stay independent, but neither will have an interest in the independence of the other. So an incentive to expand by price competition or take-overs remains. By introducing the preference for independence, we have ruled out only voluntary amalgamation, not competition or take-overs. One could go on to say that while the preference for independence may discourage amalgamation, the preference for a quiet life, as well as familiar real costs and uncertainties, will inhibit price competition and take-over battles.

III
TARIFFS AND MARKET STRUCTURE

The simplicity of the conclusions derived so far depends on the extremely simple assumptions about market structure that have been made. In the main analysis of Section I it was assumed that there was a given import supply price and a single domestic producer. The analysis could easily have been extended to encompass a collusive group of domestic producers behaving like a single producer through market-sharing arrangements. In any case, the situation was either akin to perfect competition, with a given world market price (which could also be a consequence of international oligopoly) or—once imports ceased—one of pure monopoly. In the previous section we allowed the number of domestic producers to change, but the conditions of competition were still given from outside, with the domestic price set by the given import price plus tariff. We have also been assuming (apart from a brief reference) that the domestic producer and the foreign suppliers produce essentially the same product.

Markets with these characteristics do exist so that the analysis could be applied directly in some cases. It may yield some insights even when market structures are more complex and change as a result of tariffs. Nevertheless, a more general approach, inevitably less rigorous, seems needed.

(1) *The Problem Stated*

Let us then suppose that the particular industry market under consideration consists of a number of firms, some producing domestically and some producing abroad and selling in the domestic market as well as to their own markets. The product may be differentiated, so that there is not necessarily a single price. The number of firms and their behaviour may be influenced by various barriers to entry, such as economies-of-scale, possession of know-how, skilled labour, patents or basic resources, goodwill created by advertising, consumer prejudices in favour of domestic or foreign products, and so on. The firms may behave as if they were a single monopoly or any one of a number of oligopolistic behaviour patterns may operate.

One is tempted to say that anything may happen, and refer the reader to the rich but inconclusive literature on market structure and performance to justify the view that at this stage we must either proceed to case-studies or allow the discussion to dissolve in vagueness and empty generalities. But fortunately we are not concerned with all the elements which go to determine market structure and performance.

The problem can be narrowed down considerably. Given the market as we find it before tariffs we want to know what effect the injection of a tariff will have primarily on the number of sellers, but also on the intensity of competition between them, determining the extent to which the group will exploit its joint monopoly power. A tariff is a discriminatory tax on producers, discriminatory because it taxes only the products of foreign producers, not domestic producers. But it applies to all foreign suppliers, both existing and potential (assuming there is no discrimination between countries of supply), so that it differs from those handicaps which specifically disadvantage new entrants. How then does such a discriminatory tax affect market structure?

The subsequent discussion is also applicable to import quotas, since these are also likely to handicap the foreign suppliers. If the licences are issued not to domestic traders or to firms using imported inputs but to the foreign suppliers themselves the latter may have a special opportunity to exercise monopoly power; but it is also true that they will be prevented from extending their market shares. But let us return to tariffs now.

(2) *The Dominant Firm Effect*

The main consideration is probably the dominant firm effect. There may be one or a number of dominant firms, the dominance being based on early entry, absolute cost advantages, superior efficiency, or barriers to entry of various kinds. And then there may be other, smaller firms trying to extend their market or trying to enter it for the first time. Anything that weakens the dominant firms in relation to the smaller or newer firms is likely to decrease concentration and reduce exploitation of monopoly power through price leadership or other methods. On the other hand, if the dominant firms are strengthened then concentration and monopoly power will be increased.

If the dominant firms are domestic while the smaller firms and potential entrants are foreign, a tariff will increase concentration and monopoly power. If, on the other hand, the dominant firms are foreign while the smaller firms and potential entrants are domestic a tariff will have the opposite effect. It is of course also possible that the dominant firms as well as the potential new entrants are foreign in which case the tariff represents no bias against or for the dominant firm but just contracts the market. This market contraction may indeed reduce the number of sellers, hence strengthening the dominant firm. If, finally, dominant firms as well as the smaller firms and the potential new entrants are generally domestic a tariff is not relevant: the product is (more or less) non-traded. Thus we have here a four-fold classification. The interesting case is the second one (where the dominant firms are foreign), for it yields a reason for expecting a tariff in some circumstances to increase the number of sellers and reduce monopoly.

Market behaviour depends not only on the actual sellers in the market but also on the potential sellers. Similarly, structure and behaviour depend not only on the actual tariffs but also on potential tariffs. The situation may be one of free trade and dominance by a domestic producer. If free trade were certain to be maintained foreign suppliers might enter the market, incur temporary losses, and eventually eliminate the dominance and so reduce concentration and increase competition. But it may be known that a tariff would be imposed if this appeared likely to happen. The threat of a tariff is then sufficient to preserve the monopoly situation.

Alternatively, the market may be supplied by a large number of firms, domestic and possibly also foreign, so that concentration is low and competition vigorous. If free trade were certain to be maintained a foreign producer might in time acquire a dominant position, by temporary under-cutting of prices and other familiar devices (predatory dumping). But it may be known that government policy is to preserve part or all of the market for the domestic producers, or to preserve domestic competition, so that a tariff would be imposed if a foreign supplier did acquire dominance or threatened to do so. The likelihood of this tariff will then dissuade him in the first place. The threat of a tariff in this case preserves competition.

(3) *Foreign Investment and Leaping Over Tariff Walls*

No account has so far been taken of foreign control of local production. The domestic market will be supplied not only by domestically-owned firms producing domestically and by foreign-owned firms producing abroad, but also by foreign-owned firms producing domestically. This fact does not really alter any of the preceding analysis. From the point of view of tariff protection what matters is not where a firm is owned or controlled but where it produces and sells.

The relevant point is that a foreign-owned firm can change its place of production. Instead of producing abroad and importing into our market it may begin to produce domestically. This is a natural response to tariffs: tariffs induce foreign firms to get behind the tariff walls, so that a tariff is not quite the barrier it seems. Unless it is accompanied by taxes or controls on foreign direct investment it can be leapt over.[7]

There may be a foreign dominant supplier and a tariff may be so high as to exclude imports completely. So the foreign supplier may respond by setting up his factories behind the tariff wall aiming to maintain his dominance. His success depends on the source of his power. If it is based on the low costs of production in his own country, perhaps on a skilled labour-force which would not emigrate with him, then foreign investment may be a very imperfect substitute for producing in his own country and exporting to our market; on the other hand, brand goodwill, connections with distributors, patents and know-how can all be carried with him.

While international movements of capital, technology and management make a tariff a less severe barrier than might seem at first, undoubtedly any tariff imposes some handicap on foreign suppliers of imports even if they can manage to stay in the market through moving their enterprises geographically. For if in the absence of the tariff they chose to produce in their own country, presumably that was more profitable than producing in their export markets. The tariff forces them to choose a less economic location and so forces them to raise their prices or reduce their profits. Thus the economic consequences previously discussed will follow, if in modified form. A tariff will reduce or raise seller concentration and monopoly power depending on whether those foreign firms which were originally producing abroad are dominant or not.

IV
MADE-TO-MEASURE TARIFFS

We have already referred to the concept of the *made-to-measure tariff*. This was the rate of tariff of SR/OS which was just sufficient in Figure 8.1 to protect the domestic industry without yielding it any monopoly profits. The concept of

[7] See pp. 331–5 for further discussion of the effects of tariffs on foreign investment. See also Thomas Horst, 'The Theory of the Multi-national Firm: Optimal Behaviour under Different Tariff and Tax Rates', *Journal of Political Economy*, Vol. 79, Sept/Oct. 1971, pp. 1059–1072, for an interesting theoretical analysis of the effects of tariffs on the production location decision of a multinational enterprise, allowing both for the increasing and decreasing cost case.

made-to-measure tariff-making and subsidization is of the greatest practical importance since it underlies thinking about tariff-making and the provision of subsidies and tax concessions in many countries, and helps to explain why tariff structures are often so complicated. It will play a role at various points later in this book.

(1) *The Made-to-Measure Concept*

If the made-to-measure principle were followed then the tariff structure would be carefully tailored so that no industry or product is protected more than is 'necessary'.[8] The idea is implicit in the concept of 'scientific tariff-making' which has been popular in the United States. When a tariff-making authority makes a comparison of costs of production of a product domestically and in a foreign supplying country, apparently trying to offset differences in comparative costs, it may be trying to find out—'scientifically' of course—how much protection is necessary for the domestic producer.[9] But how do we decide what is 'necessary'?

The answer in principle is given by the analysis in section I

[8] The explicit concept of 'made-to-measure tariff-making' and the term itself come from Australia, where it has long been current (but it is implicit in the policies or policy discussions of many countries). See W. M. Corden, 'Protection: Review of the Vernon Report', *Economic Record*, 42, March 1966, pp. 129–138; and Corden, 'Australian Tariff Policy', *Australian Economic Papers*, 6, December 1967, pp. 131–154.

[9] The idea in the United States of 'scientific tariff-making'—that costs of production should be equalised by tariffs—seems to stem from the beginning of this century and has been written into U.S. tariff law. It was specifically written into the Smoot-Hawley Act of 1930 that the U.S. Tariff Commission should compare U.S. with foreign costs of production and recommend tariffs to equalise the differences. Economists have frequently pointed out that as a general principle of protection it negates the law of comparative costs and indeed would kill all trade. Here we reinterpret the 'equalization of cost of production' doctrine charitably as a guide to fixing tariff-levels for those products where protection has already been decided upon, and not as a guide to which industries should be protected.

H. G. Johnson, in 'The Cost of Protection and the Scientific Tariff', *Journal of Political Economy*, 68, August 1960, pp. 327–45, interprets the term 'scientific tariff-making' as meaning the construction of an optimum tariff structure subject to constraints, which may be non-economic or specified as fixed targets. His use of the term is probably more in accord with the Oxford English Dictionary, though less with tariff history. In his sense, a large part of the present book is about 'scientific tariff-making'.

of this chapter. Given that an industry is protected one wants to minimise the burden on consumers and avoid monopoly profits. Of course this approach says nothing about why protection should be provided in the first place. The main concern is with minimizing the income distribution effects of the tariff and avoiding unnecessary rents. This is especially so—and especially logical—when the protected firms are foreign-owned. Tariff-making is used as a form of price-control for protected monopolies. In the case of intermediate goods, the consumers are other industries; the aim is then to avoid unduly adverse effects on their costs and hence profitability, and so indirectly the need for increases in their nominal tariffs.

The issue is even clearer when protection is by subsidization, whether direct or indirect, including income tax concessions. The aim is then to minimise the fiscal burden imposed by a given protective objective, thinking of the latter either as the establishment of a particular industry or the achievement of a given amount of imports replaced or exports induced. The made-to-measure principle can help to explain, for example, the details of Indian export subsidization policy.

When there is potential fragmentation of production the proper interpretation of the made-to-measure tariff would seem to be that level of tariff which prevents fragmentation without killing the industry—that is, the tariff SR in Figure 8.3. But the general philosophy of giving firms the protection they 'need' may lead to the result of ratifying existing fragmentation by providing a level of tariff which allows an existing group of firms—possibly a non-optimal number of them—to survive without any of them making significant monopoly profits. Even that, of course, may not be possible, since the tariff that is barely sufficient for the marginal firm may yield monopoly profits for intra-marginal firms. But the important distinction must here be made between a made-to-measure tariff required for survival of each firm and one required for survival of domestic production. In general we shall have in mind the latter.

So far we have been concerned with the economies-of-scale case, where the made-to-measure tariff will give the domestic producer more or less the whole domestic market. The concept is not so clear when the industry consists of several firms and has an upward-sloping supply curve.

A tariff will then yield rents for intra-marginal producers—perhaps rewards for efficiency or for a fortunate choice of location or variety of product. This is inevitable when the product is completely homogeneous: a rate that just keeps the marginal producer in business will be more than is needed for intra-marginal producers, while if it is adjusted to the needs of some sort of average producer it will lead to the decline of marginal producers. But if the product is somewhat differentiated, each producer really having his own 'product' it may be possible to split up the tariff by having separate tariff classifications for each category, and then adjusting each tariff appropriately for the needs of the relevant producer.

In any case, this sort of thinking helps to explain why many developed countries have such refined tariff schedules. It also brings out the logic behind negotiations that governments in less-developed countries carry on with international corporations that are contemplating establishing industries with the aid of subsidies and concessions of various kinds. The aim is to discover the made-to-measure subsidies—the levels just needed to attract the desired industries. The thinking has not always been explicit. But it explains why the desire to protect particular industries or products has not generally led to across-the-board tariffs or subsidies but rather to tariffs or subsidies limited rather precisely to the purpose in hand so as to minimise adverse effects on consumers, other industries or the budget. One can see here again a glimmer of the conservative social welfare function.

(2) *Disadvantages of the Made-to-Measure Method*

The made-to-measure principle involves many practical difficulties. As already pointed out, if the supply curve is upward-sloping, and if it would not be optimum to give the domestic industry the whole domestic market, there is an ambiguity about the concept. If one interprets the made-to-measure tariff as a tariff that ensures no tariff redundancy one will only be avoiding excess profits or rents for the marginal producer or unit of output, not for intra-marginal output. Even with constant or decreasing costs, it is very difficult for a tariff-making authority to find out what an industry's costs are, and in any case these costs are always changing.

Made-to-measure tariff-making is likely to reduce firms' incentives to be efficient in the *X-efficiency* sense; this is discussed in the next section. Efficiency may also be reduced in another sense by attempts to squeeze the rents of intra-marginal firms. These rents or pure profits may be the rewards of superior managerial ability. Made-to-measure tariff-making may thus reduce funds available to the nation's relatively more efficient firms, and hence is likely to discourage their expansion and reduce the efficiency of investment.

If a protected firm is the local subsidiary of a multinational corporation and imports components from another part of the corporation abroad the problem of transfer pricing arises. We discuss this more fully in Chapter 12 and here are only interested in the implications of transfer pricing within corporations for the made-to-measure method of tariff-making or subsidization. The corporation will have an inducement to fix high transfer prices for the components. This will raise the profits of the supplying subsidiary abroad. In the first instance it will lower or even eliminate the profits of the local subsidiary. This will then provide an apparent justification for the local subsidiary to require high tariffs on its final products to maintain a reasonable profit level. The essential problem, of course, is that made-to-measure tariff-making requires the tariff-making authority to know the firm's true costs, and this is not really possible unless the costs of the components-supplying subsidiary which is situated abroad are known.

A made-to-measure system can make heavy demands on the honesty of the tariff-making machinery. It depends very much on judgement, and on firms revealing cost data. It provides more opportunities for corruption than a simpler system. Hence, while one should note the advantages of a tariff system that aims to avoid excess profits and other rents, and hence tries to prevent industrialists, whether local in origin or foreign, from exploiting the country's consumers, it is not necessarily to be recommended. Similar arguments apply to the made-to-measure subsidies and tax concessions which are much used in some less-developed countries.

V
THE X-EFFICIENCY EFFECTS OF
PROTECTION AND MONOPOLY[10]

It is sometimes argued that the effects of protection and of monopoly on efficiency within firms are much more important than the effects on the allocation of resources between firms and industries. These efficiency effects are usually associated with businessmen's 'motivation'. The distinction is made between *allocative efficiency* and *X-efficiency*, the latter referring to efficiency within firms.[11] Apart from a brief reference to industry inefficiency and the preference for independence we have been concerned so far with allocative efficiency, but some systematic attention must clearly be given to *X-efficiency*.

The argument seems to be not just that firms do not always aim to maximise profits but that the *extent* to which they depart from the profit maximisation aim is related in some systematic way to trade restrictions. This has been a popular theme, for example, in writings about the effects of Britain joining the European Economic Community.[12] The body of orthodox trade theory does not incorporate these X-efficiency effects, even though they are often regarded as very important. Perhaps they can be described and analyzed rigorously, and incorporated in formal trade theory. A similar issue concerns the relevance of formal monopoly theory. It is often argued that the really important effects of monopoly are not concerned with resource allocation between firms and industries, or with income distribution, but rather with its adverse effects on X-efficiency within firms.

(1) *A Simple Argument*

In the search for a rationale for these popular but vaguely specified ideas the approach here will be comparative static.

[10] This is a revised version of W. M. Corden, 'The Efficiency Effects of Trade and Protection', in I. A. McDougall and R. H. Snape (eds.), *Studies in International Economics*, North-Holland, Amsterdam, 1970.

[11] H. Leibenstein, 'Allocative Efficiency Versus 'X-Efficiency' ', *American Economic Review*, 56, June 1966, pp. 392–415.

[12] See, for example, R. G. Lipsey, 'The Theory of Customs Unions: A General Survey', *Economic Journal*, September 1960, pp. 512–13, but also his own change of mind on this subject in J. Bhagwati (ed.), *International Trade*, Penguin Books, 1969, pp. 239–40 (footnote).

We shall consider the effects of a once-for-all change in trade policy, such as a tariff.

For any given level of the tariff the profits in the industry concerned depend on the efficiency of the firms in the industry; the lower efficiency, the lower profits. Now let efficiency itself depend on managerial effort, the cost of this effort being managerial leisure or relaxation forgone. Efficiency will also, of course, depend on many other things, including the 'state of the arts'. The argument here is that, given these other things, notably the state of knowledge, efficiency depends on effort: it takes more effort to be efficient than to be inefficient. Firms may be inefficient because of ignorance; to the extent that this ignorance is given and not responsive to an effort at managerial self-improvement, it is incorporated in the firms' 'state of knowledge'. It may seem illogical to suggest that any firm wilfully fails to use the best methods available to it, but the argument is that the search for the best method—and, in the context of a changing environment, for the best method at every point in time—requires an effort that can be regarded as having a cost.

The next step is to imagine the managerial factor to have a utility function, the arguments of which are leisure and profits, the latter representing command over goods and services. Now, for given managerial effort or efficiency, a rise in the tariff raises profits in the industry concerned and hence real income of the managerial factor. This is simply the income distribution effect of the tariff. Some of the gain in real income is likely to be taken out in extra leisure, so that managers and entrepreneurs relax and become less efficient. There is more 'managerial slack.' Conceivably there may also be a substitution effect; but as this may go either way, it will be ignored here.[13]

[13] In Corden, 'The Efficiency Effects of Trade and Protection' it was implicitly assumed that a rise in the tariff would increase profits by a constant proportion whatever the degree of managerial effort. This would mean, implausibly, that if zero profits were made with a given degree of effort at the original tariff level, zero profits would also be made with a higher tariff. If the assumption held, a rise in the tariff would then raise the marginal rate of substitution of profits for leisure and so lead to a substitution effect yielding more effort. In fact the marginal rate of substitution of profits for leisure may rise or fall, and one cannot say in general which way the substitution effect will operate. The doubts about the original treatment of the substitution effect were

The term 'efficiency' is used in a special sense here, more in accord with popular usage or with usage in the literature on *X-efficiency* than with usage in economic theory. It may indeed be efficient for managers to reduce their X-efficiency in these circumstances. We shall come back to this later. Here, in any case, we shall equate reduced managerial effort with inefficiency in the special 'X-efficiency' sense. The argument could also be generalized to allow for a variety of corporate objectives other than profit maximisation. These could be treated just like managerial leisure, being arguments in the managerial utility function, a rise in income leading to increased pursuit of these objectives.

We appear, then, to have a formal analysis supporting the argument that tariffs reduce X-efficiency. It rests, it will be noted, on four assumptions, (1) that protection has an income distribution effect which raises the real incomes of import-competing producers, (2) that the income elasticity of demand for managerial leisure is positive, (3) that there is a positive relationship between effort and efficiency, and (4) that one need only look at the industry concerned, there being no other efficiency effects elsewhere in the economy as a result of the tariff. The last assumption has been implicit.[14]

(2) *General Equilibrium Approach*

It is this implicit assumption which is the crucial weakness in the argument. Let us then retell the story in general equilibrium terms. A tariff raises the incomes of some members of the community and reduces that of others. Thinking of a uniform tariff for simplicity, and assuming policies of internal and external balance, it raises the incomes of factors intensive in the production of importables and reduces the incomes of

---

stimulated by a discussion in John Williamson, *On Estimating the Income Effects of British Entry to the EEC*, Surrey Papers in Economics, University of Surrey, 1971, pp. 10–12.

[14] The central point that a monopolist may not just seek to maximise his profits but may also vary the supply of his 'private factors' was made in J. R. Hicks, 'Annual Survey of Economic Theory: The Theory of Monopoly', *Econometrica*, 3, 1935, pp. 1–20. The implications of varying 'entrepreneurial inactivity' were thoroughly explored by T. Scitovsky, 'A Note on Profit Maximisation and its Implications', *Review of Economic Studies*, 11, 1943, pp. 57–60.

factors intensive in the production of exportables. The factors intensive in importables will become less efficient and the factors intensive in exportables more efficient.

We may prefer to assume that the opportunity to vary efficiency is open only to profit-earners and to the managerial classes the salaries of which may be closely geared to profits in the industries in which they are employed. If we assume in addition that there is no single rate of profit or of managerial remuneration throughout the economy, at least in the short-run, then we can restate the argument by saying that a tariff raises the profits of import-competing producers and reduces the profits of exporters. The former will work less hard and the latter harder.

In Britain it has been said that joining the European Economic Community would have a 'cold shower' effect on industrialists; they would give forth more effort and so presumably become more efficient. But one might ask whether this view does not see them only in their capacity of import-competitors and not exporters. A movement towards free trade, whether consisting of unilateral tariff reduction associated with exchange rate devaluation or of reciprocal tariff reduction would presumably have a relaxing 'warm sun' effect on exporters.

The story is thus clearly not as simple as was suggested earlier. There are opposing effects to take into account. First, for each gainer or loser in real income there will be an income effect (which conceivably may be offset by a substitution effect), and then the effects on gainers must be set against the effects on losers. Furthermore, one should take into account effects on the consumption side arising through the increase in the price of importables, or some importables, relative to exportables. If tastes differ, this will bring about an additional income redistribution. Those persons with consumption patterns biased towards importables rather than exportables will lose in real income as a result of the tariff, unless they happen to be the same people who are employed intensively in importables. They may then work harder as a result of the tariff so as to make up, at least to some extent, for the fall in their real capacity to consume that would otherwise result. Similarly, persons with consumption patterns biased towards exportables will tend to work less hard. A complete analysis would also allow for the effect of the tariff on the relative price of leisure-goods. If the tariff markedly raises the price of golf-clubs or alcohol the effect on managerial efficiency might be quite significant.

(3) *The Argument Rescued?*

Can one then rescue the argument that in general protection reduces efficiency?

Firstly, protection may increase the degree of monopoly and raise monopoly profits throughout the economy. For a given efficiency level the general rate of profit will then rise and income distribution will be shifted from labour to profits. 'This may be true even though, in the short-run, profits in particular exporting industries may fall. Alternatively protection may create particular pockets of monopoly profits or quasi-rents without affecting the general level of profits.

Provided that labour and other factors that have lost real income have no scope for varying efficiency-creating inputs, there will be a fall in the general level of X-efficiency or at least in efficiency in the 'pockets' of monopoly concerned. The fall in managerial efficiency will not be offset by a rise in the efficiency of labour. This approach—which is undoubtedly implicit in some popular discussions of this general subject— focuses on the effects of protection or of a movement to free trade on the general degree of monopoly or competitiveness in the economy.

Secondly, import-competing industries may be capital-intensive and there may be a general rate of profit applying throughout the economy. A tariff will raise the rate of profit at the expense of wages and rents. If efficiency variation is open only to profit-earners there will then be a fall in efficiency.

Thirdly, protection may reduce X-efficiency because the system of protection is made-to-measure. The application of this principle tends to mean that a tariff will be adjusted to avoid excess profits in the protected industry but will not be adjusted to avoid inefficiency. The system encourages import-competing firms to take out any gains in producer surpluses resulting, say, from an exogenous rise in the price of imports by reducing managerial effort rather than by taking higher profits. A made-to-measure tariff system will encourage firms to avoid profit increases resulting from extra tariffs for fear that they would lose the tariff increases; it pays then to be inefficient.

Finally, one could assume that the exchange rate is fixed, money wage rates are inflexible downwards, and effective demand is constant, or is varied to maintain external balance, not internal balance. Tariff reduction would then release factors from import-competing industries which would not be absorbed in the export industries. Efficiency in the import-competing industries and in other industries for which demand falls will improve, but there will be no reason for an offsetting rise in the export industries. In fact, tariff reduction, by switching effective demand away from the country's goods and services has then the same effect as a deliberate deflationary policy. At the cost of unemployment it may indeed increase X-efficiency.

Thus there could conceivably, but by no means certainly, be something in the view that protection reduces efficiency and increases managerial slack. All one can really say in general is that a protective structure may well have effects on X-efficiency, but the sign of these effects need not be uniform throughout the economy.

(4) *Implications of X-Efficiency Effects*

Let us now assume that there are such X-efficiency effects and consider three implications.

The first concerns the relative importance of X-efficiency effects and the orthodox resource allocation effects. It is a common intuitive view that efficiency effects are, in some sense, more important.[15] One cannot, of course, really talk about 'importance' unless one introduces the magnitudes of an actual case and an appropriate way of weighting the two effects. But consider the following simple argument. A tariff leads to a 10 per cent rise in output of the import-competing industry. The orthodox resource allocation effect focuses only

[15] See Leibenstein, 'Allocative Efficiency Versus 'X-Efficiency' '. The view can also be found in numerous journalistic discussions concerned with British entry into the E.E.C.

on this 10 per cent. But the efficiency of producing the whole output may be affected; it bears thus on the original 100 per cent as well. In this simple sense the efficiency effect is indeed always 'more important'.

The point is really that the income distribution effect of a tariff affects the whole of an industry even when the output changes resulting from the tariff are only marginal; and in our analysis here income distribution effects have become converted, in part, into efficiency effects.

The second implication is that a positive effective tariff, with all other tariffs given, may no longer lead to an increase in the output of the protected industry, and indeed, rather paradoxically, its output may fall. The rise in the tariff may have led to a decrease in the input of the managerial factor, measured in efficiency units, and this may have more than offset the inflow of factors from other industries.[16]

Thirdly, if protection *did* reduce overall efficiency, perhaps because of its effect in increasing the general degree of monopoly, or because of asymmetrical reactions of some kind, would we have an argument for free trade distinct from the ordinary static comparative cost argument? To analyse this we have to remind ourselves that the fall in efficiency would represent a redistribution of income, the fruits of which are allocated in a particular way. One might then ask first whether the income distribution effect is socially desirable and secondly whether the particular way of allocating it is. But the first question is not new, and simply raises all the familiar income distribution issues. So we are left with the second question.

One approach might be to argue that the gainers from a tariff, including the gainers of monopoly profits, know best how to maximise their utilities, and if they choose to do so by reducing managerial effort, then this is the socially optimum way; there is no social loss because an increase in producers' surplus is taken out in leisure rather than in command over

[16] For fuller discussion, see Corden, 'The Efficiency Effects of Trade and Protection', and for earlier general equilibrium analyses, V. C. Walsh, 'Leisure and International Trade', *Economica*, 23, August 1956, pp. 253–260, and M. C. Kemp and R. W. Jones, 'Variable Labour Supply and the Theory of International Trade', *Journal of Political Economy*, 70, February 1962, pp. 30–36.

goods and services for consumption and investment. From a private point of view it is optimal to become managerially less 'efficient', and there is no reason why the social welfare should require a different marginal propensity to 'consume' leisure. Any social loss results not because a sectional rise in real income is taken out in extra leisure and hence managerial inefficiency, but only because the tariff sets up a price distortion leading to the usual cost of protection.

This sounds reasonably convincing. Yet there are two qualifications. Firstly, the marginal social value of extra managerial leisure may be less than its private value because income and commodity taxes cause leisure to be underpriced relative to other goods. This, of course, applies to all leisure, not just to that taken out in managerial X-inefficiency. Secondly, 'society' or the observer may regard a spirit of slackness, a lack of competitiveness, a relaxed attitude to business as socially undesirable, as an unsatisfactory way for businessmen to consume their surpluses. One might then say that privately perceived marginal private product exceeds marginal social product in the case of increases in X-inefficiency brought about by tariffs or by monopoly.

What we have really done here is to give a new significance to the income distribution effects of a tariff and of monopoly, and especially to their effects on profits or potential profits. When orthodox international trade theory tells us that a tariff will increase profits, whether profits in general or profits in a particular industry only, we can now go on to deduce that there may also be efficiency effects with the consequences and implications discussed here. At the same time it must be stressed that the efficiency effects of protection can go either way. Quite specific assumptions must be made to lead to the result that tariffs reduce efficiency.

VI
THE RENT-SEEKING SOCIETY

There is another way in which potential monopoly profits can be used up or dissipated so that various protective policies do not lead to monetary profits to the extent suggested by orthodox theory. This is brought out in Anne Krueger's concept of 'rent-seeking'.[17] It is particularly relevant in countries where government and bureaucracy are easily subject to persuasion and influence from businessmen affected by tariffs and quotas. A simple example can explain the main idea.

An import quota is imposed on the product represented in Figure 8.4. Initially imports are *OH* and the quota limits

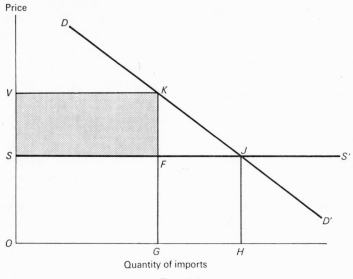

Fig. 8.4

imports to *OG*. The c.i.f. import price is *OS* and the quota will cause the price to domestic consumers or users to go up to *OV*. The familiar analysis is that there will be quota profits of *SVKF* (shaded) received by the lucky people who obtain the import licences. These quota profits are *rents* because they are not received as payments for any services, and any reduction in these profits would not affect the supply of any resource. The costs and normal profits of trading are assumed

---

[17] Anne O. Krueger, 'The Political Economy of the Rent-Seeking Society', *American Economic Review*, 64, 1974. The main point can also be found in Gordon Tullock, 'The Welfare Costs of Tariffs, Monopoly and Theft', *Western Economic Journal*, 5, June 1967, pp. 224–32.

to be included either in the import supply curve or to be subtracted from the consumers' demand curve to obtain the demand curve drawn in the diagram.

This simple analysis must be modified in at least two ways. Firstly, the potential licence-holders might pay bribes to the licence-dispensers—politicians or government officials. Hence part of the shaded area might be an income redistribution. If this sort of income came to be regular and expected it might affect the salaries and other emoluments of licence-dispensers, and could therefore not be described as rent.

Secondly, there is the cost of 'rent-seeking'. It is this that concerns us here. The potential licence-holders are likely to expend real resources in order to obtain the licenses. They will travel to the capital city—perhaps many times for one licence—use up valuable time trying to telephone officials, perhaps employ highly skilled and educated 'go-betweens', and so on. In some countries the effort to obtain licences needed to carry on any economic activity can be quite exhausting. The rent seekers will be prepared to spend on the rent-seeking process—which consists partly of payments to rent-dispensers and partly expenditure of real resources as described here—a sum of something less than $SVKF$. They will be prepared to spend up to the point where they are just left with normal profits for this activity.

Licence profits that can be obtained only through effort and expenditure of resources cannot, strictly speaking, be described as rents. But the main point is that real resources will be expended in obtaining such gross profits created by licensing, and this, in every sense, represents a real social cost. In Figure 8.4 the control of imports thus imposes not only the familiar distortion-cost of FKJ (assuming no domestic or trade divergences initially) but also a 'rent-seeking' cost which is some part, and possibly a very large part, of $SVKF$. In addition, of course, it has income distribution effects.

There is one complication. No person can be sure of obtaining the licences if he expends resources on rent-seeking. An element of risk is involved. Suppose that there are three rent-seekers, that only one can obtain the licences, that they have an equal chance, and that each knows this. If they had no risk aversion each would be willing to spend up to one-third of the maximum rent-seeking expenditure mentioned above ($SVKF$ minus normal profits) and our basic argument would be independent of the number of rent-seekers. But suppose, realistically, that they are risk averse. Hence a disutility is created through the introduction of a risk-element into the business of being allowed to import; the rent-seekers will then be prepared to spend less than otherwise on real resources. But this reduced willingness to expend resources on rent-seeking will not represent a real income gain for society since the resource-saving will be matched by the extra disutility created by the risk-element.

When the establishment of tariff and quota levels, and the allocation of licences, become discretionary and subject to influence from affected businessmen one can thus expect some dissipation of potential monopoly profits in the form of real resource costs. These are not of course the only costs of a system where the government is subject to sectional influence in the way described. Decisions are not based on a detached consideration of the national interest. In terms of our simple case of Figure 8.4, the bribes may have caused the extent of import reduction to be higher or lower than socially desirable. But here we are concerned with the direct real resource costs of rent-seeking.

Rent-seeking may be a highly specialized and skilled activity. A large firm may be able to employ people who are skilled in producing and marketing goods, and others who are skilled rent-seekers. Small firms may be handicapped since one may need several very disparate skills. This can be overcome if there are specialized rent-seeking firms (public relations consultants, private tariff advisers, and so on) that can be hired. It also helps if import licences are marketable, so that able rent-seekers obtain the licences and sell them to ordinary producers or traders.

All this is relevant for tariffs as well as import quotas. Tariffs can create monopoly profits as described in Section I of this chapter, or rents for intra-marginal firms, and it will pay firms to engage in rent-seeking expenditures designed to get such tariffs legislated. The less uniform and automatic the tariff system the more scope there is for lobbying and so the more rent-seeking expenditures might be expected.

A tariff system which is intended to be *made-to-measure*—and hence is highly complicated and depending on judgements of officials or Tariff Commissions about costs of production of protected firms—may give rise to a great deal of rent-seeking activity. It is bound to involve an element of bureaucratic discretion. We saw earlier that the aim of such a system is to avoid rents in order to minimise adverse income distribution effects. But it may perversely succeed in holding down the net incomes of protected producers not by avoiding the adverse consequences of tariffs on the incomes of consumers or users but rather by causing potential rents to be dissipated in real resource costs.

VII
DUMPING

Dumping is usually understood to mean that a product is exported at a price lower than the price at which the identical or a similar product is sold by the same producers on the exporting country's domestic market. Dumping and measures to counter it have received much attention in the commercial policies and tariff legislations of most countries and there is a G.A.T.T. Anti-Dumping Code which legitimizes anti-dumping tariffs in certain clearly defined circumstances. Here we discuss dumping because it may be either a result of monopoly or an instrument to create or strengthen monopoly.

Two questions arise. First, is it in the interests of a country to dump its own exports, perhaps encouraging a private monopolist to do so? Secondly, if foreign countries dump their exports in one's own market, does this call for anti-dumping tariffs? Before considering these two questions, let us look at various types of dumping.[18]

(1) *Private Long-term Dumping and Price Discrimination*

Dumping may be private, subsidized, or public, and if it is private, it may be short-term or long-term. In turn, short-term dumping may be sporadic or predatory. We shall look at short-term dumping more closely later. Here let us begin with

[18] The classics on the subject of dumping are Jacob Viner, *Dumping: A Problem in International Trade*, Univ. of Chicago Press, 1923, and G. v. Haberler, *The Theory of International Trade*, William Hodge & Co., London, 1936, Ch. 18. The discussion here owes much to Haberler's treatment.

private long-term dumping. This is likely to result from the profit-maximising policies of a discriminating monopolist.

Consider an export industry where there is a single producer. One might think of a manufacturing exporter producing a reasonably differentiated product. He faces less than perfectly elastic demand curves abroad and at home. He will exploit

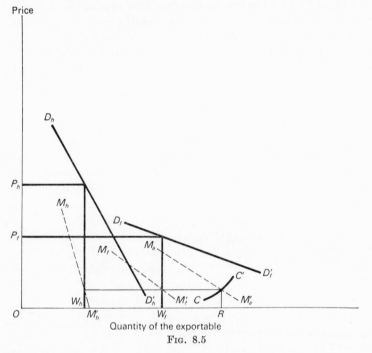

Fig. 8.5

his monopoly power by restricting supply and raising the price. If the domestic and the foreign markets can be kept separate he will choose the usual equilibrium of the discriminating monopolist, with prices relatively higher in the less elastic market. If the domestic demand is less elastic dumping, as usually defined, will then result.

This situation is represented in Figure 8.5, which makes use of a familiar diagram owed to Joan Robinson.[19] The domestic demand curve is $D_h D_h'$, the foreign demand curve is $D_f D_f'$, and the marginal revenue curves derived from these are

[19] Joan Robinson, *The Economics of Imperfect Competition*, Macmillan, London, 1948, Ch. 15.

$M_h M_h'$ and $M_f M_f'$ respectively. (Only relevant segments of these curves are drawn.) The horizontal summing of the two marginal revenue curves yields the aggregate marginal revenue curve $M_a M_a'$. The domestic producer's marginal cost curve is CC'.

The discriminating monopolist will produce at the point where his aggregate marginal revenue is equal to his marginal cost. Total output will be OR, sales at home OW_h and sales abroad OW_f. The producer has equated his marginal revenue in both markets to his marginal cost, hence satisfying the marginal conditions of his own private optimum. The price at home is OP_h and the price of exports OP_f.

In the diagram it is assumed that at the relevant points the elasticity of demand at home is lower than that abroad, so that the domestic price is higher, and thus there is dumping. Theoretically, the opposite case—with the export price higher than the domestic price—is possible. This might be called 'reverse dumping'.

In the case represented by Figure 8.5 the export price exceeds the marginal cost of production, even though it is below the price charged to domestic consumers. The limiting case is the small country situation where the exporter faces a perfectly elastic foreign demand curve. This simple but important case is represented in Figure 8.6. In that case the export price will be equal to the marginal cost of production.

It has been assumed here that the monopolist's marginal cost curve is upward-sloping. It is, of course, very likely that there are economies-of-scale and that his marginal cost curve is downward-sloping. But this does not alter the basic analysis at all, as can easily be verified by drawing the appropriate diagram. It will remain true that the producer will produce to the point where his aggregate marginal revenue curve crosses the marginal cost curve and that he will discriminate on the basis of the elasticities of the two demand curves.[20]

[20] When the marginal cost curve is downward-sloping it will never pay the monopolist to price exports at marginal cost. This is contrary to widely held beliefs. With the marginal cost curve sloping downwards the foreign demand curve must also slope downwards—for otherwise it will pay him to expand exports indefinitely. He will equate marginal cost to *marginal* revenue in the foreign market, and his foreign price (*average* revenue) must thus be above marginal cost.

It is important to stress that price discrimination is not possible without the ability to separate the two markets. If there were no tariff in the exporter's home market and transport costs were negligible, then the low-priced export product could simply be re-exported back into the domestic market and so force the price in the latter down to the export price.

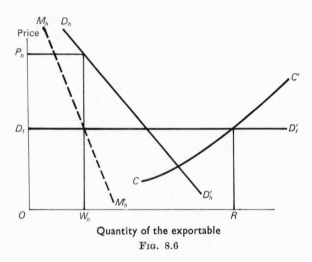

Quantity of the exportable

FIG. 8.6

Transport costs can prevent this, or at least make it possible to maintain some margin between the domestic and the export price. There will be a maximum price which the discriminating monopolist can charge domestically. The greater transport costs, the higher this price for any given export price. This maximum price set by the need to exclude imports may constrain the monopolist to charging a domestic price lower than the one represented in our diagrams.

In view of the decline in international transport costs in recent years, tariffs are probably more important causes of market separation (though tariffs in industrial countries have, of course, declined also). The c.i.f. import price of equivalent goods plus tariff sets an upper limit to the price the monopolist can charge at home. If there were no tariff in the exporter's home market, he might have very little opportunity to discriminate against domestic consumers and dump in export markets. The imposition of a tariff may thus be a cause of dumping by making it possible to raise the domestic price. Dumping could be ended or greatly modified if the tariff were removed. An import quota will have a somewhat similar effect.

It might also be noted that sometimes dumping is described as exporting below costs of production. No monopolist will choose to export for any long period below marginal cost of production since this would simply yield marginal losses. He may, of course, do so for a short period, and this will be discussed later. But it may be meant that he charges over a long period an export price below *average* cost of production. It can be shown that this is possible if his average cost curve is downward-sloping but is not possible (assuming profit maximization) if it is upward-sloping. In any case, the concern with the relationship between the export price and average cost which can sometimes be found in discussion of this subject does not seem of real value. It is better to adhere to the common definition of dumping as the price to domestic buyers exceeding the export price, as this focuses on price discrimination as the essence of private dumping.

(2) *Subsidized and Public Dumping*

Let us now turn briefly to subsidized dumping. This time we need not assume monopoly. The case of an export subsidy is represented in Figure 8.7. The domestic supply curve, which is equal to marginal costs, is $S_h S_h'$, and the domestic demand curve is $D_h D_h'$. The foreign demand curve, assumed to be perfectly elastic, is $D_f D_f'$. An export subsidy of $D_f S$ is then provided. It raises the domestic price to OS. Consumption falls and production rises, as indicated by the arrows. For the higher domestic price to be possible, there have to be either transport costs or a domestic tariff. Domestic consumers pay the price OS while the f.o.b. price paid by foreign buyers is OD_f.

Thus there is dumping in terms of the usual definition, even though the industry may be competitive. This time the export price is below the marginal cost of production, something that was not likely and would not be privately profitable on a long-term basis in the case of ordinary private dumping.

Finally there is public dumping. One might have the case of a publicly-owned exporting enterprise which is concerned with maximizing profits, but then the analysis is the same as in the case of private dumping. Here we are concerned with export marketing arrangements in which there are private competitive producers, and a marketing board or similar body intervenes with the aim of providing protection for exporters or stimulating exports rather than maximizing profits for the board or the producers. A device used in some countries to foster agricultural exports or protect agricultural

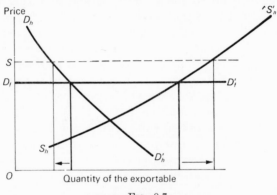

FIG. 8.7

producers is a *home-price scheme*. The aim is for the marketing board to charge a high price to domestic consumers and use the profits on domestic sales to subsidize exports. The export price will be below marginal cost of production (as in the case of an ordinary export subsidy), while the price to domestic consumers will be above it.

(3) *Is Dumping in the Interests of the Dumping Country?*

The above analysis certainly suggests that long-term dumping can be in the interests of a private monopolist who sells both domestically and exports. It might be argued that it would not be in his interest if his dumping were countered by anti-dumping duties. This may be true if he did not take the prospect of such duties into account when estimating the foreign demand curve for his exports. Essentially, if such duties are taken into account—as they should be—the demand curve facing him will become less elastic, and it may have a kink in it. Let us now disregard such duties, or alternatively suppose them to be incorporated in some way in the demand curve on which the monopolist bases his pricing policy.

The question now is whether dumping is in the national interest of the exporting country, as distinct from the private interests of the monopolist. The question is important since national policy, in the form of tariffs, can make private long-term dumping possible, and the removal of tariffs can put an end to such dumping.

Given certain simple and familiar assumptions, dumping is not in the national interest, and indeed, in certain circumstances, 'reverse dumping' may be optimal. The assumptions are those which make monopolistic exploitation by domestic producers of domestic consumers non-optimal. Domestic income distribution effects are ignored, or alternatively, income redistribution towards export producers is considered undesirable. Assuming no domestic divergences other than those created by dumping of the particular product concerned, the domestic price should be equal to the marginal cost, not above it.

If the foreign demand curve is perfectly elastic (the case of Figure 8.6) there should then be a uniform price applying to both domestic and home sales, a price equal to marginal cost. If the foreign demand curve is downward-sloping the national optimum requires marginal cost to be made equal to *marginal* revenue in the export market, but to *average* revenue (price) in the domestic market. This solution actually requires the export price to *exceed* the domestic price (reverse dumping). If the exporting industry were competitive the solution would be attained by imposing the orthodox optimum export tax.

The case is represented in Figure 8.8. The domestic demand curve is $D_h D_h'$ and the marginal revenue curve in the export market is $M_f M_f'$. The horizontal sum of these two yields $G_a G_a'$ and it is the intersection of this curve with the marginal cost curve CC' that gives us the optimum output OR' and the marginal cost at this optimum, OL. The domestic price will be OL. Marginal export revenue will also be OL, and the export price will be above it, namely OL'.

We can compare the discriminating monopolist's private optimum (originally described in Figure 8.5) with this national optimum solution. For the private optimum one draws the $M_a M_a'$ line, which is the horizontal sum of the two marginal revenue curves ($M_h M_h'$ and $M_f M_f'$), and this must be to the left of $G_a G_a'$. The private optimum requires $M_a M_a'$ to be equated with CC', so that private optimum output is OR, which is less than the socially optimal output OR'. With an upward-sloping marginal cost curve (as drawn) the private

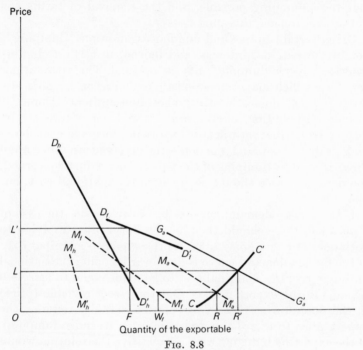

Fɪɢ. 8.8

optimum will involve more exports (OW_f) than the social optimum (OF). If the cost curve were downward-sloping it would involve less exports.[21]

[21] See S. Enke, 'Monopolistic Output and International Trade', *Quarterly Journal of Economics*, 60, February 1946, pp. 233–49, for a discussion of the theory of export monopoly, containing essentially the analysis presented in Figure 8.8. See also G. Pursell and R. H. Snape, 'Economies of Scale, Price Discrimination and Exporting', *Journal of International Economics*, 3, February 1973, pp. 85–91, for an analysis of the question whether an export industry snbject to economies of scale should be allowed to discriminate against domestic cousumers if this is necessary for its survival; they find that if no tariff is needed for discrimination then there is a gain from the existence of such an industry, but the gain could be increased by reducing discrimination. Further, they show that a tariff to permit discrimination is never optimal.

The question arises whether there is any point in a government encouraging private dumping by providing the necessary protection of the domestic market. Given our assumptions so far, from a national Pareto-efficiency point of view, the only encouragement that should be provided is not for dumping but only for the foreign market to be exploited optimally by the exporters, to ensure that they equate marginal cost to marginal rather than to average export revenue. If this is done there is no need for an orthodox optimum export tax.

This argument can be transformed by introducing some general reason for fostering exports. Perhaps the exchange rate is over-valued, and it is desired not to alter it for any one of a number of reasons: institutional rigidities, income distribution effects, ignorance, and so on. In other words, let us envisage a second-best situation where there is a marginal divergence in trade: the marginal social value of exports exceeds the private value. Disregarding the costs of financing subsidies as well as subsidy disbursement costs, the first-best optimum policy (given the second-best premise) is to provide an appropriate export subsidy. If the domestic market is not to be lost to imports this must be associated with a tariff or import quota. But an alternative policy is to provide a tariff alone, to induce the private monopolist to dump by restricting his domestic sales.

In the simple case where the export demand elasticity is infinite (Figure 8.6) the monopolist will then increase exports by the amount of the reduced domestic sales. His total production will stay constant. Thus this solution of fostering exports by encouraging private dumping with a tariff cannot achieve the same result as an export subsidy because it induces extra exports only by lowering domestic sales and, unlike the export subsidy, fails to stimulate an increase in output. But it may nevertheless be the preferable solution in circumstances where the costs of financing and disbursing subsidies are high.

This analysis is a way of rationalizing or understanding why some governments look sympathetically on dumping by their exporters. Essentially they assume that exports are more desirable than indicated by the actual domestic prices of exports, no doubt because the exchange rate is over-valued.

The same sort of analysis applies in the case of subsidized and public dumping. These may also be justified by some general case for fostering exports.

(4) *Should Foreign Dumping be Countered by Anti-dumping Duties?*

Many countries have anti-dumping regulations of some kind. One outcome of the Kennedy round of G.A.T.T. negotiations was an agreement establishing an international anti-dumping code which now governs the anti-dumping policies of signatory countries. Dumping refers to *private* dumping and is considered to occur 'if the export price of the product exported from one country to another is less than the comparable price, in the ordinary course of trade, for the like product when destined for consumption in the exporting country'. The Code provides that governments can take action against dumping only when there is 'material injury' to domestic producers. There must be 'sufficient evidence' of such injury before instituting anti-dumping proceedings.

Thus *both* the existence of dumping and the existence of injury to domestic producers must be proven. In practice this is difficult. Anti-dumping cases can yield delays, as imports are held up while investigations proceed, so there is some import-discouraging effect even when adequate evidence does not emerge, and this, indeed, seems to be the usual situation. In the United States anti-dumping duties were imposed in only eight out of 194 investigations between 1958 and 1965. In the United Kingdom, from 1958 to 1970, 139 formal applications for anti-dumping and countervailing duties were made, but finally duties were only imposed in nineteen cases.[22]

Countervailing duties must be distinguished from anti-dumping duties. While the latter apply to *private* dumping, and in addition require evidence of injury according to G.A.T.T. provisions, countervailing duties can be imposed to offset export subsidies and, according to G.A.T.T.

---

[22] For facts and general discussion, see Robert E. Baldwin, *Nontariff Distortions of International Trade*, The Brookings Institution, Washington, 1970, pp. 139–143; Gerard and Victoria Curzon, *Hidden Barriers to International Trade*, Trade Policy Research Centre, London, 1970, pp. 34–6; Brian Hindley, *Britain's Position on Non-tariff Protection*, Trade Policy Research Centre, London, 1972, pp. 13–18.

regulations, can be imposed even when there is no evidence of injury.

These are the facts. Additional facts are that the idea of anti-dumping tariffs is long-established—Viner wrote at length about the subject in 1923 and Haberler in 1936—and that dumping, or anything that conceivably looks like dumping, often provokes fearful or emotional responses from domestic industries. What then is the argument for anti-dumping duties?

A simple point has often been made by economists but has clearly made little impact among politicians and businessmen. It applies only to long-term dumping, whether private, public or subsidized. A country should take the foreign supply price as given. Whether the same or a different price is charged by the foreign producer to his own countrymen is irrelevant. It is also irrelevant whether the foreign supply price is less than the average or indeed marginal cost of production. The lower the import price the better the terms of trade. If it is desired to protect domestic production for some reason a tariff may be justified. This tariff may need to be increased if the foreign supply price falls—at least if the tariff is of the made-to-measure type—but it does not need to be higher because the foreign export price is less than the domestic sales price in the foreign country or less than the average or marginal cost of production.

What is wrong with this? Why is the simple argument that cheap imports should be welcomed by a country, and the cheaper the better, often considered 'theoretical'?

The most important qualification is that the concern is often with *short-term* dumping, which creates uncertainty and may be an instrument of monopoly. We shall consider this below. While it is doubtful that it is often important in practice, and there are problems of establishing evidence for it—as indicated by the figures quoted above—one could make a logical argument for anti-dumping duties to counter short-term dumping.

Secondly, the concern is often with producer rather than consumer interests. If the import price falls there will be a domestic income redistribution effect, which it may be desired to prevent. Implicitly the logic may be that of the conservative social welfare function. Of course, this still does not justify a concern with dumping as such—that is, with whether the import price is less than the price in the exporter's domestic market.

Finally there is some implicit concept of 'fairness'. The idea that dumping is 'unfair' seems to arise particularly when it is believed that the export price is below the foreign supplier's average cost of production, which is—as we have seen—possible in the long-term only if the average cost curve is downward-sloping.

Suppose that the domestic producer is just as efficient as the foreign producer and, indeed, has an identical downward-sloping cost curve. It would then seem 'fair' for the two to compete on equal terms, perhaps each obtaining his own domestic market, or the two splitting up the market on the basis of product differentiation. But then the foreign supplier obtains a tariff to protect him in his domestic market, and this makes it possible for him to export at a price less than average cost. Alternatively, both the foreign producer and the domestic producer may be able to discriminate because of transport costs, and the foreigner may have the 'unfair' advantage of having a larger domestic market, so that his export price can be lower, and perhaps below average cost, and he then captures the market in both countries. It remains true that consumers must benefit from cheap imports, and in the Pareto sense, assuming compensation, the country benefits. But compensation is not likely to take place, and from the producer's point of view the whole situation may appear and indeed be thoroughly 'unfair'.

(5) Short-term Dumping

Finally, let us look at short-term dumping, which we have put aside so far. It is fear of this type of dumping that motivates anti-dumping legislation and preoccupies import-competing interests in many countries.

Short-term dumping may be sporadic or predatory. *Sporadic dumping* means that the foreign supply price may fall temporarily owing to variations in production abroad, perhaps explained by harvest fluctuations or technical changes, or owing to unforeseen demand fluctuations. The fluctuations in the price may, but need not be associated with dumping as we have defined the term here; for the domestic price abroad may fall along with the export price. But it is likely to be associated with a fall of the export price, and possibly also the domestic price, below long-term average cost.

A fall in the price of imports, though temporary, is always of benefit to consumers. But it increases risks of domestic import-competing producers and arguably some offsetting tariff may be justified as second-best to an insurance policy or to avoid absolute uncompensated falls in real incomes to producers. This, of course, is a narrow sectional point of view. From an international point of view it is likely to be less costly if the consequences of unexpected fluctuations in demand and supply are spread thinly and widely around the world than if they are always confined within the originating country by the use of anti-dumping duties in other countries.

Predatory dumping is not just a manifestation of monopoly but is a technique to maintain monopoly. A foreign dominant supplier may temporarily reduce his supply price (or increase his selling expenditures) so as to force out a domestic producer trying to enter or extend his share of the market, the foreign supplier being able to sustain a price war longer than the domestic supplier. In the short-run the price reduction represents a terms of trade improvement, but if it is successful in maintaining monopoly, in the long-run it may worsen the terms of trade. The essence of this monopoly-preserving price reduction is its temporariness and its motivation.

If predatory dumping can be identified—and this is usually difficult—it may justify anti-dumping duties on anti-monopoly grounds. At the same time the possible advantages of monopoly—even foreign monopoly—through increased scale of output, hence lower costs, and possibly lower prices, should also be borne in mind.

Predatory dumping will only be effective if the foreign exporter is able to sustain a monopoly. It is not sufficient for him that he is able to force out a domestic supplier. If he has to compete with other foreign suppliers he may not subsequently be able to raise his price to reap the fruits of his predatory activities. Strictly, it is irrelevant whether the foreign monopolist charges his own countrymen a higher price than the export price, but dumping—especially if the price drops below marginal cost so that the foreign supplier is exporting at a short-term loss—may be some evidence of motivation.

9

THE INFANT INDUSTRY ARGUMENT

THIS AND the following two chapters deal with the relationships between protection and economic growth. Here we begin
with the infant industry argument because this has in the
past been the main growth-related argument for protection.

The argument was endorsed by John Stuart Mill in a
famous and oft-quoted passage in his *Principles*.

The only case in which, on mere principles of political economy,
protecting duties can be defensible, is when they are imposed
temporarily (especially in a young and rising nation) in hopes of
naturalizing a foreign industry, in itself perfectly suitable to the
circumstances of the country. The superiority of one country over
another in a branch of production often arises only from having
begun it sooner. There may be no inherent advantage on one part, or
disadvantage on the other, but only a present superiority of acquired
skill and experience. A country which has this skill and experience
yet to acquire, may in other respects be better adapted to the
production than those which were earlier in the field: and besides, it
is a just remark of Mr. Rae, that nothing has a greater tendency to
promote improvements in any branch of production than its trial
under a new set of conditions. But it cannot be expected that
individuals should, at their own risk, or rather to their certain loss,
introduce a new manufacture, and bear the burden of carrying it on
until the producers have been educated up to the level of those with
whom the processes are traditional. A protecting duty, continued
for a reasonable time, might sometimes be the least inconvenient
mode in which the nation can tax itself for the support of such an
experiment.[1]

Haberler has written that 'Since Mill gave it his approval,
the infant-industry argument has been accepted in principle

---

[1] John Stuart Mill, *Principles of Political Economy*, Book V, Ch. 10, *Collected
Works of John Stuart Mill*, Vol. III, University of Toronto Press, 1965, pp.
918–19.

by many Free Trade economists'.[2] It is undoubtedly now one of the most widely accepted arguments for protection in less-developed countries, just as it has in the past been used to justify protection in the United States, Germany, Canada and Australia. Its origin is usually associated with Alexander Hamilton whose *Report on Manufactures* (to the Secretary of the United States Treasury) was published in 1791,[3] and Friedrich List, whose *Das nationale System der politischen Oekonomie* was published in Germany in 1841.[4]

In order to analyse this argument clearly, two distinctions must here be stressed. The first is between *economies of scale*—essentially a static concept—and *economies of time*. The former result in falling costs as the scale of output at any point in time increases (with given factor prices) while the latter result in falling costs as the length of time over which output has proceeded increases, and are sometimes called *dynamic* economies. The second distinction is between *internal* and *external* economies, the question being whether the economies, whether of scale or of time, are internal or external to the decision-making unit, namely *the firm*.

The essence of the infant industry argument is that it is an argument for temporary protection. Hence (a) *time* must enter the argument in some essential way; it cannot rest solely on static economies of scale, whether internal or external. Furthermore, (b) it is an argument for intervention to alter the pattern of production, and so will require some kind of distortion, imperfection or externality in the system somewhere. These two requirements of (a) a time element and (b) an imperfection or externality element govern the subsequent discussion.

It will be shown that an argument for temporary protection can be built in two main ways: one resting on dynamic

---

[2] Gottfried von Haberler, *The Theory of International Trade*, William Hodge & Co., London, 1936, p. 281.

[3] Alexander Hamilton, *Report on Manufactures* in *Papers on Public Credit, Commerce and Finance by Alexander Hamilton*, ed. Samuel McKee Jr., Columbia University Press, New York, 1934.

[4] Friedrich List, *The National System of Political Economy*, Longmans, Green, and Co., New York, 1904. For an excellent discussion and summary of List's influential ideas, see Bert F. Hoselitz in Hoselitz et al., *Theories of Economic Growth*, The Free Press, Glencoe, 1960, pp. 195–204.

internal economies and the other on dynamic external economies. But some other approaches, concerned in turn with investment coordination and with static economies of scale, will also be mentioned.[5]

I
DYNAMIC INTERNAL ECONOMIES

We shall now see whether an infant industry argument for protection can be built with dynamic internal economies (or internal economies of time) as the main ingredient. The average costs of a firm are assumed to fall the longer its output has continued; it learns from experience. This is to be distinguished not only from costs falling with scale of output in the static sense but also from costs falling over time for exogenous reasons. Sometimes these dynamic economies are called *irreversible* economies, since experience of production in one year causes costs to fall once and for all for later years.

But there is a preferable way of looking at this. When factors of production are engaged in producing output in a particular year two products really result, *visible* current output, saleable currently on the market, and the *invisible* accumulation of experience and knowledge, in fact the creation of human capital. The process of learning is a form of investment which builds up a stock of productive capital. Insofar as the invisible capital does not depreciate the effects of learning are indeed irreversible. If current output stopped the stock of capital would remain. But no new capital would be created, so that the process of learning—that is, investment in human capital—can be stopped by reducing or ending current output.

Given certain assumptions, internal dynamic economies do not provide an argument for protection. If finance is freely

---

[5] See Haberler, *The Theory of International Trade*, pp. 278–85, for extensive discussion of the main issues of this chapter and references to early writers. Valuable recent contributions which stress some of the main themes of this chapter are *Trade and Welfare*, Ch. 16; Kemp, *The Pure Theory of International Trade*, Prentice-Hall, Englewood Cliffs, 1964, Ch. 12; Johnson, 'Optimal Trade Intervention in the Presence of Domestic Distortions'; and H. G. Grubel, 'The Anatomy of Classical and Modern Infant Industry Arguments', *Weltwirtschaftliches Archiv*, 97, December 1966, pp. 325–42. The latter is a notably comprehensive survey of the issues.

available to the firm at a rate of interest that correctly indicates the social discount rate, if the firm has correct expectations about the fruits of the learning process, and if there are no uncorrected divergences of any kind in the economy, there is no case for intervention through tariffs or subsidies. Since the firm reaps its own rewards from its investment in learning it does not need to be subsidized to encourage it to go through the infancy period. This is a familiar argument much stressed by economists.[6]

The learning may be on the part of the management of the firm or of its workers. If the firm's employees gain in experience and so become more valuable to it then for the analysis here to apply their wages and salaries must not rise so much that the firm does not profit at all. The suppliers of inputs to the firm may be the learners; if this is to lead to internal economies to the firm concerned the prices of the products of the suppliers must fall (or the quality of the products must improve). In the small country model this is possible in the case of traded inputs only if these are subject to made-to-measure tariffs; otherwise this effect requires the inputs to be traded or to become non-traded as a result of domestic learning.

The learning of the firm concerned may lead to improved quality of product, including improved consistency of the product (an important effect in less developed countries). It may consist in part of *market familiarization*—learning to give the customers the type of product and the service they really want—hence becoming more efficient at product differentiation. And the improved efficiency may be as much in distribution as in production. The quality improvement that results from continued production and sales may be only in the minds of customers; as they become more familiar with the product they may be prepared to pay more for it, or, alternatively, less advertising is needed to sustain sales at a given price and given physical quality.

An infant industry argument can be constructed from dynamic internal economies in three possible ways, the third version being rather tenuous. In all cases the first-best argument is for some form of direct subsidization.

(1) *Imperfection of Private Information*

The first approach rests on an imperfection of private information or of private ability to assess information. Investment in learning capital may be very long-term investment. One may have to produce for many years before significant fruits emerge. It could be argued that private enterprise simply does not look so far ahead, and that the state, in the form of its civil servants or planners, has a longer view and sees a more favourable learning curve than the firm's owners or managers do.

The crucial point of difference may concern not the prospective fall in costs but rather demand prospects. Perhaps an expansion of the economy leading to a general growth in demand is envisaged by the government's planners; only if this demand eventuates will an investment in an infant industry pay. Private enterprise, owing to lack of imagination or information, or to excessive caution, may not expect this growth in demand. There is no divergence between private and social interests but the government or its agents has a better view of what these interests are than the firms concerned.

One could argue that long-term investment takes two main forms: infrastructure investment—which is normally left to governments—and learning investment—which governments may have to subsidize private firms to make. This is about as strong a case as one can make along these lines, and appears to provide some support for infant industry protection when there are dynamic internal economies.

But the case is not really very strong. First-best policy is for the state to spread more information. If private entrepreneurs are not aware of the prospects for expansion of demand that will result from planning or other developments, there are well-known ways of making them aware. 'Indicative planning' is a possible method of spreading information about market prospects. But do government, planning authority or tariff board really have better information?

Why should the private firm (or state enterprise) concerned have less information about the prospects for its own cost curves than a central state authority? The planners or civil servants may be more optimistic than private firms and more ready to speculate about the future because they will not personally have to meet the losses if the risks do not come off. A private firm, if small, might go bankrupt if its investments are misplaced and so has more at stake. Private enterprise may also be justified in being sceptical about ambitious growth targets and forecasts of demand expansion contained in national plans when past experience suggests that target-setting is easier than target fulfilment. Why should a government or tariff board see the long-term profit prospects of particular industries more clearly than the firms concerned? If the enterprise concerned is one of the large international companies it is certainly very likely to have better information and understanding of the future development of its costs and markets than the civil servants of host governments.

Often the applicants for protection supply the relevant information to the tariff board or Ministry of Commerce; in the course of their requests for protection they are liable to paint optimistic pictures, indicating that they have every chance of eventually growing out of the infancy stage. Any public expectations can then be based only on professed private expectations and are hardly likely to be more optimistic. The true divergence may be between those hard-headed private expectations on which investment decisions would be made in the absence of protection and the *publicly professed* private expectations designed to induce public authorities to provide protection.

These doubts are stated here in very general terms and expressed strongly. They certainly apply to protection in developed countries, and to protection of the branches and subsidiaries of international corporations anywhere. In particular cases, mainly applying to indigenous firms in less-developed countries, there may indeed be a divergence of information for which a tariff is a second or third-best remedy.

(2) *Imperfection of Capital Market*

The second approach to building an infant industry argument out of dynamic internal economies depends on imperfection of the capital market. The potential infant may be unable to obtain finance to cover initial losses at a rate of interest which correctly indicates the social discount rate. There will then be under-investment, or failure to invest at all, in the creation of long-term learning capital. (We assume now that there is no divergence between the private firm's information or expectations and the government's.)

The argument can be developed along four possible lines.

(1) Investment in human capital or in goodwill is not embodied in physical goods and hence is more difficult to finance; the capital market is biased against 'invisible' investment.

(2) The capital market may be fragmented, most financing being perhaps self-financing. While existing enterprises have no trouble in obtaining funds new enterprises find it difficult, especially when their investment is not embodied in tangible assets; there is a bias against infant firms. This may be important in less-developed countries, and not only in these. But many countries have special development banks designed to provide finance for new or small enterprises, and this is obviously the best approach.

(3) The rate of interest for all long-term investment may be too high, social and private time-preference diverging. Society (with the government acting on its behalf) is less 'myopic' than private enterprise is when using its own funds or than private lenders are on the capital market. Furthermore the private discount for risk on long-term investment may be higher. Private investors may not be prepared to endure the suspense of the years of infancy. Thus there is a general bias against long-term investment. Partly this is overcome by government investment in the infrastructure and in educational facilities—and partly it would be overcome by assistance for infant industries.

(4) Finally, there may be some indivisibility in the capital needed to finance infant industries. The minimum time-period required for any significant learning to result may be high, so that initial losses would inevitably be substantial. The difficulty then may be in obtaining large amounts of long-term finance on the capital market.

Taking capital markets as they are, one can comment that in many countries it has certainly been possible for new ventures to start even though they had to incur initial losses;

if they produce non-traded goods they are very unlikely to receive assistance. When new industries developed in California they did not receive infant industry protection against the established industries of New England (though transport costs, of course, provided some natural protection). Indeed the financing problem only arises in the case of firms which require outside finance; large firms, such as the multinational corporations, or firms producing many products but confined to one country, can finance new activities out of their own resources, and—as research and development expenditures of many large corporations bear out—are often not hesitant to take a long view. Probably argument (3)—that the rate of interest for all very long-term investment may be too high—is the weightiest one, at least with respect to firms other than the very large ones. The effect of infant industry protection may then be to ensure that all very long-term private investment is not confined to very large firms.

In all cases devotees of first-best solutions can argue that the proper solution is to improve the capital market. McKinnon has made a strong case along these lines. He stresses that 'Tariff-setting machinery has not the discipline of even a moderately efficient capital market in identifying those activities which really are socially profitable'.[7] Nevertheless, one has to face the fact that capital markets are imperfect, especially in less-developed countries, and it is often easier or cheaper to impose tariffs than to create an effective capital market. There seems little doubt that, in spite of many qualifications, a valid, practically relevant infant industry argument for subsidization of new manufacturing industries resting on capital market imperfections can be made for many less-developed countries.

---

[7] Ronald I. McKinnon, 'On Misunderstanding the Capital Constraint in Underdeveloped Countries', in J. Bhagwati et al. (eds.), *Trade, Balance of Payments and Growth: Papers in Honor of Charles P. Kindleberger*, North-Holland, Amsterdam, 1971. He has developed the argument at length, and very convincingly, in *Money and Capital in Economic Development*, The Brookings Institution, Washington, 1973, stressing that 'Tax-subsidies, tailored to a degree of fineness beyond the knowledge and administrative capacity of the government, cannot substitute for a financial system where borrowing and lending are undertaken freely at high rates of interest.' (p. 34). This argument he would apply even more to tariffs.

(3) *Why Should Protection be Temporary?*

Let us pause for a moment to consider why protection to foster internal dynamic economies should or could be *temporary*. If the protection were provided in the form of direct production subsidies, if learning were continuously associated with production, and not just for some initial period, and if there were continuous imperfection of information or the capital market, as discussed above, then the case would be for permanent subsidization even if the import-competing industry turned into an export industry as its costs fell over time.

The temporary element can enter in three ways. (1) The learning may itself be temporary, being a characteristic of the firm's infancy period. (2) The imperfection of information or of the capital market, as these apply to the firm concerned, may be temporary: as the firm expands and its costs fall it may find it easier to finance further investments, whether in visible or invisible capital. (3) We may be constrained to the use of a tariff as a method of protection (the fiscal constraint ruling out direct or indirect export subsidization), so that the tariff could end once imports of the product have been completely replaced, and *should* end if the firm has monopoly power and above-normal profits are to be avoided.

(4) *Pecuniary External Economies*

A third approach to getting an infant industry argument out of dynamic internal economies is owed to Negishi and rests on the so-called *pecuniary external economies* which result from indivisibilities.[8]

Any indivisible lump of investment will generate pecuniary external economies (producers' and consumers' surpluses) outside the investing firm: the extra investment is likely to lead to an increase in the wages of the workers employed in the industry concerned, to higher incomes in industries

---

[8] Takashi Negishi, 'Protection of the Infant Industry and Dynamic Internal Economies', *Economic Record*, 44, March 1968, pp. 56–67. Negishi emphasizes consumers' surpluses generated by an investment, and the discussion here has extended his analysis. He does not explicitly use the concept of pecuniary external economies. He is concerned with cosmopolitan welfare and allows the terms of trade to vary as the result of the expansion of an industry, and also for the possibility that the protected industry becomes an export industry.

supplying non-traded inputs and to extra output and so a lower product price, unless the latter is set by a fixed import price. Thus the gains from the investment will not stay completely within the firm making it.

An investment in learning is likely to be 'lumpy' so that we are not just concerned with a marginal output decision. A significant rise in demand for labour and in output will result from the learning. The rewards to the investing firm will then fall short of the social gain by the producers' and consumers' surpluses generated outside the firm, and hence an investment that would pay socially might not be privately profitable. It seems then that internal economies of time—which express the effects of learning—have led to pecuniary external economies and so to an argument for protection. It should be noted that we are not referring to the ordinary externalities which by-pass the price system but rather to a source of market failure that arises when goods and factors are priced, when prices indicate *marginal* valuations and costs, but when a more-than-marginal investment is involved.

The weakness in the argument is that all investment, if it is indivisible, gives rise to pecuniary external economies; investment in learning is not unusual. One would have to show that the indivisibilities in learning are in some sense greater than in alternative investment, whether within the same firm or other firms. There appears to be no logical reason derived from pecuniary external economies for subsidizing this particular form of investment in preference to others. Furthermore, in the small country case, if imports of the product concerned do not cease the price to consumers will not change. So no consumers' surplus will be generated, though there will still be producers' surplus to workers.

II
DYNAMIC EXTERNAL ECONOMIES

When the process of production by firm X creates an invisible capital asset the benefits of which go in later years to other firms, and other firms are not charged for it, there are dynamic external economies or external economies of time. These provide some basis for an infant industry argument since they contain both the externality and the time element, the two requirements for an argument for temporary protection. But in addition it is necessary that the dynamic external economies be *reciprocal*.

(1) *Reciprocal External Economies*

If firm X's production creates a capital asset the product of which goes to firm Y, and Y does not pay X for it, then X should be subsidized to encourage it to produce an amount of the asset that takes into account the latter's marginal product to Y. But it is not an argument for temporary subsidization, since production by X may add to this asset at all times. One might then introduce the assumption that production by X adds to this asset only for a limited period; then we have an argument for temporary subsidization, but not an 'infant industry' argument since it cannot be said that firm X 'grows up' in any sense.

So we must assume *reciprocal* dynamic external economies. Production by X creates an asset the benefits of which go in later years at least partly to Y, while at the same time production by Y creates an asset the benefits of which go partly to X. Now there is a case for subsidizing both firms to encourage them to produce these assets. The result of the accumulation of the two assets will be that in later years the costs of both firms fall. Initial subsidization of the two firms will expand their production, perhaps getting both going for the first time, and eventually the subsidies can be removed without their outputs falling as a result.

The considerations determining whether protection should be permanent or temporary are essentially the same as in the case of dynamic internal economies discussed earlier. If external economies are being continually created and assistance can be given in the form of direct subsidies, then subsidization should be permanent, the firms possibly turning eventually from import-competitors into exporters. But either if the externalities are only being created for a limited 'infancy' or learning period, or if the method of assistance has to be by tariff, then protection should be temporary, ending in the latter case once the tariff has eliminated imports of the goods concerned.

The external economies may be generated not through *production* by the firms concerned but rather by inputs of particular factors of production, say labour, into these firms. It is then obvious that the first-best argument is for subsidization of these inputs, production subsidies being second-best and tariffs third-best.

Now let us generalize this analysis. There are a number of firms, X, Y, Z .... Production by any of them leads not only to current output but also to accumulation of an invisible asset, K, which we can think of as Knowledge. All of the firms contribute to production of K. The benefits of K spread indiscriminately to them all, lowering the average cost curve for each of them. K might be thought of as a public good, so that the amount of it that is used by one firm does not reduce the amount available for other firms (though this is not essential to the argument). When production by X adds to the common stock of K, X itself will benefit; to that extent there is an internal dynamic economy. But the greater part of the benefit will go to other firms, so in the main we have an external dynamic economy.

There are two classic versions of the infant protection argument: one might be called Marshallian, while the other was developed by Friedrich List.[9] It is possible to show that the two versions have the same basic structure. If the group of firms which generate the reciprocal external economies for each other is called an 'industry' the external dynamic economies are then internal to the industry. As time passes the costs of the firms in the industry fall. This is the Marshallian case, where economies are external to firms but internal to the industry. Alternatively we might call the group of firms the 'manufacturing sector'. The economies are then internal to the manufacturing sector. This is the case on which List's *infant economy* argument rests—an argument widely used

[9] List, in *The National System of Political Economy*, was, of course not rigorous. Alfred Marshall, in his *Principles of Economics*, developed a concept of static economies of scale that are external to firms but internal to the industry; the concept made it possible for him to reconcile industry economies of scale and perfect competition (a purpose of no interest to us here). His concept was not dynamic and he did not apply it to the infant industry argument. But many expositions of the infant industry argument since Marshall have had an essentially Marshallian structure.

today which is much broader than the Marshallian infant industry argument since it provides an argument for general protection of the whole manufacturing sector of an economy.

An error to which partial equilibrium analysis can lead should be avoided. Economic activity anywhere might add to the nation's experience and so raise indiscriminately the general level of productivity. Since protection means favouring one sector of the economy at the expense of others one cannot protect the whole economy; hence there can only be an argument for protection if the external economies are limited to some groups of firms within the economy or if they differ between groups. The assumption so far has been that there are no actual or potential external economies outside the group under discussion.

Conceivably all firms in the economy might belong either to group A or group B, and within each there could be a structure of reciprocal external dynamic economies, production within group A creating asset K_a and within group B, asset K_b. There may then be no case for encouraging one group rather than another.

(2) *Labour Training*

What might give rise to dynamic external economies? Most commonly the training of labour is cited, the accumulation of human skills through 'on-the-job' training being the 'K' of our story. If a firm trains labour specifically for use only in that firm no external benefits will accrue to other firms. We are here concerned with more general training, whether in technical skills appropriate to an industry or perhaps in the habits of working regularly in a factory environment, a skill required for all manufacturing industry. Firm X trains labour in this way, and this experienced or trained labour may then go off to other firms. An essential element in the argument is the inability to tie the workers to the firms where they obtained their training.

The argument is that there is an externality because firms are creating assets from which other firms benefit even though the other firms do not pay for them. The assets in this case are not 'public goods' because the more labour one firm takes out of the common pool of skilled labour the less is available for others; rather they are, in Meade's language, 'unpaid factors'.[10] There are implicit assumptions in this argument.[11] It must be borne in mind that the asset K—the Human Capital—is being embodied in specific workers and they will reap the rewards once they are trained by being able to obtain appropriately higher wages. Insofar as there are indivisibilities, pecuniary external economies will accrue to the firms that employ them in the future, but the gains will go principally to the trained labour, at least at the margin.

The crucial question is who pays for the training. If the workers will seek higher wages from firm X as soon as they are trained, or if they are liable to leave the firm, then it may not pay firm X to train them at its own expense. The workers will have to pay for their training themselves by accepting especially low wages during the training period. This is, of course, achieved by the apprenticeship system. If the workers do not wish to drop their consumption standards to this low wage they will have to borrow on the capital market to finance the training.

Here we come to a possible argument for intervention or protection. Capital markets are not usually well organized for such purposes. In practice, if reliance were placed wholly on the capital market, under-investment in this form of invisible capital creation would be likely, especially in less-developed countries. If workers have to finance their training by accepting low consumption standards during the training period—as in practice they have to—then their own rate of discount will govern the amount of training received—but, in view of their low incomes, this rate may be much higher than the social rate of discount as expressed by the rate of interest in the capital market.

[10] J. E. Meade, 'External Economies and Diseconomies in a Competitive Situation', *Economic Journal*, 62, March 1952, pp. 54–67.

[11] See Gary S. Becker, *Human Capital*, National Bureau of Economic Research, Columbia University Press, New York, 1964. Chapter 2 of this book contains a rigorous analysis of the economics of on-the-job training, and shows how training will be financed and who will obtain the rewards in an economy which is perfectly competitive and which has a perfect capital market. See also V. K. Ramaswami, 'Optimal Policies to Promote Industrialisation in Less Developed Countries', in Streeten (ed.), *Unfashionable Economics*, Weidenfeld and Nicolson, London, 1970, which discusses the issues with reference to less developed countries.

First-best policy is to improve the capital market (if that is feasible) and second-best policy is to provide finance or subsidization specifically for labour training. A reasonable third-best policy might be to subsidize the employment of labour by those firms or industries that contribute in a significant way—more than average—to the pool of trained labour available to firms other than themselves. Subsidizing their output might be a convenient fourth-best and tariffs fifth-best.

This version of the dynamic external economies argument for infant industry protection rests on capital market imperfection just as the principal version of the dynamic internal economies argument does. In the present case the problem is that labour cannot borrow adequately at appropriate rates while in the other case it is assumed that private entrepreneurs have difficulty in doing so. Furthermore, even if workers could borrow on the capital market to finance training, ignorance, fear of 'mortgaging the future' or simply lack of foresight may be inhibiting factors, just as ignorance or myopic expectations might explain a failure of private entrepreneurs to exploit internal dynamic economies.

(3) *Knowledge Diffusion*

Dynamic external economies may also arise in other ways. Knowledge may spread from firm to firm. To some extent firms can keep new knowledge and ideas secret, patent them or market them. If they can, then there is no external effect. But there is an inevitable diffusion process. The more new knowledge there is to diffuse the more external economies there are likely to be; hence there may be a case for providing protection for those industries which are in a stage of experimentation and where techniques are changing or knowledge is advancing.

Again, there are more direct ways of fostering the creation and diffusion of knowledge and new ideas, and perhaps of assisting specifically those firms which contribute most to the diffusion process. Baldwin[12] has stressed that protection

---

[12] Robert E. Baldwin, 'The Case against Infant-Industry Tariff Protection', *Journal of Political Economy*, 77, May/June 1969, pp. 295–305.

provides no incentive for a firm to acquire more knowledge than it would have otherwise even though the social gains exceed the private gains. All protection can do is to favour firms which are known to be knowledge-creators and diffusors. A subsidy related in some way to knowledge creation, such as research, or subsidizing training of researchers, will be preferable to a subsidy related to the production of the final product, and even more, to a tariff. This is our familiar argument that a policy that is applied as close as possible to the point of the relevant divergence is always first-best.

Furthermore, while these methods may encourage knowledge creation, they will not encourage knowledge diffusion. Johnson has stressed that once knowledge has been created the marginal cost of spreading it outside the firm that has created it is very low or zero. Hence it is a public good, so that, once it exists, it should be free. But firms have incentives to prevent diffusion or to charge for it, a problem which protection, whether through subsidy or through tariff, cannot overcome. The extent to which new knowledge does get diffused depends on the ability of private firms to appropriate the benefits to themselves through patenting or commerical secrecy.[13]

Thus the infant industry argument based on knowledge diffusion has been much criticized. But the criticisms have not really destroyed it, though they have underlined its possible third or fourth-best nature. A criticism that has not been generally made is that knowledge can, these days, leap across countries, especially if governments have agencies designed to channel information and research results to their citizens. There is thus less need to foster domestic production, at

[13] Harry G. Johnson, 'A New View of the Infant Industry Argument' in McDougall and Snape, *Studies in International Economics*, North-Holland, Amsterdam, 1971. In this paper Johnson seeks to confine the term 'infant industry argument' narrowly to arguments relating to investment in the acquisition of knowledge by firms, hence excluding arguments concerned with factor-market (capital or labour-market) imperfections. As Snape pointed out in a comment to the article, this is a much narrower definition than the one accepted by Johnson earlier (in 'Optimal Trade Intervention ...'), and in any case the choice of term is just a matter of semantics. It is more in conformity with general usage and tradition to define the infant industry argument to include any argument, however based, for temporary protection for a new venture or set of ventures (as in this chapter).

high cost to the Treasury or the consumers, for external benefits to filter through to other domestic firms.

(4) *Atmosphere Creation*

Another form of dynamic external economy is the creation of a general atmosphere that is conducive, say, to factory work or organized economic activity, or to the development of mechanical or scientific interests. Here the training and experience of workers in firm X may affect the quality of other workers; thus the new attitudes and knowledge leap from worker to worker; the gains will then go in the main to workers but will no doubt also spill over to firms.[14] This atmosphere-creating effect may well provide a strong basis—possibly the strongest of all—for an 'infant economy' argument for generalized protection of manufacturing in countries inexperienced in manufacturing production and yet containing the human potentialities for it—the sort of country Friedrich List was writing about.

(5) *Selling Externalities and Goodwill*

Finally, successful production and selling by firm X, whether in the home or the export market, may increase goodwill for all similar products produced in the country. Production by firms X, Y, and Z adds to the asset Goodwill from which they all subsequently benefit, an external dynamic economy internal to the industry. If the sales are in the domestic market this will represent a social gain only if consumers make better-informed choices as a consequence of the experience of buying domestically-produced products.

Since the creation of domestic Goodwill may not create a social gain, even though it creates a private gain, the argument may be mainly relevant to exporting. But the possibility of external diseconomies cannot be ignored. Firm X, exporting poor quality products under misleading labels, will create Illwill not just for itself but possibly for all products from its country.

---

[14] In this case the knowledge leaps from worker to worker; in the previous (knowledge-diffusion) case it leapt from firm to firm; and in the labour-training case it leapt from firm to worker, though in that case the market would not necessarily be by-passed.

III
THE MILL-BASTABLE TEST

Mill wrote: 'But it is essential that the protection should be confined to cases in which there is ground of assurance that the industry which it fosters will after a time be able to dispense with it'.[15] As Kemp has stressed, 'Bastable rightly remarked that the mere prospect of overcoming a historical handicap is not enough to warrant assistance. It is necessary further that the ultimate saving in cost should compensate the community for the high costs of the protected learning period. It is necessary, in particular, that when a suitable time-discount is applied to the early excess costs and to the eventual cost savings, the commodity should still be worth producing'.[16]

In general terms this is unexceptional. Consider a simple case of internal dynamic economies, with initial average cost $100, later cost $78, a fixed import price of $80 and a social rate of interest of 10 per cent. Assume that (a) the potential infant industry consists of a single firm, (b) it is quite unable to finance any of its investment in learning itself, (c) the scale of output is fixed by technical possibilities, and (d) the learning is limited to the initial period. The Mill test in this example would only require the average cost to fall to $80; Bastable's contribution was to stress the need for it to fall to $78. This also applies to an industry consisting of several firms which are unwilling to finance their initial learning themselves because the dynamic economies are external to the firms, though internal to the industry.

Nevertheless, the Mill-Bastable test requires careful interpretation and could be misunderstood. The nature of the initial costs and later returns must be carefully defined. The Mill test does not necessarily mean that an initial tariff (or subsidy) must eventually be eliminated for protection to have been justified and the Bastable test does not necessarily mean that the price to consumers must fall below the price of imports.

---

[15] *Principles of Political Economy*, op. cit., p. 919.
[16] *The Pure Theory of International Trade*, p. 186. See also C. F. Bastable, *The Commerce of Nations*, 10th edition, Macmillan and Co., London, 1921.

First of all, suppose that, in the example just given, the subsidy was not eliminated at all even though learning ceased after the initial year. In later years there would then be excess profits of $22 per unit (possibly shared with the firms' labour in higher wages), of which the subsidy of $20 would be a straightforward static redistribution from the Treasury to the protected industry, while $2 would be the return on the original learning investment, a return received by the producers even though the investment was originally made by the Treasury. If we are prepared to disregard income distribution considerations we can argue that the investment had paid off. This does not conflict with the Mill-Bastable test, since the subsidy *could* have been ended, and the producers *could* have paid a return of $2 a year per unit to the Treasury.

Secondly, the industry may be able to finance its own learning investments but may require a rate of return of 20 per cent, compared with the social rate of 10 per cent. A subsidy of $10 per unit in the initial year will then induce the investment, after which the subsidy could be taken off. But this time no return *could* come to the Treasury unless producers' expectations are to be falsified; the whole $2 per unit is required to achieve the 20 per cent private rate of return.

Thirdly, if a tariff rather than a subsidy is used for protection, a consumption-distortion cost will be added to the initial production-distortion cost, so that to achieve the same 10 per cent social rate of return as before the cost of production would have to fall to below $78. If the tariff were kept on this consumption-distortion cost would be continued, and so the cost of production would have to fall even more to justify protection.

Fourthly, suppose that the scale of domestic output is variable (upward-sloping static supply curve): the higher the tariff or production subsidy the greater the output will then be. If learning varies with scale of output, the extent of the downward shift in the supply curve in later years will depend on the initial level of the tariff. For each initial tariff there is then a particular initial cost, representing the cost of the investment in learning, and this must be set against the later gain. This later gain will depend on whether the tariff or subsidy is removed or is kept on. If protection is removed, the gain will be greater. If it is kept on, the gain could disappear, though some gain could still remain, so that the investment in learning would still have been profitable.[17]

It is worth noting that the optimum policy will be to remove the tariff in the later period, and so allow the import price to set the domestic price. It will not be optimal from a Pareto-efficiency point of view to force the price received by the

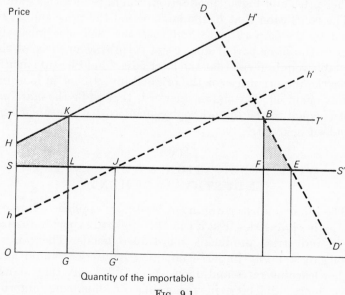

Quantity of the importable

FIG. 9.1

producer to a level below the import price so as to give the consumer a return on his investment, because such a lower price would lead to under-production. The reason is that the cost of the investment in learning incurred in the early period is a fixed cost, independent of later output.

This is illustrated in Figure 9.1. The initial supply curve is HH'. A tariff of ST brings about domestic production of OG. The excess cost of this production over the cost of imports is $SHKL$. In addition, the tariff inflicts the consumption cost FBE. These two costs together (shaded in Figure 9.1) represent the costs of the investment in learning. This

[17] In 'Optimal Trade Intervention in the Presence of Domestic Distortions' Johnson has a geometric proof of this in terms of the two-sector model.

investment causes the supply curve (marginal cost curve) to shift in later periods to hh'.

Given this new supply curve, optimum output becomes OG', where marginal cost is equal to the cost of imports. If free trade rules in the later period the gain from the investment is the area hSJ, all of which goes to producers. A lower price to producers, and hence lower production, or alternatively, a higher price and higher production, would reduce this gain.

The conclusion that in the later period the price should be equal to the import price and that this will maximize the gain in the later period may seem surprising in view of the Bastable requirement that the later cost should be sufficiently *below* the import price for the original investment in learning to have paid off. The simple answer is that while the *marginal* cost is equal to the import replacement cost, the *average* cost is indeed below it.

IV
WIDE APPLICATION OF THE INFANT
INDUSTRY ARGUMENT

The infant-industry argument is usually applied to import-competing industries. But it can also apply to export industries or to industries producing non-traded goods. There is no strong reason why import-competing industries should have higher learning rates and, more specifically, why they should have more difficulty in foreseeing or financing internal dynamic economies or generate more external economies.

Conceivably one could argue that learning takes place mainly in new industries or in the production of products new to the country and, while production might be for export right from the start, it is more likely to be for the home market. Perhaps it is also easier for a tariff board or planning authority to foresee the future costs of producing a new product that is already being produced abroad and for which the domestic market has already been reconnoitred by imports, than one for which an export market has to be created. But all this is a little tenuous.

Many industries and countries have experimented in exporting new products. Indeed, there is a special learning problem in breaking for the first time into foreign markets, so that there could be an *infant marketing argument* additional to the usual infant industry argument concerned with production. Here again, an argument for *protection* must rest additionally on capital market imperfection, inadequacy of private information or judgement, or externalities.

The infant industry argument is also usually associated with manufacturing. This may be because historically many countries have deliberately sought to shift the output pattern from agriculture to manufacturing, and inevitably there is more scope for learning in any new field of activity. Perhaps manufacturing is more difficult, requires a higher stage of techniques and knowledge, and so inevitably generates more learning economies.

But one must remember that a protection argument requires not just learning, but in addition either capital market or information imperfection, or otherwise externalities. In less-developed countries private agriculture usually finds capital harder to obtain and more costly than industry (as stressed in Chapter 6), and surely a lack of information or foresight is a greater weakness of farmers than of urban industrialists. Furthermore, the scope for learning in agriculture is often immense. Governments usually deal with this directly by providing free agricultural extension services, but it is doubtful whether in any country the subsidization implicit in this matches the subsidy equivalent of manufacturing protection.

Finally, learning can potentially take place in many activities. If it is desired to foster the learning process either because of information or capital market imperfections believed to be associated with the process, or because it has external effects, then one must protect the industries where learning is relatively higher. In a general equilibrium framework all externalities and arguments for protection must be seen in relative terms. If one fosters industries where there is learning but does this by reducing the output of, or perhaps completely ending, other industries where there are much greater potentialities for learning (measuring learning in terms of its later fruits) then total learning for the whole economy may actually decline as a result of protection.[18]

[18] See also pp. 309–19 below.

V
THE PROBLEM OF
INVESTMENT CO-ORDINATION

Various industries or activities may be *complementary* with each other. The more expansion there is in one, the more profitable it is to expand the other.[19] It is possible to argue that in such a case temporary subsidization of the various complementary activities may be justified.

Consider first a closed economy. Two products, A and B, are complementary. If capital is invested to produce or to expand A with no production or expansion of B, the return on capital would be only 2 per cent. But if B were expanded at the same time the return on capital in A would leap to 10 per cent. Similarly, expansion of B alone would mean a very low return, but expansion in combination with A would yield a 10 per cent return to investment in B. There are mutual benefits which might be described as reciprocal *pecuniary* external economies because they operate through the price system, as distinct from ordinary externalities which by-pass the price system. They are in part potential since industry B will not benefit fully from the expansion by A until it, B, actually responds itself by expanding output.

The argument runs that, in the absence of any co-ordination, investment will not take place, though socially it should possibly take place. If capital and processes were divisible there would be no problem. There would be a little expansion of one, then a little of another, and so on, the two outputs expanding in step with each other. It would not be necessary to make a big investment in one before knowing whether there would be complementary investment in the other.

To some extent we have a familiar information problem. The information A requires is not information about what B is planning to do independently, but rather about its reaction to a move by A. In our simple example it seems obvious that the two industries should co-ordinate their activities,

---

[19] This is the problem of investment co-ordination which underlies the well-known arguments for and against 'balanced growth'. See Tibor Scitovsky, 'Two Concepts of External Economies', *Journal of Political Economy*, 62, April 1954, pp. 143–151.

a planning authority might arrange the necessary co-ordination, or the pecuniary externalities might be 'internalized' by one firm taking over the other. But when many firms would have to expand together it may be more reasonable to assume decentralized decision-making with no collaboration or co-ordination. The solution is then to subsidize temporarily some or all of the activities so as to get them going. Even if the guaranteed rate of return were provided simultaneously to all the industries concerned they would inevitably not all start at the same time. There would be varying time-lags in the investment responses and different gestation periods for the initial investment; thus there would be a period when a subsidy would actually have to be paid. The cost of this subsidy may reflect the cost of failing to co-ordinate the expansion of the industries properly from the beginning. It is a form of infant industry subsidization and needs to be subjected to the usual Mill-Bastable test.

The argument has been made as convincing as possible, though in fact there are plenty of qualifications. Deliberate co-ordination, consultation, indicative planning, and so on seem to be the best methods to use for the purpose, and are in fact used all the time. Furthermore, indivisibilities are often not so large that they prevent the development of complementary industries by a piecemeal step-by-step approach. Probably the major complementarity in an economy is between the infra-structure and manufacturing industry. The former is usually publicly owned, is established to some extent in advance of the latter, and is run at a loss. If we think of the infrastructure as industry A and manufacturing as industry B, then usually A leads the way; if it did not, B would not get established unless heavily subsidized.

Introducing trade weakens the whole argument considerably. Suppose that all goods are and stay traded, that world prices are given, and that there are no non-traded goods and services. Then the whole problem disappears. When industry A expands, and this increases the demand for product B, perhaps because it is an input into A, it is no longer necessary for industry B to expand; the extra B can be imported. If we abandon the small country assumption world prices are no longer given, and an expansion of weaving, for example, would be likely to raise the price of yarn. Nevertheless, when there is the possibility of trade extreme complementarities disappear. The problem really only returns when non-traded goods are introduced.[20]

VI
ECONOMIES OF SCALE AND THE PSEUDO-INFANT INDUSTRY ARGUMENT

It is common for an import-competing industry to be established with the help of protection, but gradually for its costs to fall because domestic demand for its products expands, so that eventually it no longer needs protection. Can one say that such an industry is an infant that has grown up, and that the fact that protection is no longer needed has justified the original protection? This type of argument is often used to justify protection in economies with small markets that are expected to grow. The essential element is internal economies of scale.

(1) *The Pseudo-Infant Industry Argument*

The issues are expounded in Figure 9.2. AA' is the private and social average cost curve (including normal profits) of the single actual or potential domestic producer, and SS' is the import supply curve. The demand curve moves to the right over time, perhaps because of population growth or growth in real income per head. Curve D_1D_1' refers to time t_1 and D_2D_2' to t_2. In the absence of a tariff or subsidy production would commence at time t_2. A tariff of ST would cause production to commence at t_1. As the demand curve moved to the right the scale of output could expand, costs would fall and the tariff could be gradually reduced until at t_2 it could be

---

[20] Wolfe has based an infant industry argument on a case where an input is potentially non-traded but actually traded. Domestic yarn production would only start if it were assured of a domestic market, and, owing to economies of scale it could then be produced at less than import cost. But cloth production would only be profitable if it were assured of yarn supplies from domestic producers; if the import price has to be paid for yarn, cloth production would not start. Thus cloth producers wait on yarn production, and yarn producers on cloth production. The solution is to get production going with temporary protection. See J. N. Wolfe, 'A Generalisation of the Infant Industry Argument,' in Tullio Bagiotti (ed.), *Essays in Honour of Marco Fanno*, Vol. II, Padua 1966.

removed. The industry has apparently been an 'infant' that has grown up.

Yet no argument for tariff protection has been produced. It is a *pseudo-infant-industry argument*. To establish the industry prematurely at an output less than *OH* inflicts a social cost. Static economies of scale combined with a moving demand curve provide no argument for protection as such. Something else must be added.

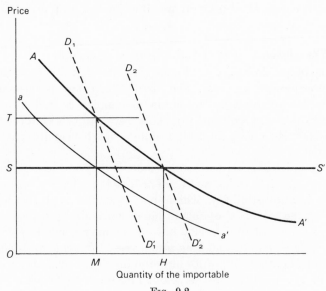

Fig. 9.2

Now introduce external economies generated by domestic production so that the average social cost of production *aa'* is below the private cost curve *AA'*, the marginal divergence at output *OM* being *ST*. Domestic production should now commence when output *OM* can be sold. So a tariff *ST* should be imposed to bring the industry into existence at t_1 causing it to produce *OM*. As the demand curve moves to the right the tariff could be reduced until at t_2 it could be removed completely. If we were only concerned with the industry being established at the right time it would be sufficient to impose the tariff *ST* from the beginning and keep it on. But if it is to be a made-to-measure tariff designed to avoid monopoly profits it should be reduced gradually until at t_2 it reaches zero. In the period from t_1 to t_2 the industry would not exist privately but should exist socially so that it is in this stage that the tariff is crucial.

The argument for protection stems from the externalities which have been assumed to cause the marginal divergence in domestic production. The contribution of static internal economies of scale is only that, in combination with a moving demand curve, they make it possible for the protection to be temporary.

(2) *The Doubtful Role of Static Economies of Scale*

The infant industry argument is often muddled up with static economies of scale. The distinction between these and economies of time or dynamic economies is not always clearly drawn. In fact most of the analysis in this chapter has not depended on static scale economies, whether internal or external.

Economies of scale have a crucial role only in the argument concerned with investment co-ordination. That argument rests on indivisibilities, which may take an absolute form (so that the total cost of output below a certain capacity is the same as the cost of capacity output) or may result in significant scale economies making low levels of output, and hence piece-meal expansion, quite uneconomic. They also play some role in the research-and-development issue to be discussed below.

While scale economies are not crucial to the main forms of the infant industry argument it is true that the precise amount and time-scale of protection required when there are internal or external *dynamic* economies may depend on static scale effects. Various relationships between scale of output and learning are possible. (a) Learning per unit of output may vary in some way with the scale of current output. Thus learning may increase with both scale and time, and hence rise with *cumulative* output. This assumption is made in some theoretical models. (b) The productivity of learning may depend on the scale of future output. Thus experience may only be needed if one wants to produce at large-scale. Hence the rewards of learning may increase with scale of output in later years. If this is so it may pay to start with small-scale production and, as the rewards of learning emerge, gradually expand scale, clearly a common situation. Finally, (c) the fruits of learning may be specific to the relevant technique of production, technique varying with scale. These considerations do not alter the fact that static economies of scale cannot be a sufficient basis for an infant industry argument.

Case (b), where it pays to start with small scale and gradually expand, is an interesting one. It is often assumed that the attainment of large-scale production takes time, not because demand increases over time (as in our pseudo-argument above) but because there is some learning element. A firm must walk before it runs. If this were so the learning effect and the increase in scale would be simultaneous, and hence dynamic learning economies and static scale economies would be realised at the same time. The need for learning would intro-duce the crucial time-element, but learning would be required because large-scale production is desired, and this in turn may be so because, once learning has taken place, average costs are lower at large-scale than at small. If one is thinking of plants and plant economies one would expect a plant of appropriate size to be set up in the first place, rather than starting with a small plant and gradually expanding. But in the case of a firm with many plants (possibly a conglomerate) gradual expansion because of learning may be optimal.

VII
RESEARCH AND DEVELOPMENT, AND THE PROTECTION OF ADVANCED-TECHNOLOGY INDUSTRIES

It may seem surprising that the infant-industry argument is currently used to advocate and justify protection of certain industries in some of the most developed countries, including the country that had the first industrial revolution. The industries concerned are the so-called 'advanced-technology' or 'science-based' industries—principally the computer and aircraft industries—where vast and speculative expenditures on research and development are required. This is one of the main ways in which protectionist thinking manifests itself in Europe at present.

(1) *The Special Features of Research and Development*

Research and development (R & D) expenditures are expenditures on investment in learning. This form of investment can be distinguished from the investment in learning discussed so far in this chapter, which is a by-product of the process of production. The distinction is, in fact, not a rigid one since some production might take place at a very early stage, and R & D expenditures might be associated with production, learning being a joint result of both. But in the case of the advanced-technology industries one can broadly distinguish production of earlier models on a production line from the development of new generations of computers or a new type of aircraft (i.e., from R & D).

Furthermore, R & D can be distinguished from ordinary investment, say in capital equipment. One can conceive of an ordinary plant being installed in a very short period and then being ready to bear its fruits. R & D is like a plant with a very long installation and gestation period, with costs having to be incurred for many years before significant fruits emerge. Thus time is of the essence with this form of investment. A static model is clearly inappropriate. Of course ordinary capital also takes some time to instal and bear fruit, so it is a matter of degree.

R & D has three additional features. First, it is subject to great economies-of-scale. Generally, significant increasing returns are likely to be yielded by increases in cumulative expenditures, at least up to some very high point and especially if X-efficiency is being maintained. These economies could be thought of as static scale economies, but it must be assumed that any given level of expenditure at a point in time is roughly maintained for a considerable period. It follows that, for such expenditures to be at all effective, very large sums may be required.

Secondly, there is usually great uncertainty about what will come out of R & D.

Thirdly, some of the knowledge generated may be useful for other projects, so there may be considerable external effects.

The extent of these external benefits may also be very speculative, just as the internal benefits are.

(2) *Is There a Case for Government Assistance?*

Where does the possible need for government assistance or protection come in? Firstly, the very fact that the investment expenditures required are so vast may mean that an ordinary private capital market would not be willing to supply the funds. This argument has been used, for example, in the case of the Anglo-French supersonic plane, the *Concorde*. One might then suggest that the capital market is imperfect, but in the case of the British aircraft industry one would be suggesting that the world's second largest and perhaps the most sophisticated market—the City of London—is 'imperfect'.

In any case, if a capital market is unable to cope with a very large demand for funds and yet these funds are available to the government, the government should only invest them if it expects rates of return, discounted for risk, that are comparable with returns on other government funds and with returns on private investment. After all, these funds could always be channelled on to the private capital market. Thus in this case there may be an argument for government intervention in the form of public investment, but not for government subsidization.

Secondly, there is the familiar point that the benefits are often likely to be well in the future, and public and private foresight, as well as time-preference may differ.

Thirdly, private risk aversion may be greater than social risk aversion as it affects particular projects. If this were so, then the advanced-technology industries, with their long-term investments and considerable uncertainties about the outcome, would be candidates for some kind of assistance. But obviously this does not justify ignoring the degree of uncertainty. Furthermore, really large firms are certainly willing to engage in quite speculative expenditures, so that this argument cannot apply to assisting the very large multinational firms.

Fourthly, assistance may be justified if there are externalities. Three cases must here be distinguished.

(1) It has been argued that there may be externalities of

some kind through the employment of scientific labour in a country. The scientific labour market seems to work quite well, and is probably much less imperfect than most other segments of a country's labour market. But perhaps one can conceive of some 'cultural overspill' resulting from a lot of exceptionally intelligent people residing in a particular country rather than abroad.

(2) The externalities may take the form of new ideas or innovations seeping out of the firms engaged in research and benefiting other firms. This is the 'knowledge externality' discussed earlier. Such externalities undoubtedly do exist, although they do not appear to be exceptionally important in computers or aircraft, and there seem to be difficulties about predicting them. A country can conceivably benefit here even when the research and development are being carried on in another country: if knowledge can leap across the boundaries of a firm it can also leap across national borders.

(3) There are externalities external to a project but internal to a firm. These sorts of 'externalities' would not be classified traditionally as externalities at all, and do not justify government intervention or subsidization in themselves, but they are relevant for making correct cost-benefit calculations for particular projects, and may justify 'project subsidization'.

In the case of all three kinds of externalities the resources used—notably the high I.Q. manpower—are likely to have yielded externalities in their alternative uses. An argument for protection on these grounds exists only if the value of the externalities is significantly greater in the project or industry under consideration.

(3) *Monopoly and Infant-Firm Protection*

It has been suggested that the European Economic Community should foster a single European computer industry to compete with the American company International Business Machines (I.B.M.) which presently dominates the world computer market.[21] One argument on which this proposal

[21] The case for the development of a unified European computer industry to meet the 'American challenge' is put vigorously in J.-J. Servan Schreiber's best-selling *The American Challenge*, Hamish Hamilton, London, 1968. See also Select Committee on Science and Technology of the House of Commons, *The Prospects for the United Kingdom Computer Industry in the 1970s, Report,*

rests is familiar. Economies-of-scale in this industry—not only in R & D and in production but perhaps even more in marketing and servicing—are so large that private enterprise may not have the resources to make the vast investment required. The other argument is more interesting since it implies an infant-firm argument based on an anti-monopoly argument with a touch of the orthodox optimum tariff argument.

A world monopoly is able to exploit its customers, and when it is faced with competition it may be compelled to reduce the prices or raise the quality of the goods and services it supplies. A large European computer firm which forces I.B.M. to lower its prices will have created benefits for computer users which are external to the firm, and such externalities may justify government subsidization. Insofar as Europe is and remains a net importer, its terms of trade will have improved. The time-element enters because of the learning (R & D) element in developing the new competitor.

On the other hand, it is arguable that Europe would get its computers cheaper and better if I.B.M. were allowed to dominate world supply. Customers may benefit more from the lower costs of the monopoly owing to the economies-of-scale that its world monopoly position makes possible than from the reduced profits and possibly higher X-efficiency in a market with two dominant firms. In any case, the object here is not to discuss all aspects of the question whether the domestically-owned computer industry should be subsidized in the E.E.C., but only to disentangle some strands in a popular argument.[22]

Minutes of Evidence and Appendices, H.M.S.O., London, 1971, especially Christopher Layton's Memorandum, 'Theme: Elements of a European Policy', in Vol. II, pp. 249–254.

[22] There are other arguments. It is said that computers are likely to become so crucial to the operation of government and the economy that no government can permit a monopoly, especially one that is foreign-owned. This is mixed up with the general idea that Europe should be 'independent' of the United States. It is also argued that I.B.M. only attained its dominant position thanks to U.S. government finance of R & D and U.S. government procurement, mainly for military purposes. So Europe should surely do the same. This is a familiar fallacy. U.S. subsidies for computer research have benefited I.B.M.'s other customers, whether American or European. For Europe it is a datum: the U.S. government has indirectly improved Europe's terms of trade. It is a form of export subsidisation or long-term dumping (the U.S. government has paid more than European customers), and provides no case for Europe to do likewise.

10

PROTECTION AND CAPITAL ACCUMULATION

PROTECTION can affect the rate of growth in various ways, one being through effects on the rate of capital accumulation. This then raises two questions. First, what precisely are the effects of protection on capital accumulation? Secondly, if protection, or a particular protective structure, raises the rate of capital accumulation, is this an argument for imposing or altering the structure? It will be assumed here that internal and external balance are consistently maintained. In addition, we assume absence of capital inflow or outflow (see Chapter 12).

I
THE REAL INCOME EFFECT OF PROTECTION ON SAVINGS AND INVESTMENT

Protection may raise or lower savings, hence investment, hence the rate of capital accumulation and hence the rate of growth of real income. These simple relationships will now be spelt out in more detail.[1]

Suppose for a moment that a country's aggregate savings depend only on its total real income, and not on income distribution or on relative prices, except insofar as the latter affect real income. Provided the marginal propensity to save is positive a rise in real income will cause savings to go up and a fall, to go down. For all the reasons discussed in earlier chapters, protection may raise or lower a country's static real income. If there are no domestic divergences and the country cannot influence its terms of trade, protection is likely to lower real income. In that case it will cause savings to fall. On the

---

[1] For a more rigorous development of this argument in a fully-specified neo-classical growth model context, see W. M. Corden, 'The Effects of Trade on the Rate of Growth', in Bhagwati et al. (eds.), *Trade, Balance of Payments and Growth*, North-Holland Publishing Co., Amsterdam, 1971.

other hand, if protection improves the terms of trade real income may be raised and some part of the gains in real income will then be saved.

The next step is to assume that investment is determined by intended savings out of full employment income. We could suppose that the rate of interest is so adjusted as to achieve this, or alternatively that public investment is varied so as to absorb the savings available. Hence investment will rise when real income rises and will fall when real income falls. This remains true if savings are positively related to the rate of interest. A rise in income, leading to a rise in saving at a given interest rate, will lead to a fall in the rate of interest, and a positive relationship between savings and the interest rate then means that the final increase in savings and investment will not be as great as it would have been if savings had been interest inelastic. (Complications resulting from changes in the relative prices of investment-goods are ignored for the moment.) Thus a protective structure that imposes a cost of protection (a fall in real income) will reduce investment.

The change in investment, in turn, will alter the rate of capital accumulation. This is, perhaps, not so obvious. If the rate of capital accumulation is k, net investment is I, the stock of capital at any one point in time is K, and real income is Y, then

$$k = \frac{I}{Y} \cdot \frac{Y}{K}.$$

The propensity to invest (= propensity to save) is I/Y, and is given, and so at a point in time is the stock of capital K. It follows that a fall in Y—which is a fall in static real income—will lower k. Alternatively, one need not mention the propensity to save and invest at all, but can just point out that a fall in real income will lower I and hence must lower I/K.

It is not always understood that protection will affect the *rate* of capital accumulation even when the average propensity to save and invest is constant. Yet it is not an original point. Adam Smith wrote:

The industry of the society can augment only in proportion as its capital augments, and its capital can augment only in proportion to what can be gradually saved out of its revenue. But the immediate effect of every such regulation is to diminish its revenue, and what diminishes its revenue is certainly not very likely to augment its capital faster than it would have augmented of its own accord had both capital and industry been left to find out their natural employments.[2]

If for 'revenue' we read 'real income' and bear in mind that not 'every such' regulation may reduce real income, we see that Adam Smith has anticipated the simple argument here.

The final step in the argument is to note that the change in the rate of capital accumulation is likely to change the rate of growth of real income and of real income per head in the same direction. We thus have the result that a policy which has the once-for-all effect of reducing static real income is also likely to reduce the rate of growth of real income.[3]

If protection reduces static real income by 10 per cent and if the propensity to invest is constant, the rate of capital accumulation will also fall by 10 per cent—say from 4 per cent to 3·6 per cent. The rate of growth of real income may well fall by less than 10 per cent in that case. For example, if the rate of growth of real income is a weighted average of the rate of capital accumulation and the rate of growth of labour, and if the latter is constant, it will certainly fall by less than

[2] Adam Smith, *The Wealth of Nations*, Everyman edition, Vol. I, J. M. Dent and Sons, London, 1910, p. 402.

[3] It is *not* assumed here that the economy is in a steady state, with capital and output growing at the same rate, since there is no particular reason to make such an assumption, popular though it is in the theory of neo-classical growth. In the simplest neo-classical growth model, with two factors, labour and capital, constant returns to scale, no technical progress, and labour growing at a given rate, the steady state rate of capital accumulation and rate of growth of real income will be equal to the given labour growth rate, and even protection could not alter them. Starting in steady state, protection would jolt the economy away from the steady state, say lowering the rates of growth of capital and of real income, along the lines of the argument given above. The proportionate fall in real income would be *less* than the proportionate fall in the rate of capital accumulation. But gradually the two growth rates would recover, to approach again the given steady state rate.

This process of the return of the growth rate to its original level might take a very long time during which it would be below its steady state level. Hence, even if we start in steady state and always tend to return there, the point made here about the effect of protection on the rate of accumulation can be important. See Corden, 'The Effects of Trade on the Rate of Growth' for a fuller analysis.

10 per cent. [4] This is even more so if there is some given rate of technical progress which is not directly related to capital accumulation (and hence is disembodied). On the other hand, if the average propensity to save and invest falls (the marginal propensity being higher than the average propensity) the rate of capital accumulation will fall by more than 10 per cent. But it is clear that one should not overrate the effect under discussion. It may be significant only if protection has a significant static effect in the first place.

It might also be noted that one could short-circuit the analysis here by focusing directly on the Harrod-Domar equation:

$$g = \frac{I}{Y} \cdot \frac{\mathrm{d}Y}{I}$$

in which g is the rate of growth of real income and $I/\mathrm{d}Y$ is the marginal capital-output ratio. Protection which reduces static real income, and so makes the economy less efficient, raises the marginal capital-output ratio.

It should be stressed that the effect of protection on capital accumulation is not additional to the static effects of protection. Rather, it is an implication of the static effects. When

---

[4] Let $y =$ rate of growth of income, $k =$ rate of capital accumulation, $n =$ rate of growth of labour, and assume that $y = \alpha k + \beta n$, where $\alpha + \beta = 1$. Then

$$\frac{\mathrm{d}y}{\mathrm{d}k} = \alpha = \, < 1$$

and

$$\frac{\mathrm{d}y}{\mathrm{d}k} \cdot \frac{k}{y} = \alpha \frac{k}{y} = 1 - \beta \frac{n}{y} = \, < 1$$

Thus the rate of growth of income falls by less than the rate of growth of capital accumulation, and also (provided $n > 0$) *proportionately* by less.

Furthermore, considering the rate of growth of income per head, $y - n = \alpha k + \beta n - n$, and hence

$$\frac{\mathrm{d}(y-n)}{\mathrm{d}k} = \alpha = \, < 1$$

and

$$\frac{\mathrm{d}(y-n)}{\mathrm{d}k} \cdot \frac{k}{y-n} = \frac{y - \beta n}{y-n} = \, > 1$$

Hence the rate of growth of income per head falls to the same extent as the rate of growth of income, but *proportionately* it falls by *more* than the rate of capital accumulation (provided $n > 0$).

protection inflicts a static cost, and so reduces static real income, consumption and investment are reduced. The lower consumption inflicts a current welfare loss while the lower investment reduces the rate of growth and so inflicts a future welfare loss. The cost of protection is the sum of these two losses.

II
SAVINGS AND THE INCOME DISTRIBUTION EFFECTS OF PROTECTION

(1) *Different Propensities to Save*

Protection can affect savings and investment through its effects on income distribution. If the marginal propensity to save differs between different sections of the community the ratio of total savings to total incomes will alter. This effect of trade and protection on growth via its effects on income distribution has often appeared in the literature, first in the writings of the classical economists and then in the recent economic development literature.

The marginal propensity to save out of profits is usually higher than that out of other incomes, especially wages. Hence any redistribution towards profits will raise the overall savings propensity. In Ricardo's model it was assumed that all savings were out of profits and that a movement towards free trade in the form of abolition of the Corn Laws (restrictions on imports of wheat) would shift income distribution away from rural rents towards profits. It followed that free trade would raise the rate of growth.

In less-developed countries today the effect is often claimed to be the other way: protection of manufacturing industry is likely to raise profits and so increase savings. It is of course an empirical question whether the marginal propensity to save out of rural incomes is less than out of urban incomes derived from manufacturing. The difficulty of measuring saving and investment in agriculture, especially by peasant smallholders, may sometimes create the impression that the savings propensity in agriculture is lower than it really is.

Protection may also reduce the profits of the commercial sector through reducing the volume of trade, and an interesting question then is whether the marginal propensity to save out of commercial profits is greater or less than that out of the profits of the protected manufacturing sector. On the other hand, if protection takes the form of quantitative import restrictions it may increase the profits of the commercial class, or at least of those elements in this or another class that are lucky enough to obtain the scarce import licence privileges. If a tariff is the instrument of protection one must also take into account whether the revenue raised will be saved by the government; we shall come to this shortly.

It cannot be assumed automatically that protection which redistributes incomes towards a high-savings sector must cause total savings and hence investment and the rate of growth to increase. The overall average propensity to save will rise only if the redistribution is towards the sector with the higher *marginal* propensity to save; conceivably this might not be the sector with the higher average propensity. More important, total savings may fall if protection reduces real income through reducing the productivity of the economy. It was stressed earlier that a fall in real income will reduce savings if the marginal propensity to save is positive; this might be combined with the income redistribution effect. For example, if protection reduces wages by $100 and raises profits by $70, and if the marginal propensity to save out of wages is 15 per cent and out of profits is 20 per cent, then total savings and the rate of growth will fall, even though there has been a redistribution in favour of profits and hence a rise in the overall propensity to save.

The normative analysis of protection designed to increase savings via income distribution effects should follow the principles set out in Chapter 2. There is a hierarchy of policies. Let us assume that the savings generated by the free trade situation—this time, in particular, by the free trade pattern of real incomes—are below the 'social optimum'. We leave aside the question of how the 'social optimum' is determined. Hence the pattern of expenditure diverges from the socially-desirable pattern and there is a domestic *expenditure* divergence.   First-best policy is directly to subsidize or supplement savings or investment. This could be done by raising taxes in a minimum-distortion way, and using the revenue to feed loanable funds on the capital market, so bringing the rate of

interest down and the availability of credit up until investment is at the desired level. The tax revenue minus the
reduction in private savings by taxpayers themselves would
represent the desired increase in the savings of the community.
It might be suggested that, instead of feeding the revenue on
to the capital market, it could be used to subsidize uniformly
the purchase of investment-goods. But there are great
practical obstacles to achieving complete uniformity. Another
possibility is to use the revenue to finance public investment,
so that the inadequate private investment would be supplemented by public investment. At the same time the tax-
subsidy system should be used to attain whatever income
distribution is desired, as discussed in Chapter 3. This means
that the pattern of taxes imposed to finance extra investment
needs to take into account both the usual by-product distortion effects and the need to get closer to the desired income
distribution.

It is hardly necessary to go down the hierarchy of policies in
any detail. Second-best policy would be to use direct taxes and
subsidies to alter income distribution in favour of the sections
with relatively high marginal propensities to save. A by-
product income distribution distortion would then be created.
Third-best policy would be to subsidize the production of those
products that are intensive in the factors that have the relatively high marginal propensities to save. This would add a
by-product production distortion. A tariff would add a by-
product consumption distortion.

(2) *Savings and the Fiscal Effects of Protection*

Protection can affect savings and investment through its
effects on government revenue, this being a particular type of
income redistribution effect. The marginal propensity to save
and invest of the government may be higher or lower than that
of the sections of the community that pay taxes.

There are three types of revenue effect. Firstly, subsidies of
various kinds cost revenue. In order to conserve government
revenue to finance investment it may thus be appropriate to
refrain from providing protection through subsidies, or at
least to protect less in this form than otherwise. We shall
come back to this shortly.

Secondly, tariffs raise revenue. It is not difficult to show that, if the quantity of investment is sub-optimal to a considerable extent and provided extra government revenue is used to finance investment, the benefits from a tariff that raises revenue may far outweigh the by-product distortions created by the tariff.[5] The question (discussed in Chapter 4) still remains whether a tariff is a first-best way of raising any given amount of revenue, or, more generally, what the role of tariffs in an optimum tax package might be. The greater the by-product distortion cost associated with raising any given amount of revenue, the lower the optimum level of publicly-financed investment, and hence the lower the optimum rate of growth. The by-product distortions created by taxation reduce the efficiency with which savings (defined as consumption forgone) are converted into investment.

The third revenue effect of protection is that it may affect income tax collections. For each sector there is a marginal propensity to pay income tax, depending on levels of income and collection costs, just as there is a marginal propensity to save. Thus total income tax collections may rise or fall, depending on whether protection shifts incomes towards those sectors prone to pay high marginal taxes (perhaps the corporate sector) or towards sectors paying low or no taxes (perhaps the unincorporated services sector in some less-developed countries).

(3) *The Shadow-Wage*

The idea that the choice of techniques, projects and industries should be influenced by the need to generate a pattern of incomes that will yield appropriate savings and reinvestment is one that has a key place in the literature on investment criteria and social cost-benefit analysis.[6] It is

[5] Jaroslav Vanek, 'Tariffs, Economic Welfare and Development Potential', *Economic Journal*, 81, December 1971, pp. 904–913.

[6] A. K. Sen, *Choice of Techniques*, Blackwell, Oxford, 3rd ed. 1968; S. A. Marglin, *Public Investment Criteria*, Allen and Unwin, London 1967; I. M. D. Little and J. A. Mirrlees, *Manual of Industrial Project Analysis in Developing Countries*, Vol. II: *Social Cost Benefit Analysis*, Development Centre of the O.E.C.D., Paris, 1969; P. Dasgupta, S. Marglin and A. K. Sen (for U.N.I.D.O.), *Guidelines for Project Evaluation*, United Nations, New York 1972. See also 'Symposium on the Little-Mirrlees Manual of Industrial Project Analysis in Developing Countries', *Bulletin of the Oxford University Institute of Economics and Statistics*, 34, Feb. 1972.

closely related to the previous discussion, and indeed raises the
same issues. We shall be discussing the relation between the
theory of protection and of social cost-benefit analysis in
more detail in Chapter 14. Here one aspect only will be
considered.

The basic approach is second-best, and hinges on the belief
that in some countries it is not possible to obtain the socially
necessary extra savings directly or explicitly through the

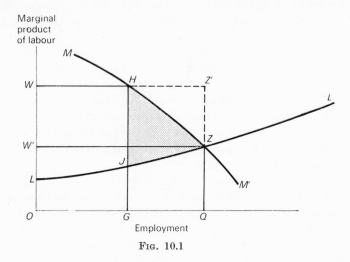

Fɪɢ. 10.1

taxation system so that, to some extent at least, it must be
done by fostering projects and industries with high savings
and investment-generating capacities. Furthermore, it is
generally assumed that wage income and the incomes of the
rural sector go wholly or mostly in consumption while
government revenue is used for investment. These are all very
controversial assumptions.

Figure 10.1 refers to employment in the industrial sector of
a less-developed country and is similar to various diagrams in
Chapter 6. MM' is the curve showing the value of the
marginal product of labour in that sector. LL' shows the
social opportunity cost of labour to the industrial sector.
It can be thought of as indicating the value of the marginal
product of labour in agriculture, and could take into account
the effects of extra industrial employment on the pool of urban

unemployment. Finally, OW is the given real wage that the industrial sector has to pay.

If the given wage OW were to determine investment and employment decisions, employment in the industrial sector would be OG. On the other hand, if employment were carried to the point where the marginal product of labour is equal to its marginal social cost, employment would have to be increased to OQ. This increase would generate extra manufacturing output of $GHZQ$ at the cost of reduced agricultural output $GJZQ$. The excess of the former over the latter is the shaded area and, in the Pareto sense, represents a social gain. Employment beyond OQ would reduce this gain, so that OQ is the Pareto-efficient employment point.

All this is familiar. It suggests that the *shadow-wage*—that is, the wage used for planning or investment purposes— should be OW'. Calculations made with such a wage-rate would show all projects up to Q as being profitable, and additional projects as unprofitable.

But there are complications. If extra employment of GQ is to be provided the extra industrial wage payments will be determined by the ruling wage OW, and not the shadow-wage, so that extra wage payments will be $GHZ'Q$. Since the value of the extra output is only $GHZQ$ there will be a loss to be made up. Some kind of protection has to be provided. Now there are two possibilities.

Tariffs or import restrictions might be imposed so as to raise the domestic prices of the industrial products concerned. A financial loss can then be avoided and consumers and users of the products will in fact be paying the subsidy. There will also be some gain in profits to the producers. In terms of Figure 10.1, the value of the marginal product curve facing producers will be raised so as to intersect the point Z'. The value of the private marginal product will then exceed the value of the social product.

Alternatively, the loss may be made up out of government revenue. If the enterprises are publicly-owned, public profits will go down by $HZ'Z$. If the enterprises are private, they might be subsidized, directly or indirectly, out of public funds. The modern theorists of investment criteria for less-developed countries seem generally to have in mind the case of public enterprise. While in most non-communist less-developed countries import-competing industries tend to be privately-owned, they are sometimes subsidized out of public funds, and in India—the source and inspiration of much of this theorizing—there is a substantial publicly-owned manufacturing sector. We shall now assume that the loss is financed out of public funds or public profits forgone.

The position then is as follows. It is illustrated in Figure 10.2. The extra employment has reduced agricultural output

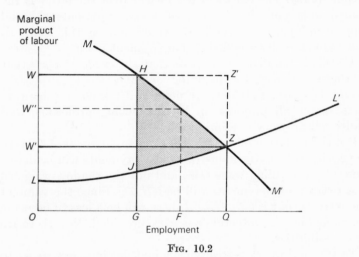

Fig. 10.2

by $GJZQ$. We can assume that this forgone output would have been consumed. On the other hand, extra wage payments of $GHZ'Q$ are being made, and all these wages are also used for consumption. It follows that extra consumption is $JHZ'Z$. This is the sum of two areas: the shaded area JHZ which represents the net social gain in the Pareto sense, and the loss in public revenue $HZ'Z$. The loss in revenue is assumed to lead to an equal loss in savings and investment, so that total consumption has increased while savings have decreased by a lesser amount.

If it were possible to tax wage income without significant collection or distortion costs and without providing a compensating rise in the money wage-rate so as to maintain the after-tax real wage at its original level of OW, there would be no need to reduce savings. Tax revenue of $HZ'Z$ would compensate the Treasury for the cost of the subsidy, and wage-earners would still be left with an increase in income equal to the (shaded) Pareto gain. Even if taxation did involve collection and distortion costs there might still be a gain, but it would be less than the shaded area.

In the literature referred to above it is assumed that such taxation is not possible, or at least not likely. The problem then is to weigh the consumption gain to wage-earners against the loss of public revenue and hence savings. If no value at all were placed upon extra consumption by wage-earners when this is at the cost of savings, it will be optimal not to increase employment beyond the point H where public profits are maximized. The actual wage OW will then be the shadow wage. On the other hand, if an equal value at the margin is placed upon wage-earners' consumption and government revenue (= savings) it will be optimal to increase employment to OQ, the Pareto-efficient point. The shadow-wage will then be OW'. This will apply if it is thought that the split-up between consumption and saving is optimal.

The interesting case is where the marginal value attaching to wage-earners' consumption is positive but is less than the social value of savings. It will then be necessary to trade-off consumption against savings and to choose an employment level somewhere between OG and OQ. It will result from making decisions on the basis of a shadow-wage between OW and OW', say OW'', yielding employment OF (Figure 10.2).

Figure 10.3 shows how the second-best optimum is obtained, and also brings out some of the central issues.[7] The horizontal axis shows wage-earners' income (including income in agriculture), assumed to be equal to consumption, and the vertical axis shows government income, assumed to be equal to savings. Both are valued at ruling free trade prices—that is, at prices of imports and exports at the borders of the country. The points H and Z correspond to the points H and Z in Figure

[7] This diagram is a slight variant of one in Vijay Joshi, 'The Rationale and Relevance of the Little-Mirrlees Criterion', *Bulletin of the Oxford University Institute of Economics and Statistics*, 34, Feb. 1972, pp. 3–32. The whole of this discussion is much influenced by Joshi's article.

10.2. H shows the combination of consumption and savings when investment decisions are determined by the ruling real wage and the point Z the combination at the Pareto-efficient point. The line HZ (not necessarily a straight line) shows how the combination changes as the shadow wage is lowered from the ruling wage to the Pareto-efficient wage.

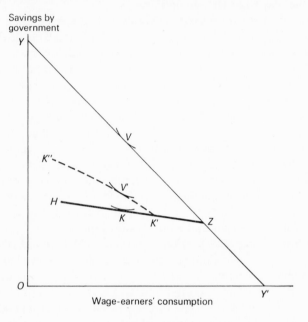

Fig. 10.3

There is a map of indifference curves indicating the community's (or its 'social welfare function's') preferences between consumption and savings. The optimal point is at K where an indifference curve is tangential to HZ.[8]

---

[8] The use of an indifference curve map is a useful heuristic device in the present case, and brings out the main points. Given expected future production possibilities, including expected future divergences or constraints, and given an intertemporal social utility function which tells us the community's preferences between present and future consumption, there will be such a map at any one point in time. The complication that is ignored here is that expected future production possibilities, as well as divergences and constraints, will depend, among other things, on present production, divergences and constraints.

If it were possible to tax wage-earners without distortion and collection costs the national income at Z could be redistributed between consumption and savings along the line YY' $(OY = OY')$ as desired. The first-best optimum would then be attained at V where an indifference curve is tangential to YY'. Thus first-best policy is to attain the Pareto-efficient point Z and then tax optimally to get to V. This is clearly superior to the second-best optimum at K, attained by biasing government decisions towards less labour-intensive projects.

If extra taxation is feasible but there are collection and distortion costs of taxation the first-best optimum policy will consist of moving to some point such as K' between K and Z (which implies a shadow wage in Figure 10.2 between W'' and W') and then transforming consumption into savings along a line such as $K'K''$ (flatter than YY'), attaining the optimum at V'. This follows from the analysis of Chapter 3.

Some of the characteristics and limitations of this approach should be borne in mind. It takes into account the problem discussed in Chapter 6 of the wage-rate facing the industrial sector being above the opportunity cost of labour and justifies protection for this reason. Its special feature is that it is also concerned with the question of optimum savings, but in an indirect way, through the effects of investment decisions on the distribution of income, especially the distribution of income between wage earners and the government. Furthermore, it assumes that potential losses are financed from government revenue and that government revenue is always used for productive investment, or at least that the government is behaving rationally and using its revenue in such a way as to equate at the margin the value of investment and public consumption. It is surely more common for the burden to be imposed on consumers rather than on the government, through industries being protected by tariffs or import quotas. Furthermore, government revenue may sometimes be used for public consumption or military purposes, rather than productive investment. A crucial second-best assumption is that, though aggregate investment is below its optimal level, there are constraints which prevent the government from attaining the optimum by taxation policies.

III
CHEAP IMPORTED CAPITAL-GOODS, INVESTMENT AND EMPLOYMENT

So far we have assumed that savings and investment vary only with real income and its distribution. We now disregard these effects and concern ourselves with an important *relative-price* effect. Protection may reduce the prices of capital-goods ruling within a country in relation to the prices of consumption-goods. Many countries deliberately discriminate in their tariff or import control systems in favour of imports of capital-goods. There are high tariffs or severe restrictions on imports of manufactured consumer goods and low tariffs on imports of capital-goods, and ready availability of import licences for them. Even though the unprotected exportables may also be consumer goods the net result is then to lower the domestic relative prices of capital-goods.

This type of policy raises a number of questions. The first is whether the policy is likely to succeed in increasing the amount of investment, as it clearly is intended to do. Another question is whether it increases or decreases employment in the advanced sector, bearing in mind that it may encourage the use of more capital-intensive techniques. Finally, the various by-product distortions it gives rise to need to be sorted out.

(1) *Do Cheap Capital-Goods Increase Investment?*

If the foreign currency prices of imported capital-goods decreased, total investment in real terms would increase. A given amount of expenditure on capital-goods, representing a given amount of consumption forgone, would buy a larger quantity of capital-goods. The extra investment would be financed out of improved terms of trade. If the capital-goods were financed or subsidized through tied aid, then the extra investment would be aid-financed. Expenditure on capital-goods need not stay constant; it might well decrease somewhat, but it is certainly likely that investment in real terms would rise. At any rate, consumption would not fall, and if expenditure on capital-goods declined, would actually rise.

But here we are concerned with a slightly different issue.

We assume given foreign prices, no change in foreign aid, and consider the consequences of altering the domestic price structure. If real investment is to increase there will have to be a reduction in consumption by someone in the country concerned. What then will determine the increase in investment, if any, and whose consumption will be reduced?

Suppose that all capital-goods used in the country are imported, with no domestic capital-goods production at all, not even construction. A uniform tariff is then imposed on all imports of consumption-goods combined with a uniform subsidy at the same rate on all exports. The exchange rate is appreciated appropriately to maintain balance of payments equilibrium. This exchange rate adjustment will reduce the domestic prices of capital-goods and modify somewhat the rise in the prices of consumption-goods facing domestic consumers brought about by the tariff and the export subsidy. Hence the prices of consumption-goods rise relative to the prices of capital-goods. Any gap between the revenue from the tariff and the cost of the export subsidy is covered by non-distorting payments to or taxes upon the general public. The result is the same as if a tax had been imposed on all consumption-goods, its revenue being used to subsidize the purchase of capital-goods.

Will more capital-goods be purchased at the expense of consumption-goods just because their relative domestic prices have changed?[9] It is not inevitable. One possibility is that the demand for consumption-goods is completely inelastic. When prices of consumption-goods rise consumers reduce their savings sufficiently for purchases of consumption-goods to stay constant. The reduced flow of savings onto the capital market raises the rate of interest. This rise in the rate of interest will offset the effect of the lower price of capital-goods, and the cost per annum to a firm of a given capital-good will remain unchanged. Nothing will in fact have changed; real savings and investment will be the same as before.

The story could also be told as follows. The higher prices of consumption-goods in the first instance lower real wages. So

---

[9] The main lines of the following argument are presented geometrically in terms of a two-sector neo-classical model in Corden, 'The Effects of Trade on the Rate of Growth', pp. 126–131.

money wages rise to restore the real wage level. The higher money wages reduce the money profits of firms and thus the funds they have available for investment. But the lower prices of capital-goods compensate them for this, so on balance they purchase just as many capital-goods as before. The subsidization of capital-goods has failed to increase investment.

All this has assumed that the tariff system which has created the relative price change has no protective effect for domestic consumption-goods production. Hence there is no by-product production-distortion and no fall in total real income at world prices. When consumption stays unchanged as a result of the appropriate reduction in money savings or rise in the money wage, real investment will also, finally, stay unchanged.

By contrast, if there is a production-distortion effect, total real income will fall and hence, if consumption stays unchanged, real investment will actually decline. A policy which was intended to raise investment will actually have lowered it! This is by no means improbable. A device that has significantly reduced the efficiency of the economy through the distortions it has created is quite likely to lower investment even though it has lowered the relative price of capital-goods to consumption-goods.

Of course, real consumption need not actually stay constant; that is just the limiting case. People may reduce their savings somewhat when prices of consumption-goods rise but not so much as to maintain constant consumption levels. Similarly, money wages may rise, but not so much as to restore the original wage-level. Hence consumption will fall and, finally, investment may have risen—though, because of the by-product production-distortion effects, it is also possible that finally both have fallen somewhat. The point to stress is that, in assessing the effects of a changed relative price pattern on real savings and investment, one must not neglect the possible effects on the rate of interest and the availability of investment funds. Comparing the effects of a tariff system which discriminates in favour of capital-goods imports with the effects of the free trade alternative, real investment will not necessarily be higher to the full extent indicated by the

lower price of capital-goods, and indeed it may not be higher
at all.

(2) *Effects on Employment*

Will the subsidization of capital in the advanced sector
increase employment in that sector? On the one hand, one
might expect increased investment in a sector to provide
increased employment opportunities. After all, it is often
argued that a cause of unemployment in less-developed
countries is shortage of capital. On the other hand, it is also
frequently argued that cheap capital leads to the use of
excessively capital-intensive techniques and the over-en-
couragement of capital-intensive industries, with the implication that the resultant substitution against labour will
reduce employment.

Simple analysis soon brings out the source of the confusion.
Provided that the subsidization of capital is not financed
directly or indirectly by a tax on wage earners in the advanced sector and that it does lead to increased investment
(in spite of the considerations discussed above), it is likely
both to increase employment in the advanced sector *and* to
raise the ratio of capital to labour in that sector. In the
special case of an infinitely elastic supply of labour to the
sector, it will increase employment without raising the capital-
labour ratio. A crucial assumption is that the subsidy is not
financed by urban wage earners, whether directly through
the tax system or through tariff-imposed higher prices of
consumption-goods. This important case will be considered
shortly. Here we must assume that the subsidy comes from
elsewhere—say the urban middle class, rural landlords or
foreign aid.

Figures 10.4 and 10.5 illustrate this argument. The vertical
axis in Figure 10.4 shows the quantity of capital available to
the advanced sector and the horizontal axis the quantity of
labour. The quadrant contains a map of isoquants (not
drawn) showing outputs resulting from various combinations
of the two factor inputs. It is assumed that there are constant
returns to scale, and factor prices are equal to marginal
products.

Hence factor prices depend solely on the ratios in which the

factors are employed and are independent of scale. The higher
the capital-labour ratio, the higher the marginal product of
labour and hence the real wage, and the lower the return to
capital ('rental' on capital-goods). Various rays from the
origin represent different ratios of the factors and hence
different factor prices. Ray OR_1 shows a low capital-labour

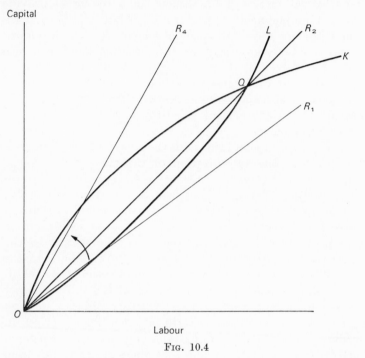

Fig. 10.4

ratio and hence a low real wage and high return to capital,
while ray OR_4 represents a high capital-labour ratio, and hence
high real wage and low return to capital. A movement in the
direction of the arrow represents a rise in the real wage and
fall in the return to capital.

Next we draw in the supply curves for labour and capital.
Curve OL shows the quantities of labour supplied to the sector
at various capital-labour ratios, and hence various levels of the
real wage. We assume that the higher the wage the more
labour is supplied. Similarly, curve OK is the capital supply
curve: the higher the return on capital the more capital is

forthcoming to the sector. Equilibrium is at the intersection of the two curves, yielding a capital-labour ratio indicated by OR_2.

When capital is subsidized its price to the sector for any given quantity supplied falls. Thus the sector faces a new capital supply curve, OK' (Figure 10.5). Employment increases from ON to ON' and the capital-labour ratio rises

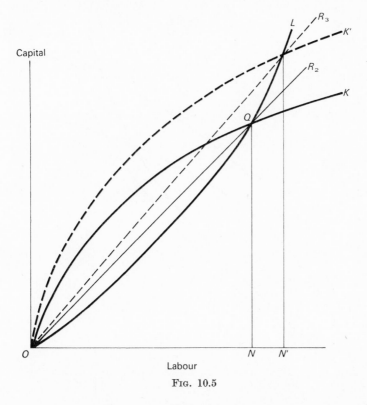

Fig. 10.5

from OR_2 to OR_3. The real wage rises, the cost of capital (= marginal product of capital) to the sector falls, but the supply price of capital excluding the subsidy rises. Thus we see that total employment has increased absolutely but the labour-capital ratio has fallen.

The fact that the subsidization of capital increases employment in the advanced sector is obviously not a justification for it. As said so often already, if the opportunity cost of labour to the sector is below labour's supply price, first-best policy is to subsidize labour, not capital. In Figure 10.6 the curve OL' shows the supply curve of labour facing the advanced sector if the employment of labour were subsidized optimally. It is assumed that it would yield the same amount

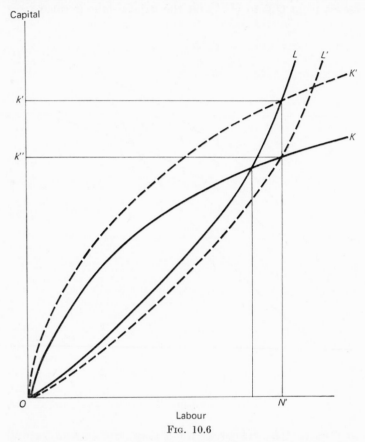

Fig. 10.6

of employment, namely ON', that the subsidization of capital represented by OK' yielded. But it involves less use of capital, Ok'' instead of Ok'. Subsidizing capital in order to stimulate employment wastes capital because it increases unnecessarily the capital-intensity of production.

This approach assumes that the subsidy to capital is financed neither by the capitalists in the advanced sector nor by labour in that sector but comes, in some sense, from outside the system. If profits in the advanced sector were taxed in order to subsidize the use of capital in that sector, then, in terms of this simple analysis, the state would be taking with one hand what it hands out to the same people with another, and there would be no inducement to increase the use of capital and hence to increase employment.

The interesting case is the one in which labour in the advanced sector is taxed to subsidize capital, perhaps through a tariff system that affects the urban but not the rural cost-of-living. Employment would then actually fall. Economists who argue against artificially making production more capital-intensive have this case in mind. The real wage payable in the advanced sector would rise and the cost of capital would fall; the substitution effect would then make production more capital-intensive. But since there would be no net subsidization of the advanced sector—as there was when the subsidy came from outside the sector—there would be no income effect, and it is this that raised the demand for labour and the real wage in the earlier case.

(3) *Distortions Created by Subsidizing Imported Capital-Goods*

It may well be true that savings and investment are suboptimal, and government action is needed to increase them. Furthermore, subsidizing imported capital-goods through the tariff or import control system may well increase real investment to some extent—though this is by no means certain, especially bearing in mind the adverse income effect induced by the inefficiencies created by this system. But let us grant these assumptions.

It will certainly not be first-best to foster savings and private investment through trade intervention, as is done so often. As already pointed out, first-best policy is to impose a minimum distortion set of taxes, possibly a uniform consumption tax, and feed the funds obtained on to the capital market, or make them available to private industry through a development bank. A genuine uniform subsidy on *all* investment-goods whether home-produced or imported, and whether used by the advanced sector or the subsistence sector, could also attain the first-best result, but such a completely uniform subsidy to embrace absolutely all investment-goods is hardly practicable. If investment were better done by the public sector, then it is obvious that straightforward taxation is required.

The systems generally used for making imported capital-goods artificially cheap create three types of distortions.[10]

The most important distortion is that investment in the advanced sector is subsidized but investment in the subsistence sector is not. Imported fixed capital equipment is used mainly in the advanced sector and not in the subsistence or unorganized sectors. To the extent that the unorganized sector has some need for imported capital-goods, if only of a modest kind, it is backward in the technique of extracting import licences from the bureaucracy compared with the influential industrialists sitting in the big cities. Furthermore, investment in the subsistence sector consists mainly of labour services, particularly for construction work, and this is not subsidized.

For all these reasons the system of artificially cheap imported capital-goods causes the nation's scarce savings to be diverted away from the subsistence towards the advanced sector. As already mentioned in Chapter 6 it is one cause of the dual economy, along with an imperfect capital market and the wage differential. The subsistence sector becomes excessively labour-intensive and the advanced sector excessively capital-intensive.

A second distortion is in the use of capital within the advanced sector. Fixed capital equipment is cheap relative to building and construction and to working capital, and indeed also relative to investment in human capital. This leads to waste of fixed capital in the form of excess capacity, to relatively less investment in building, to unduly low working capital, and possibly also to under-investment in human or 'learning' capital. It may also bias methods towards those associated with imported capital-goods suited for the factor proportions of developed countries. This may give a special advantage to foreign-owned firms.

[10] See Ronald I. McKinnon, 'On misunderstanding the capital constraint in LDCs: the consequences for trade policy', in Bhagwati et al. (eds.), *Trade, Balance of Payments and Growth*, North-Holland, Amsterdam, 1971. See also his *Money and Capital in Economic Development*, Brookings Institution, Washington 1973.

A third distortion is that domestic production of capital-goods is artificially discouraged in relation to production of consumer goods. This is the familiar by-product production distortion effect of tariffs and import restrictions, and explains why some less-developed countries have highly developed import-competing consumer good industries but have been slow to move into the capital-goods area.

IV
CONSUMPTION POSTPONEMENT AND THE DEMONSTRATION EFFECT

Protection may affect savings and hence investment in a way that is quite distinct from those so far discussed. If quantitative import restrictions or tariffs are imposed on imports of some consumption goods, we might expect consumers to substitute other consumption goods, whether unprotected importables or exportables. But it is also possible that they substitute savings instead, in the expectation that eventually the import restrictions or tariffs will be removed. There would be no sense in substituting savings if the restrictions were believed to be permanent. But if the restrictions are believed to be temporary a consumer may optimize his utility over time not by substituting the purchase of other goods for the restricted goods but rather by saving with the intention of buying the same goods later.

This analysis is related to Nurkse's 'demonstration effect'.[11] Nurkse argued that the availability of cheap and attractive imports produced in advanced countries causes consumers in less-developed countries to reduce their savings, or to be reluctant to increase them, since they wish to use their incomes to live as far as possible at the standards of living the advanced countries have 'demonstrated' to them. If the imports were not available, it is suggested, there would be less pressure to consume, and hence more savings. This argument clearly requires that import restrictions which are designed to reduce the 'demonstration effect' are expected to be temporary. Even then it must be remembered that the real income

---

[11] Ragnar Nurkse, *Problems of Capital Formation in Underdeveloped Countries*, Blackwell, Oxford 1953.

effect of restriction may cause savings to fall, so that on balance an increase in savings is by no means certain.

There is another, related, reason why reducing the availability of foreign goods may reduce savings rather than, as Nurkse suggested, increase them. This follows from the effect of the availability of foreign goods on the incentive to work. The availability of cheap imports may lead to a substitution effect between leisure and goods, including future goods; one might expect people in less-developed countries to work harder to obtain these desirable goods, both to consume more now and to save more in order to consume more later, and conversely for a restriction of imports to reduce savings for this reason.

V
THE INDUCEMENT TO INVEST AND BACKWARD LINKAGE

In an influential book published in 1958, *The Strategy of Economic Development*, Albert Hirschman presented some stimulating and apparently paradoxical ideas which have never been sorted out properly. Some of these are relevant to the present discussion. He emphasized the need to 'maximize "induced" investment decisions' in a less-developed country. He also stressed the need to maximize 'linkage effects' in order to induce investment and so foster the industrialization process.

(1) *Does the Inducement to Invest Matter?*

Protection may raise or lower the marginal efficiency of capital schedule for the economy as a whole. It thus affects the 'inducement to invest'. It will obviously affect the schedules of particular industries, and direct resources into some industries and out of others. But here we are concerned with the effect on total investment.

The first question is whether the 'inducement to invest' matters. In our analysis so far it has not mattered. Total investment was determined by savings, and an increase in the private inducement to invest would do no more than raise the equilibrium rate of interest or reduce the ratio of public to private investment. It was of course assumed that there is no capital inflow or outflow, an assumption we shall continue to make here. Let us then see through what channels a change in the inducement to invest could affect total investment.

The familiar Keynesian problem is that there may be a floor to the rate of interest, or alternatively that investment may not be interest-elastic, so that either the rate of interest could not fall sufficiently or, if it did fall, it might not succeed in bringing investment to the level of ex-ante savings generated at full employment. Hence the marginal efficiency of capital schedule would have to be shifted sufficiently to maintain full employment. But in most less-developed countries this is probably not a problem. The problem is that demand is too great relative to available productive capacity, not too little: investment is limited by the availability of savings, not the lack of desire to invest.

If lack of investment and demand were a problem then, as Keynes pointed out, one way would be to increase public investment sufficiently, financed by credit from the central bank. Public and private investment combined can always be brought to equality with ex-ante full employment savings. But while this method could certainly deal with the employment problem it would not necessarily lead to the most efficient pattern of investment and hence the highest rate of growth given the available savings: private investment may be socially more productive at the margin than public investment so that there may sometimes be a case for trying to raise the marginal private efficiency of private capital in order to raise the private content in total investment.

It is also possible that to some extent investment generates its own savings. This is probably what Hirschman had in mind. If investment becomes more profitable extra resources may become available to finance it, either through reduced consumption or through reduced leisure, entrepreneurs and others working harder so as to finance the investment. If an increase in the inducement to invest generates higher savings then in fact savings are profit- or interest-elastic. Resources for investment will increase if incentives are sufficient. Hence raising the inducement to invest might indeed raise total private investment, and thus the rate of growth, even in the absence of capital inflow or outflow. This may be an

11

important effect in some, and possibly most, less-developed countries, including some of the Latin American countries with which Hirschman was concerned.

In any case, assuming now that the inducement to invest is relevant for the rate of capital accumulation, let us see how protection could affect it.

(2) *How Protection Affects the Inducement to Invest*

The question is simply how protection affects expected profits, which can be assumed to depend on current profits. If protection reduced aggregate real income it may also reduce the rate of profit, at least if there is no great difference in factor-intensities between industries. If there is, then protection will raise the rate of profit provided relatively capital-intensive products are protected.

More realistically, capital may be somewhat sector-specific, with separate supply curves for different types of capital, each type of capital combined with entrepreneurship responding to its own profit rate. Protection may then raise some profit rates and reduce others. Increases in profits will induce investment and decreases will discourage it. If protection raises profits in those industries where the elasticity of supply of new capital is high at the expense of profits in industries where the elasticity of supply is low, as well as at the expense of other factor incomes, then total investment will increase.

There may be a *threshold effect*, this being an element in Hirschman's analysis. New investment into an industry will only be induced if the rate of profit increases above a certain threshold; small increases or decreases in profits would have no effect. If it is desired to induce an increase in total investment, protection must then concentrate profits, raising profits in some industries substantially, even though this is at the expense of other industries. Large profit increases in a few industries associated with modest profit decreases in many other industries will then raise total investment.

Finally, given that intervention through protection or other devices can raise the overall inducement to invest and that an increase in the inducement can actually raise investment and the rate of growth, the question still remains whether such intervention is justified. Should a static optimization policy be supplemented by a deliberate investment-inducing policy which will lead to some departure from the static optimum?

Essentially this depends on considerations previously discussed, namely whether private and social time-preference diverge, and whether private expectations of the returns on investment fall short of the correct expectations—or, at least, fall short of expectations based on the (presumably) superior information of the policy-makers. There may also be externalities associated with investment; these would not enter the private calculus but should enter the social calculus.

If it is socially desirable to foster new investment then first-best policy would be to subsidize new investment directly. A method which subsidizes an industry so as to induce new investment also draws existing capital and labour into the industry, and hence causes some misallocation of existing resources, apart from possibly undesirable income redistribution effects. Hence subsidization of output of an industry into which new investment is expected to go is second-best. A tariff to foster the industry adds a by-product consumption distortion, and so is third-best.

(3) *Backward Linkage Examined*

Hirschman's idea that 'linkage effects' should be maximized is worth examining with some care in view of the influence this idea has had on many people's thinking.

When a particular industry is protected its own actual or potential profit rate will be increased. Now the question is how actual or potential profitability of investment in other activities will be affected. Hirschman focuses on backward linkage.[12] Profits or potential profits in activities which produce inputs for the protected activity will increase. These must be non-traded inputs, or at least inputs which, because of economies of scale, would be wholly domestically produced and hence non-traded if the demand for them were sufficient.

If cloth were protected, so that the weaving industry expanded, the demand for yarn (which can be produced domestically if the scale of output is large enough) would also expand and a backward linkage effect would be generated.

[12] A. O. Hirschman, *The Strategy of Economic Development*, Yale University Press, New Haven, 1958, Ch. 6.

This might be contrasted with protecting another product which does not require produced inputs that could be produced domestically, or where there may be some domestic production of the inputs but any extra demand for them would be supplied from imports available at given world prices. The argument is that protection of cloth should be favoured relative to the latter type of industry.

Hirschman is not specifically concerned with protection, this being just one device for 'rearranging and concentrating the pattern of imports' so as to induce sufficient demand for those inputs where there is a possibility of a backward linkage effect. Protection of cloth, in this example, raises profits in weaving and spinning and so induces investment in these two activities. Assuming that this is a net addition to the stock of capital, and is not just capital diverted from going somewhere else, there may then be some case of fostering the cloth industry.

Yet this is only half the story. Protecting cloth may reduce profits and hence investment in other industries. Labour may be drawn into weaving and spinning from other industries, and these two activities may use non-traded inputs, including the services of public utilities, which are also used by other industries. While spinning is complementary with weaving, there will be activities directly or indirectly competitive with weaving. The backward linkage effect is no doubt the main complementarity effect, but it is by no means the only effect to take into account. Even if the supply of unskilled labour were perfectly elastic, other activities may be competitive in the use of skilled labour and some non-traded inputs. Investment in these industries will become *less* attractive.

If tariffs are the instruments of protection, one must bear in mind that the effective protection of the using industries will be reduced unless they are compensated by appropriate increases in their nominal tariff rates, while if the using industries are export industries they will suffer negative effective protection from tariffs on inputs unless they benefit from 'export drawback' or similar arrangements, or obtain compensating export subsidies.

Finally, if a tariff is seen as a method of indirectly subsidizing a using industry so as to induce investment via the backward linkage effect into other industries which supply inputs to it, it is a very indirect approach and sets up a whole series of by-product distortions. If indeed there is any inducement-to-invest problem, it would be more sensible to subsidize directly investment into the input-supplying industries concerned.

VI
LEARNING AND THE RATE OF GROWTH

To conclude, let us introduce *invisible* capital accumulation· As we saw in the previous chapter, when an industry's cost curve falls over time because it is learning by experience one could say that the learning process represents the accumulation of *invisible* capital. An infant industry is an industry that is undergoing such a learning process. If we make use of this simple concept of invisible capital accumulation we can relate the analysis of Section I of this chapter to the analysis of infant industry protection. The latter usually takes a partial equilibrium form—as essentially in the previous chapter—but here we consider the implications of infant industry protection for the rate of growth in a simple general equilibrium model.[13]

We consider an economy with only two industries, A and B, in both of which learning is related to the scale of current output. Thus production of A generates invisible capital K_a and production of B invisible capital K_b which will yield fruits in later years. If the current 'visible' outputs of A and B are X_a and X_b respectively, then true current output is $X_a + X_b + dK_a + dK_b$. A part of current income will be saved and invested in the ordinary way. Hence total investment consists of ordinary investment plus invisible investment.

Now suppose that it is desired to increase the rate of growth by fostering industries with learning. Protection then shifts the output pattern from one industry to another, say from A to B. Learning in A will go down and in B go up. Let us

---

[13] A rigorous general equilibrium two-industry model with learning-by-doing that is external to firms but internal to the learning industry is in Ch. 7 of Pranab K. Bardhan, *Economic Growth, Development, and Foreign Trade*, John Wiley and Sons, New York, 1970. Bardhan presents a mathematical solution of the optimizing problem.

assume that, measured in terms of the productivity of learning—that is, in terms of the value of the increment to future output (measured at world prices)—this raises total learning. We might then say that invisible capital creation has increased and so the rate of capital accumulation *on this account* has increased. But it must be borne in mind that if the shift in output imposes a current cost of protection, and hence lowers current real incomes, it is likely to reduce ordinary saving and investment out of current income for the reasons set out earlier. Thus against the increase in invisible capital creation must be set a decrease in ordinary investment.

It is perfectly possible that protection of genuine infant industries reduces the sum of ordinary investment and invisible investment—and hence the rate of growth—even though total learning or invisible investment on its own has clearly increased. This could happen if a large static cost of protection is required to bring about a shift to industries with relatively high learning rates and if the propensity to save out of current income is high.

Finally, it should be noted—though no formal proof is attempted here—that if all learning were internal to firms and firms were able to finance current losses by borrowing on the capital market or out of their own resources, profit-maximizing firms would forgo current profits or even incur losses for the sake of obtaining the benefits of learning only to an extent required to equate the expected marginal productivity of learning with that of ordinary investment. They would not invest in invisible capital to an extent that reduced expected future output and hence the rate of growth. Of course their expectations may be mistaken; this may lead them to over-invest in learning, or alternatively and more likely, they may undervalue the rewards of learning and so may under-invest in activities which require relatively long loss-making periods while experience is being gained.

11

SOME DYNAMIC ASPECTS OF TRADE POLICY

It is very common for critics of international trade theory, and especially of the comparative-advantage argument for free trade, to point out that it is static, and therefore of little value. What is needed, it is often said, is a 'dynamic' theory. But the fact that much of formal international trade theory is normally presented in terms of a purely static, timeless model does not mean that the main insights of the theory cease to be relevant once time is introduced explicitly, and once growth is allowed.

In writings concerned with the less-developed countries it is often suggested that while orthodox trade theory might well make a case for free trade, modified by just a few arguments for protection such as the orthodox optimum tariff argument, once 'dynamic' considerations are introduced the situation is transformed. 'Dynamics' seem to make a case against free trade. But the actual dynamic arguments are rarely specified with any rigour. By contrast, in discussions in Britain concerned with the possible gains and losses of her entry to the E.E.C. it was common to find the advocates of British entry conceding the static costs (mainly resulting from the Common Agricultural Policy) but arguing that these were likely to be outweighed by the 'dynamic' advantages. Thus in this case the dynamic arguments were assumed to make a case *for* (intra-European) free trade. Sometimes it was suggested that, for quite unspecified reasons, the British growth rate might rise as a result of entry into the E.E.C., and then it was not difficult to show that even quite a small rise in the growth rate would in time outweigh the adverse and more easily calculable static effects on Britain's real national income. The rise in the growth rate was implicitly assumed to be costless, and therefore not the result of increased savings.[1]

In listing the dynamic effects in this British discussion, economies of scale and efficiency effects induced by increased competitiveness were usually included. Economies of scale are indeed sources of gains from trade and from the formation of a customs union, and can be analysed in a comparative static way.[2] It is a well-known proposition of static trade theory that countries can mutually gain from trade even when their production functions and relative factor-endowments are identical if they can realise scale economies as a result. To realize these gains, adjustments may well have to be structural, or large, rather than marginal, but this does not introduce any essentially dynamic element. The effects of trade and of increased competitiveness on X-efficiency were discussed in Chapter 8. Again, the essential ideas can be analysed in static terms.

Growth and various dynamic effects (involving time in some essential way) have already made their appearance in the previous two chapters. In addition, the concept of the conservative social welfare function is essentially dynamic. In the present chapter we analyse briefly some arguments for trade intervention or for free trade which have appeared in the literature or in popular discussion and which cannot easily be analysed in purely static terms, and most of which involve changes over time or rates of growth. In some cases second-third- or fourth-best arguments for protection can be established. Section VI discusses some arguments in favour of trade which are often described as dynamic but can in fact be analysed in static terms.

[1] The 'dynamic' approach to deriving expected gains from U.K. entry into the E.E.C. could be found in a great deal of journalistic writings (for example, in *The Economist*) and in popular books. The Government's White Paper, *Britain and the European Communities* (Cmnd. 4239, London, H.M.S.O., 1970), used this approach and contained some imprecise arguments, but did not really invent them. For a serious discussion see *National Institute Economic Review*, No. 57, August 1971, pp. 46–50, 'The Dynamic Effects'. 'Dynamic' effects of trade liberalization are discussed in Chapter 5 of Bela Balassa, *Trade Liberalization Among Industrial Countries*, McGraw-Hill, New York, 1967.

[2] See W. M. Corden, 'Economies of Scale and Customs Union Theory', *Journal of Political Economy*, 80, May/June 1972, pp. 465–475.

I
THE ANTICIPATION OF PRICE CHANGES

In the late nineteen-fifties and in the nineteen-sixties, the view was popular that there was a long-term movement of the commodity terms of trade against less-developed countries, that this trend required the less-developed countries gradually to reallocate their resources out of primary product exports and that protection was necessary to induce this reallocation, or at least to induce an appropriate bias towards manufacturing in their growth patterns. These terms of trade expectations were based on extrapolations of alleged past trends. There has been dispute about these past trends; and even if there were a clear trend in the past, it does not follow that it will continue into the long-term future. But assuming that it is reasonable to expect a downward movement in the terms of trade, three general issues arise.

Firstly, is it necessary to take expectations about prices in the future into account when making current resource allocation decisions? If export prices will fall in the future, why not wait until they actually do fall? This simple-minded question directs attention to a possible loss in real income when reallocation of production and consumption patterns is premature. If factors of production were instantaneously and costlessly malleable from one use to another there would be no need to anticipate price changes. But since capital turns from 'putty' into 'clay' once installed, and since it takes time and resources to retrain labour, the future must indeed be anticipated. This is sometimes put by saying that resource allocation must be governed not by static but by 'dynamic' comparative advantage.

Secondly, why should governments or their planning authorities make better forecasts than private decision-makers? Clearly the private sector *could* anticipate a decline in export prices and adjust appropriately. The only general presumption is that large enterprises—such as companies owning mines or estates—are as likely to make well-informed forecasts as governments do, and will indeed be well aware of any government forecasts, but that small firms and peasant small holders are unlikely to be well-informed, and might tend to assume that recent prices will continue into the future.

Thirdly, what is the optimum way of correcting private decision-making, given that it is based on erroneous expectations derived from past or present prices? First-best policy appears to be to spread better information, subject to the costs of information dissemination. Second-best policy is to change present prices themselves through taxes, subsidies or the protective structure, hence anticipating future price changes affecting export and import competing producers by making them occur now. But the income distribution implications of such a policy are disturbing. Exporters are in any case expected to incur an income loss in the future; this loss will now be inflicted on them earlier.

II
FLUCTUATIONS IN EXPORT EARNINGS

It is a common argument that countries should reduce their dependence on trade and also diversify the pattern of their exports because of fluctuations in export earnings. We will assume at this stage that the fluctuations are predictable and certain. These fluctuations should be distinguished from expected *trends*, just discussed.

(1) *First-Best Policy*

The source of export income fluctuations may be either fluctuations in foreign demand conditions, reflecting perhaps fluctuations in supply conditions abroad, or in domestic supply conditions, perhaps climatic in origin, or some combination of the two. These will not only cause incomes of producers to fluctuate but will filter right through the economy. Suppliers of inputs and of goods consumed by producers will be affected. Revenue from taxation will fluctuate; in countries where taxes on trade are important sources of revenue this may be one of the main effects. The supply of funds to the capital market may fluctuate and hence the users of funds will be affected through variations in credit conditions and perhaps in the structure of interest rates.

Now fluctuations in consumption and investment levels are inconvenient to people, and similarly it is inconvenient and

costly for governments to have to vary expenditures in response to fluctuations in revenue. If everybody behaved with
self-interested good sense and foresight—and we have
assumed that the fluctuations are predictable—people would
save in good times and dissave in bad. The government would
run a surplus when export tax revenue was high and a deficit
when it was low. If fluctuations in investment are inconvenient
to producers, and especially if the rate of interest was determined by the world market rate, investment would not
fluctuate even though incomes fluctuate. The net result would
be that the country would accumulate foreign exchange
reserves in good times and run them down in bad, these
changes being matched partly by changes in government debt.
It should be noted particularly that changes in foreign
exchanges reserves would be explained only partly by changes
in the government net budgetary position; partly, they would
result automatically from private adjustments to income
fluctuations.

Let us now assume that private people lack foresight and
believe that every boom will last forever or is it the herald of an
upward trend. There is thus a domestic divergence, leading to
a misallocation of resources and calling for intervention.
First-best policy would be for the government to raise taxes
or special levies in good times and remit taxes or make actual
repayments in bad times. Indeed, highly progressive income
or export taxes build in such a mechanism. The government
will then save and dissave on behalf of the private sector and
hence will ensure that foreign exchange reserves fluctuate
appropriately. This requires politicians or government
officials to have foresight, unlike their private compatriots.
If *neither* have foresight—so that one cannot rely on governments increasing taxes and showing expenditure restraint
in times of export boom—there will be some case for a second-
best policy.

(2) *Second-Best Policy*

Reducing export income fluctuations by deliberately
inducing a movement out of those industries with exceptional
fluctuations will clearly be second-best if it has the net result
of lowering average incomes over the period of a full cycle.

Fluctuations for the economy as a whole can be reduced either by shifting resources out of industries where income fluctuations are above some kind of average, or by diversifying into products which have offsetting fluctuations.

Let us now analyse this second-best policy (fiscal policy in all cases being first-best). Three cases must be distinguished.

Firstly, the source of fluctuations may be in demand conditions abroad. If there is a fluctuation in general demand conditions abroad and this affects all or most of a country's actual and potential exports the case is for making the country less dependent on exports, hence shifting resources out of export into import-competing industries. The movement should be particularly out of exports where demand fluctuates more, such as the more income-elastic exports. Appropriate export taxes would be second-best, and tariffs third-best. If the fluctuations affect particular exports only the case is, at least partly, for export diversification towards products with less fluctuations or with fluctuations that are not synchronized.

Secondly, the source of fluctuations may be on the supply side, and may affect export products but not import-competing products. Second-best policy is then to discourage production of these products, possibly through a production tax. If there is also significant domestic consumption of exportables, an export tax would only be third-best.

In both cases—whether the source of fluctuations is in foreign demand or in domestic supply—the resource shifts should be either towards products where demand or supply fluctuations are less or where they are not synchronized (not co-varying).

Thirdly, the source of fluctuations may be on the supply side but may affect the country's production quite generally, not just the export sector. Production in the economy as a whole may be affected by fluctuations in political conditions. The fluctuations may be climatic in origin and may affect both export and import-competing industries, with both being mainly agricultural. This is an important case applying to many smaller less-developed countries.

In this general type of case, trade has the virtue of cutting or modifying the links between domestic consumers and producers. The more domestically-consumed products are imported, the less domestic production fluctuations affect domestic consumers. This gain from trade is distinct from the usual gain derived from comparative costs. One would expect consumers themselves to diversify their sources of supply by relying on imports as well as on domestic production, so the government would not necessarily be justified in deliberately fostering or increasing trade. In the limiting small country case, with export demand and import supply both perfectly elastic, there will be no problem provided trade has been opened up and quota-free trade is permitted, since trade will automatically absorb changes in the pattern of production.

The crucial empirical questions are whether fluctuations in export receipts tend to originate abroad or at home, and if they originate at home, whether the causes are limited to the export sector or are general to the economy. The popular policy issue arises principally for less-developed countries. The empirical literature suggests that fluctuations may be more domestic (supply) than foreign (demand) in origin and that the extent of export income fluctuations and of GNP fluctuations tend to be less the more open the economy.[3] If these supply fluctuations are general to the economy and not just limited to export products, it seems to follow that in general one cannot even make a second-best case for trade restrictions designed to reduce economic instability.

[3] See Alasdair MacBean, *Export Instability and Economic Development*, Harvard University Press, Cambridge, 1966, which was based on data covering 1950–58, and Donald Mathieson and Ronald McKinnon, 'Instability in Underdeveloped Countries: The Impact of the International Economy', in M. Reder and P. David (eds.). *Essays in Honor of Moses Abramovitz*, Academic Press, New York, 1973, which reviews the literature subsequent to MacBean's book and also shows that openness as indicated by the export-GNP ratio may be a significant explanatory variable of the degree of export and GNP fluctuations in less-developed countries.

MacBean's main conclusion, based on extensive empirical work (both cross-sectional and special country studies) for the period 1950–58 is that income fluctuations in less-developed countries are not closely related to export-earnings fluctuations and that there is no evidence that export instability reduces capital formation. Thus there is no instability problem, except in special cases where the source of the trouble is usually on the supply side. His results for this period have been confirmed by the cross-sectional results of P. B. Kenen and C. S. Voivodas, in 'Export Instability and Economic Growth', *Kyklos*, 25, 1972—Fasc. 4, pp. 791–803, but for 1956–67 the latter find some indication that export instability has reduced the level of investment.

III
RISK AVOIDANCE

So far we have assumed that expected changes or fluctuations in export earnings are certain. Let us now allow for
risk.[4] The logic underlying risk aversion is well-known.
Gamblers may place as much value, or even more, upon a
50 per cent chance of making a fortune as upon the other
50 per cent chance of going bankrupt or starving, but most
people and nations will wish to avoid this sort of risk. A
discount against risk will only emerge if the marginal utility of
income changes significantly over the range of the possible
outcomes. Hence, if a particular project or source of income
affects only a small part of a person's total income (and if
variations in the income derived from it are independent of
variations of income derived from other sources) the discount
against risk will also be very low, and for practical purposes
can be treated as being zero. It follows that if the risk from an
activity is spread over a large number of persons, so that the
outcome of the activity has a negligible effect on the income of
any one person, there will be a negligible utility loss from
risk.

It is often argued that exporting is especially risky owing
to uncertainties in foreign demand, and that this justifies
government intervention to restrict dependence on exports.
This is not self-evident, since risks may be lowest when sales
are spread over many markets, and not just the domestic
market. Similarly, in the case of consumers of importable
goods, risks would probably be lowest if they purchased from
many sources, less dependence on imports not necessarily
reducing their risks. On the supply side, in so far as risks for

---

[4] The theoretical literature on uncertainty and trade policy is rather sparse
and the only significant paper appears to be William C. Brainard and Richard
N. Cooper, 'Uncertainty and Diversification in International Trade', *Studies
in Agricultural Economics, Trade and Development*, Vol. VIII, No. 3, 1968
(Yale Economic Growth Center Paper No. 145). It does raise the crucial issue
of why there might be a divergence between social and private risk perception
or ability to adjust to risk, and suggests (rather unconvincingly) that private
investors in primary exports are not likely to diversify enough (public intervention should be *anti*-risk).

The present discussion has been much influenced by various essays in
Kenneth J. Arrow, *Essays in the Theory of Risk-Bearing*, North-Holland
Publishing Company, Amsterdam, 1971.

different products are independent of each other, risk will indeed be increased through the tendency to production specialization induced by trade. Perhaps foreign trade is—or appears—especially risky because private traders and producers have less information about foreign than about domestic markets and sources of supply.

Let us now grant the proposition that reducing trade will reduce the nation's risks and focus on the main issue. Is social intervention required to shift resources out of more risky activities, or combinations of activities, chosen by private decision-makers into less risky activities or combinations of them? If peasants choose to grow rubber for export even though rubber demand is very risky rather than to grow rice for home consumption where there is much less risk involved, is this likely to be non-optimal from a social point of view, calling perhaps for an export tax or production tax on rubber or a subsidy to rice production.

A strong argument can be made that there is no case for social intervention at all. Each private decision-maker will make his own assessment of risks of various activities. Let us assume that his assessment is correct, or at least is the best that is possible given the knowledge and experience available. We assume that he is risk-averse so he is willing to trade-off expected income against risk reduction.

He can reduce his risks in various ways: he can shift out of more risky activities—for example out of export crops into subsistence crops—he can diversify into activities with low co-variance (this being one aim of conglomerate corporations) and he can buy risk reduction for a price, whether by issuing shares in his firm on the stock market or through taking out some kind of insurance policy. In practice there are costs associated with all these means of risk reduction, and these costs will, in effect, reduce his expected net income, so it will not pay him to dispose of his risks completely. Indeed it may not be practicable to do so because all activities have some risks associated with them and there are limitations to insurance.

All this seems to lead to the conclusion that the degree of risk that private decision-makers assume will be socially-optimal. This is true even though, if the risks could be spread over the nation, it might be socially optimal to ignore risks completely and just aim to maximize expected incomes. The costs of insurance and of devices having similar effects prevent the risks from being spread completely, or make it uneconomic to do so and, given this, it will be socially optimal to allow private risk aversion to determine resource allocation and consumption decisions in the private sector.

Let us now consider possible qualifications to this basic approach. Firstly, private decision-makers' information or ability to assess information may be poorer than that of the government and its advisers. Thus rubber-growing may in the past have been a fairly safe activity, but it is becoming more risky. The government may realize this before private producers, and some intervention in the interests of the private producer may be justified. Of course the argument could go the other way, private producers over-estimating the risks of unfamiliar activities—this being a basis for an infant-industry argument. It is plausible to argue that private producers and consumers are likely to have less information about economic conditions abroad than the government, so they are likely to over-estimate the risks through trade. It follows that the very argument which underlies the general view that trade with foreigners is more risky than trade with locals provides a basis for *pro-risk* intervention by the government.

Secondly, the effects of taxation on risk-bearing must be taken into account. This is rather a complex subject on which there is a large literature. The main point is the following. Normally profits are taxed but losses cannot be wholly offset, if at all. If a risky activity turns out favourably the Treasury shares the gains, but if it turns out badly it does not share the losses. This introduces a bias against risk-taking. This is also true when losses are not made but the income tax is progressive and the averaging of incomes over several years is not allowed. It follows that government policy should be offsetting—that is, it should be *pro-risk*. In so far as trade increases risks—which, of course, is not certain—it would seem again that pro-trade intervention is called for, at least as a second—or third-best policy.

Thirdly, it may be usual for the government to come to the

rescue of sections of the community which are liable to lose income or employment as the result of external events, such as declines in export prices. This is an application of the conservative social welfare function, which, as pointed out in Chapter 5, is a form of social insurance. Hence government action tends to reduce the risks faced by the private sector. So one might argue that since it is the government through its taxpayers which is carrying the risk it should aim to reduce risks through trade, at least marginally, since private people will no longer have an incentive to do so.

This parallels the preceding taxation argument. The focus this time is on government assistance when losses are being made rather than on taxation when there are profits. One must really analyse it together with the taxation argument. The government may take over risk in the sense of not only helping an industry when it is in difficulties but also taxing it when it is doing well, hence not altering its expected income. In that case the argument would be valid, provided social risk aversion were positive when risk is borne by the nation as a whole. But the act of spreading the risks reduces the utility loss from risk, and the discount for risk for the nation as a whole may be minimal. When this is so there would be little or no case for intervention to discourage risky activities.

On the other hand, when the utility loss from risk is not extinguished by being transferred to the taxpayer—for example, when the risks of a major export industry accounting for a large part of the national product are transferred to the government—there may conceivably be some case for *anti-risk* intervention. It is not necessary to assume that income from the industry concerned is large in relation to national income for risk aversion to remain; it would be sufficient if the industry's income is co-variant with other important sources of government income.

Furthermore, it is possible that risky industries can expect to get much more help when they are in trouble than they are likely to pay in taxes when they do well. If this is so, their expected incomes will have been raised by the social insurance system created by the conservative social welfare function, and there will be a tendency on this account for over-expansion of risky industries. We have then another possible argument for *anti-risk* intervention.

IV
PROTECTION TO EMPLOY
A GROWING POPULATION

The following argument has in the past been advanced in some less-developed countries, and also in Australia and Canada. We have ahead of us a large growth in the labour-force. The extra labour will have to be employed in import-competing manufacturing since the labour-absorptive capacities of the agricultural export industries are very low, and even if exports could be increased they could be sold only at worse terms of trade. So we must expand manufacturing, and this can only be done with the help of protection.

It is implied that, given free trade and a fixed exchange rate, extra labour and capital could not automatically find employment in exportable and import-competing industries.

Extra labour cannot be employed in the export industries because the money wage is given, these industries are subject to diminishing returns, so that extra output would require their domestic prices to rise sufficiently, which in turn requires a sufficient expansion in the foreign demand for exports that is not forthcoming. Extra labour cannot be employed in import-competing industries for essentially the same reason. One might imagine an array of import-competing (manufacturing) industries, each with constant cost, but with industries ranked in order of costs, so that for the import-competing sector as a whole costs rise with output expansion.

Existing import-competing industries can indeed expand insofar as the domestic market (total expenditure) expands, but the development of new industries would require a devaluation or tariffs. The extra labour-force could be employed in the non-traded sector provided fiscal and monetary policies generated extra demand for non-traded goods and services, but some of this extra demand would spill over on to exportables and importables and so worsen the balance of payments.

As pointed out in Chapter 7, the reluctance to devalue

could reflect certain arguments for protection. It might be said that devaluation would worsen the terms of trade. Further, relatively low income distribution weights might be attached to those members of the exporting community, notable land-owners, who would be the principal gainers from higher domestic prices of exports. Devaluation may also be avoided because of the indiscriminate income distribution effects on import-competing producers and consumers of importables: a made-to-measure approach, possible only with subsidies or tariffs, may be preferred.

If protection is needed to employ a growing population it may well raise the rate of growth of real output. Suppose the labour-force is growing at some given rate, say 3 per cent per annum. Given no increase in protection and a constant output per person employed, aggregate demand must also expand at 3 per cent to maintain full employment. Assume that the average and marginal propensities to import (including the imported materials content in domestic production) are fixed, so that imports would also expand at 3 per cent. Now we assume the value of exports to grow at a slower rate, say 2 per cent. The growth of export income is assumed to be quite independent of domestic supply conditions and hence of labour-force growth. Hence full employment growth without protection would involve a growing balance of payments deficit.

To avoid this deficit there are then two extreme possibilities:

(1) Aggregate demand may be increased only at a 2 per cent rate. The growth rate of the economy will then be 2 per cent, there will be growing unemployment, but balance of payments equilibrium will be maintained.

(2) Tariffs or other devices may be used to reduce the ratio of imports to the gross national product over time. Aggregate demand can then expand at more than 2 per cent, so maintaining full employment, while imports grow only at 2 per cent. In real terms the economy will grow at something between 2 per cent and 3 per cent, with income per head declining.

Comparing the second alternative with the first, protection has raised the rate of growth by avoiding the growing unemployment that would otherwise be necessary. Protection has brought the *actual* growth rate up to the *potential* (full

employment) one. Of course, in the absence of the exchange
rate constraint the potential growth rate might be higher.

The real issue concerns the rigidity of the exchange rate and
of factor prices. One can argue that this model is as extreme
and unrealistic as the Harrod-Domar growth model, with its
fixed capital/output and capital/labour ratios. It is perhaps
more plausible as a short-run model. In most countries in the
short-run there is some downward rigidity of money factor
prices and of the exchange rate. Even if the exchange rate is
varied, this is normally done only at discrete intervals, and
in the intervening periods there might conceivably be some
role for protection. But it must be remembered that measures
that make short-run adjustments in factor prices or the exchange rate unnecessary are likely to reduce the incentives to
make long-run adjustments. Thus protection may yield a
gain in real income in the short-run through making full
employment possible; but in the long-run, when full employment would have been attained even with free trade, it will
only impose a cost.[5]

V

THE COST OF PROTECTION AND
THE RATE OF GROWTH

Protection can turn a positive rate of growth into a negative
one. This is a case of 'tariff-induced immiserizing growth'.

[5] A variant of this argument, applied in Canada by Dales, and also assuming
a fixed exchange rate, is that failure to expand employment opportunities
through protection would dry up the flow of immigrants. See J. H. Dales,
The Protective Tariff in Canada's Economic Development, University of
Toronto Press, 1966. This is distinct from a well-known argument for protection
in Australia and Canada which assumes that immigration depends on the real
wage-level, that protection raises the real wage on Stolper-Samuelson lines
(import-competing industries being labour-intensive and exportables land-
intensive) and that growth in population is a desirable (non-economic)
objective. See J. B. Brigden et al., *The Australian Tariff—An Economic
Enquiry*, Melbourne University Press, 1929 (which originated this argument).
See also Jacob Viner, 'The Australian Tariff', *Economic Record*, November
1929 (reprinted in Viner, *International Economics*, The Free Press, Glencoe,
1951); A. J. Reitsma, 'Trade and Redistribution of Income: Is there Still an
Australian Case?', *Economic Record*, 34, August 1958, pp. 172–188; W. M.
Corden, 'The Logic of Australian Tariff Policy', *Economic Papers*, No. 15,
1962 (published by Econ. Soc. of Aust. and N.Z., Sydney); C. L. Barber,
'Canadian Tariff Policy', *Canadian Journal of Economics and Political
Science*, 21, November 1955, pp. 513–30; John H. Young, *Canadian Commercial
Policy*, Royal Commission on Canada's Economic Prospects, Ottawa, 1957,
pp. 89–93.

Capital accumulation or technical progress may expand the production possibilities but the cost of protection may rise so much that real income actually falls.[6] This is represented in Figure 11.1 which could be interpreted in partial or general equilibrium terms.

The import supply curve is SS' and the given tariff is ST. The given demand curve is DD'. The supply curve of import-competing production, indicating marginal costs, is HH' in

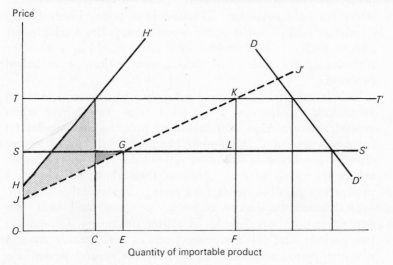

Price

Quantity of importable product

FIG. 11.1

period I and JJ' in period II. In period I the tariff inflicts a production and a consumption distortion cost. In period II the domestic import-competing supply curve has shifted downwards so that output has increased from OC to OF. The consumption-distortion cost remains the same.

On the original output of OC there is a gain owing to the reduced costs. On that part of new output which would

[6] The possibility was first uncovered by J. H. Dales. See J. H. Dales, *The Protective Tariff in Canada's Economic Development*, Ch. 3. It was presented in terms of two-sector geometry in H. G. Johnson, 'The Possibility of Income Losses from Increased Efficiency or Factor Accumulation in the Presence of Tariffs', *Economic Journal*, 77, March 1967, pp. 151–154, and earlier Johnson had also presented a partial equilibrium exposition (Johnson, *Aspects of the Theory of Tariffs*, pp. 113–114).

continue even without a tariff (CE) there is also a gain. These two gains are shaded in Figure 11.1. On the remaining new output EF there is a loss, being the production-distortion cost of protection of period II (the triangle GKL), If the latter exceeds the two gains there is immiserizing growth.

While this immiserizing case is of some intellectual interest, its policy significance should not be overrated. If there is genuinely a cost of protection, with no adequate argument for protection, the tariff should be removed irrespective of whether it creates immiserization. The case is of policy interest only if—rather oddly—tariff policy were not a policy variable but growth policy were, in which case there would be a case for restraining the rate of capital accumulation or technical progress.

This immiserization case is actually just a special case. In general, protection may affect the rate of growth even when capital accumulation and technical progress are unaffected by protection, and full employment and full capacity production are being maintained. (We thus rule out considerations discussed earlier). Assume then that the country's production possibilities expand owing to population growth, capital accumulation or technical progress, and that this expansion does not depend on protection policy. Comparing two periods, and assuming free trade in both, real income in the first period might be Y_1 and in the second period Y_2, with Y_2 greater than Y_1. Now, suppose that, instead of free trade, protection is imposed in both periods. Protection could inflict a static cost or a gain. Suppose it inflicts a cost. Let this cost be C_1 in the first period and C_2 in the second. Hence real income under protection is $(Y_1 - C_1)$ in the first period and $(Y_2 - C_2)$ in the second period. The protection rate of growth will then be a weighted average of the rate of growth of Y and of C.

If Y and C grew at the same rate then the protection rate of growth would be the same as the free trade rate of growth. In general, the cost of protection could grow faster or slower than free trade real income. If it grew more slowly, the rate of growth of real income with protection would be higher than the rate of growth under free trade. This must not be misunderstood to mean that protection is desirable because it raises the rate of growth. At every point in time a cost of protection would be incurred, but proportionately, the cost would decline over time. It is also possible that C grows more rapidly than Y, in which case protection will lower the rate of growth, immiserization being an extreme case of this. [Free trade yields $Y_2 > Y_1$, and protection yields $(Y_2-C_2) < (Y_1-C_1)$.][7]

VI
THE 'DYNAMIC' EFFECTS OF TRADE ON KNOWLEDGE

In the literature one finds references to 'dynamic' effects of trade which seem to be concerned, broadly, with changes in knowledge and in techniques resulting from trade. They appear on first sight to be arguments in favour of trade that have nothing to do with comparative costs as analysed in static models, or even with capital accumulation in the usual sense. All of them can in fact be analysed perfectly adequately in comparative static terms.

(1) Trade makes new goods available to a country and so, it would appear, changes tastes and expands wants. In fact there is nothing analytically new here. The essential nature of the static gains from trade is that trade widens choice by presenting a country with a consumption-possibility frontier that differs from the closed economy one. The country can still choose the closed economy bundle of goods, but it can now choose from many more bundles than it was able to in the closed economy. If it is argued that the changes in tastes are undesirable, it is implied that the opening or expansion of trade may have adverse effects in terms of some kind of social welfare function that does not accept the primacy of individual choices.

[7] The special case of free trade immiserizing growth caused by adverse terms of trade effects was mentioned in Chapter 7 (pp. 178–9). In that case, imposition of the orthodox optimum tariff turned a negative into a positive growth rate, the cost of protection itself being negative.
[i.e. $Y_2 < Y_1$ and $(Y_2-C_2) > (Y_1-C_1)$.]

Johnson has explored the circumstances—in a two-sector model with given terms of trade—in which protection raises or lowers the rate of growth on these grounds. No simple answers seem to emerge. See H. G. Johnson, 'A Note on Distortions and the Rate of Growth of an Open Economy', *Economic Journal*, 80, December 1970, pp. 990–92.

The availability of new goods may stimulate people to work harder so as to be able to buy the new goods that have become available. In fact, when the price of imports falls from infinity there is a two-fold substitution effect: consumers substitute imports for domestically-produced goods *and for leisure*. But the gains from trade will also yield an income effect, leading possibly to an increase in leisure. John Stuart Mill, in a famous passage, implied the substitution effect, arguing that the availability of new goods leads to a substitution against leisure in favour both of consumption now and of savings so as to increase consumption later.

A people may be in a quiescent, indolent, uncultivated state, with all their tastes either fully satisfied or entirely underdeveloped, and they may fail to put forth the whole of their productive energies for want of any sufficient object of desire. The opening of a foreign trade, by making them acquainted with new objects, or tempting them by the easier acquisition of things which they had not previously thought attainable, sometimes works a sort of industrial revolution in a country whose resources were previously undeveloped for want of energy and ambition in the people: inducing those who were satisfied with scanty comforts and little work, to work harder for the gratification of their new tastes, and even to save, and accumulate capital, for the still more complete satisfaction of their tastes at a future time.[8]

(2) Trade brings contact with foreigners, with favourable effects on methods of production, tastes, and so on. A capital-good, Knowledge, is being created or augmented. To some extent it is a public good, though some of the benefits of traders going abroad will be purely internal, the traders themselves gaining. If greater knowledge leads to improved methods and choices, there is a gain additional to the gain from the availability of more or cheaper goods. This general effect of exposure to foreign contacts will raise the rate of growth since foreign contact is the channel through which continuous technical progress abroad passes to the country concerned. Of course, contact with foreigners could also have adverse effects. The gains and losses from imperialism may have to be regarded as part of the gains or losses from trade,

---

[8] John Stuart Mill, *Principles of Political Economy*, Book III, Ch. XVIII, Sec. 5.

especially trade that is primarily carried on by foreigners and leads to their settlement in the country.

(3) The availability of imports carves out a market for new goods and so shows domestic producers what it is possible to sell in the country, reducing the risk element in developing new products.[9] Imports create the asset Knowledge for the local producers—knowledge that there really is a domestic market. This is an external economy (public good) effect. The implication is that the development of a domestic industry may first require the free entry of imports, and when the necessary market has been 'reconnoitred' infant industries will be set up, possibly with the help of protection.

(4) The availability of imports enables them to be copied, each imported good carrying some potential knowledge-creation with it.[10] It shows potential domestic producers what it is possible to produce. Domestic firms will often not develop new products or new processes until they have proof that such products can be made at a given price. It might be argued that the greatest cost of research is the cost of finding out which problems are soluble. This means that if free trade merely results in goods of established types and made by well-known techniques coming in cheap because foreign labour is cheap, the efficiency-stimulating effect of foreign trade will be small. However, if free trade opens the door to the entry of goods from countries where factor costs are not too different from domestic factor costs, the new products or processes will in principle be imitable. This knowledge is also a public good, benefiting potential import-competing producers.

[9] Albert Hirschman, *The Strategy of Economic Development*, Yale University Press, New Haven, 1958, Ch. 7.

[10] I owe this point to Professor Harold Lydall.

12

PROTECTION AND FOREIGN INVESTMENT

THE RELATIONSHIP between protection and foreign investment has often been observed, it being generally believed that tariffs and import restrictions can attract foreign capital. In a number of countries, notably in Latin America, as well as Canada and Australia, a great deal of foreign capital is in manufacturing industries which are dependent on high rates of effective protection, and much evidence has been advanced that protection has induced the inflow of this capital.

Johns[1] has found that of eighty-six U.K. companies which first invested in Australian manufacturing in the decade 1950-60, forty-six stated that import controls or tariffs were a 'very important' motive for investment. Brash, referring to American investment, writes: 'One half of all respondent companies mentioned the desire to by-pass tariff barriers as a motive in their establishment in Australia, and such companies were widely dispersed through industry'.[2]

In other countries also, foreign investment in manufacturing has been explained by tariffs, exchange control, and similar devices. Barlow and Wender conclude from their thorough study of the motivation of American capital outflow that 'manufacturing companies invest abroad primarily to maintain a market that has been established by export but which is in danger of being lost (through tariffs, exchange control etc.)'.[3] Brecher and Reisman write that the 'Canadian tariff has been an important factor historically—and in some periods the

---

[1] B. L. Johns, 'Private Overseas Investment in Australia: Profitability and Motivation', *Economic Record*, 43, June 1967, pp. 233–61.

[2] D. T. Brash, *American Investment in Australian Industry*, Harvard University Press, Cambridge, 1966, p. 36.

[3] E. R. Barlow and I. T. Wender, *Foreign Investment and Taxation*, Prentice-Hall, Englewood Cliffs, 1955, p. 160.

dominant factor—in encouraging foreign companies to locate in Canada'.[4]

These observations raise a number of questions. Firstly, does protection induce more foreign investment than would enter otherwise, taking into account not only the direct but also the indirect effects? It will be shown that the answer is not as obvious as seems on first sight. Secondly, does the foreign investment effect of tariffs represent an additional argument for tariffs over and above the usual arguments or, on the contrary, are tariff increases particularly reprehensible when they benefit mainly foreign-owned companies? In order to consider this latter question it is first necessary to examine various arguments for directly taxing or subsidizing foreign investment. The third question is whether there are particular reasons for encouraging or restricting foreign capital inflow because a country has a given tariff system. At the end of this chapter the implications of multinational corporations for the analysis will be considered.[5]

I

THE EFFECT OF PROTECTION ON
FOREIGN CAPITAL INFLOW

(1) *The Direct Effect of Protection*

If an industry is protected, its profits or potential profits increase. This should naturally lead to a movement of capital and labour into this industry; the tendency would be for both domestic and foreign capital and labour to move in, the actual movements depending on the supply elasticities of the factors. The domestic factors would come out of the country's other industries while the foreign factors would be an addition to the country's total factor supply. Thus the observations of Brash and of Barlow and Wender that protection induces foreign capital inflow appear to make sense.

[4] I. Brecher and S. S. Reisman, *Canada-United States Economic Relations*, Royal Commission on Canada's Economic Prospects, Ottawa, 1957, p. 117.

[5] This chapter is a substantially revised version of Corden, 'Protection and Foreign Investment', *Economic Record*, 43, June 1967, pp. 209–232. Sections V and VI are new, and account has been taken of *pecuniary externalities*, arising from indivisibilities, which were neglected in the article.

Furthermore, protection not only raises profits in the domestic industry but also reduces profits in that foreign industry the exports of which have been excluded by protection. This foreign industry will then naturally seek to restore its profits by maintaining its market in the protecting country through investing in the newly profitable domestic industry. This is 'defensive investment'. The foreign industry will have a particular reason for investing in that country and in that same industry rather than in other countries or other industries because its trade will have made it familiar with the market and may have led to investment in distribution facilities and in goodwill that it can make use of as a domestic producer. This is the logic behind Barlow and Wender's observation, that manufacturing companies invest abroad primarily to maintain a market. It strengthens the conclusion that protection encourages capital inflow, though it is not essential to it.

(2) *The Indirect Effects of Protection*

While this sounds convincing it is not the whole story. There are three qualifications.

Firstly, a tariff on a product which is an input in another industry will reduce effective protection in the other industry even though it increases protection directly. If the protected product is an input in the production of non-protected exports, the tariff yields a negative effective rate for these exports—in fact just like an export tax. The decline in protection elsewhere through the reduction in effective rates can be expected to lead there to less capital inflow than would have taken place otherwise and conceivably even in time to capital outflow.

Secondly, some foreign investment is complementary with imports, so that, if imports are discouraged by a tariff, such foreign investment will be also. The main example is investment in distribution facilities for imported goods. In a sense this is a similar effect to the one just discussed. We might think of a product at the point of final retail sale as being the final product, the imported good being just an input into this final good. The value-added element within the country consists essentially of the service of distribution. A tariff then reduces effective protection for the final good, indeed making effective protection for it negative, and naturally discourages investment in it.

The third qualification seems the most important. Protection of an industry draws, as we have seen, domestic as well as foreign capital and labour into the industry. But where do the domestic factors come from? There may initially be unemployed capital and labour, in which case there are no adverse repercussions on other industries. But if there is full employment, the extra demand for domestic factors will raise their prices. The factors required by the protected industry are drawn out of those industries where product prices do not or cannot change—whether import-competing or exporting. The increase in costs causes export industries and other import-competing industries to become less profitable.

If we think of general protection of import-competing industries, the familiar point is that protection creates a bias against exports. While it will lead to an inflow of foreign capital into import-competing industries it might also be expected to lead to reduced capital inflow into export industries. This assumes that in the export industries the domestic factors the prices of which have risen are complementary, not competitive, with foreign capital. If they were competitive, then protection might actually *attract* foreign capital into the export industries.

What, then, is the total effect of protection on capital inflow? In analysing this issue, it will be assumed here that foreign capital is 'sector-specific'. This seems appropriate for direct investment, the form in which most countries now get most of their foreign capital. Direct investment consists of a package of capital, technology and managerial know-how, and the package and its availability are different for different countries. As Brash says: 'foreign capital of the direct investment variety tends to move from a particular sector in the capital-exporting country to the *corresponding* sector in the capital-importing country, induced by a profitable investment opportunity in that sector'.[6] Thus one can think of separate foreign supply curves for chemical-capital,

[6] Brash, *American Investment in Australian Industry*, p. 276.

motor-car-capital, mineral-capital, and so on. Perhaps a single supply curve may apply to a group of associated industries.

Now when protection reshuffles the pattern of actual or potential factor prices, raising the profits of some import-competing industries, and reducing the profits of export industries (and also of actual or potential non-protected or low-protected import-competing industries), one must ask what the relevant supply elasticities of the various types of foreign capital may be as well as how the profit pattern will be affected. The matter is quite complex, for the inflow of capital also depends on the technical possibilities of increasing production in the relevant industries and on the elasticity of supply of domestic complementary factors. Yet the general point is clear. If the country did not protect, or protected less, would not the export industries and other industries which receive little or no protection be more profitable and so attract more capital? And can one be sure that the capital gained by protection on the one hand is not lost on the other? Hence, taking into account indirect effects it is by no means as obvious as appears on first sight that protection generally increases total capital inflow.

It might be noted that the models of the pure theory of international trade offer simpler answers. If one assumes that capital is homogeneous and mobile as between different industries and that there is a single upward-sloping supply curve of foreign capital (which may move to the right over time), one must ask whether protection raises or lowers 'the' rate of profit. If it raises it foreign capital flows in, while if it lowers it foreign capital flows out (or flows in less than it would otherwise). The question then becomes whether relatively capital-intensive industries are being protected and so the rate of profit is being raised. One thus obtains the simple answer that general protection through import tariffs induces foreign capital inflow if importables are on the whole capital-intensive.

Furthermore, in the simple Heckscher-Ohlin model, assuming, among other things no 'factor reversals', a country that is a net capital importer must have capital-intensive importables so that protection will indeed induce capital inflow. The point is simply that a country is likely to import that product which is intensive in the country's relatively scarce factor, and if the country imports capital then, presumably, its scarce factor is capital.[7]

II
TAXATION OF FOREIGN INVESTMENT

What are the effects of foreign investment on optimum tariff policy? To answer this question a related issue must first be sorted out at some length. Should foreign investment be subsidized or taxed, and if so, how much? We shall be concerned here only with the interests of the capital-receiving country.[8]

Simple Case: No Tax or Subsidy Required

As a starting-point for the analysis the following nine assumptions, to be subsequently removed one by one, will be made. Assumption (9) will only be removed in Section V.

(1) Capital is paid its marginal social product, so that the market gives a correct indication of social values. There are no domestic divergences. This means, among other things, that there are no non-market external economies and diseconomies.

(2) Foreign investment projects are perfectly divisible. This means that they will be expanded up to the point where the return to capital is equal to its private marginal product, which will, by assumption (1), be equal to its social marginal product.

(3) Foreign capital is available at a given price, the supply curve being infinitely elastic. This means that the incidence of a tax cannot be borne by the foreign investors.

[7] The classic discussion of the relationship between protection and factor mobility in terms of the two-country two-factor trade model is in R. A. Mundell, 'International Trade and Factor Mobility', *American Economic Review*, 47, June 1957, pp. 321–35; this article leads to the rather fantastic conclusion that the smallest tariff will stimulate a capital movement which so alters factor proportions as to eliminate all trade. This result depends on the assumptions that the foreign supply curve of capital is affected by the protection-induced trade reduction and that production functions in the two countries are the same.

[8] This discussion has been much influenced by G. D. A. MacDougall, 'The Benefits and Costs of Private Investment from Abroad: A Theoretical Approach', *Economic Record*, 36, March 1960, pp. 13–35 (reprinted in Jagdish Bhagwati (ed.), *International Trade*, Penguin Modern Economics, 1969).

(4) Additional revenue raised from taxes on foreign capital would be redistributed to the public in the form of reductions of other taxes. Thus the object of taxing foreign capital is not to raise revenue, but rather to affect resource allocation and the price at which foreign capital is obtained.

(5) There are no taxes on domestic capital.

(6) There are no taxes on exported capital in the capital-supplying countries. Thus the profits on foreign capital are taxed only by the capital-receiving country.

(7) Foreign capital is homogeneous, with a single supply curve. This is contrary to the 'sector-specific' approach used earlier. This assumption, like the others, will be removed in due course. But the arguments to be developed below *before it is removed* could all be applied separately to the various types of heterogeneous capital.

(8) Foreign capital does not earn monopoly profits.

(9) The country cannot affect its terms of trade (small country assumption).

If we grant these assumptions, then we must conclude that there is no case for taxing or subsidizing foreign capital. The incidence of the tax cannot be borne by the foreign investors. A tax would certainly raise revenue and channel some of the gains from foreign investment to the Treasury, but it would nevertheless inflict a social loss. A particular tax rate will maximise revenue, but there is no virtue in such a tax if it is not revenue but social welfare that needs to be maximised.

Figure 12.1 illustrates the point. The quantity of foreign capital (measured in value units) is shown along the horizontal axis and its price and marginal product along the vertical axis. SS' is the supply curve. DD' is the social marginal product curve; given our assumptions, it is also the demand or average revenue curve. The marginal revenue curve is mm'.

In the absence of tax or subsidy, OR of capital will come in yielding DQS of surplus (consumers' surplus and producers' surplus to complementary factors). If a tax at the rate WS/OW were imposed, revenue would be the shaded area $WVMS$; this is the rate which maximises this area, that is, the *maximum revenue tax*. It is given by the intersection of mm' with SS'. The social gain will then be DVW of surplus plus

WVMS of revenue, the sum of which falls below the surplus in the absence of tax by *VQM*. The triangle *VQM* is the social loss because of the tax. Capital *TR*, on which the social return is greater than the cost, is lost.

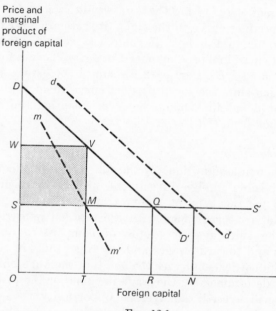

Fɪɢ. 12.1

(1) *External Economies and Diseconomies*

Now remove assumption (1) and introduce external economies associated with foreign capital inflow. These may be of many kinds. New techniques of production and management may be brought into the country. If these then spread to local firms without cost, they are external economies. It is often forgotten that if they stay within the foreign firms or if the foreign firms sell them at a market price, they are not external economies. Bringing new methods into the country may yield external economies; but in itself the import of the new methods is not an external economy.

In addition there are external diseconomies, such as the danger to the balance of payments from its dependence

12

on new capital inflow, the possible threat to political independence resulting from a large foreign-owned sector, and the loss of initiative and the sense of dependence which may result when much of the country's managerial class is responsible to overseas superiors.

In terms of the diagram these externalities shift the social marginal product curve to the right (if there are net external economies) or to the left (if there are net diseconomies). The private marginal product curve, or demand curve for capital, remains at DD'. Figure 12.1 assumes net external economies, which shift the social product curve to dd' and which mean that a subsidy should be given to bring the foreign capital stock from OR to ON.

(2) Indivisibilities

Now remove assumption (2) and allow for the effects of a large indivisible foreign investment project in raising the wage-rate for the type of labour employed in the project. We are not concerned here with the internal redistribution effects—the shift against profits of competitive domestic industries and towards wages and towards profits of complementary industries—but with the fact that the wage bill of the project is determined by the wage payable for the last man, but this will exceed the average opportunity cost of the labour employed. Intra-marginal producers' surpluses or *pecuniary externalities* are generated.

This point is illustrated in Figure 12.2. The indivisible increase in foreign capital is TR. The marginal product curve DD' slopes downwards because of the increase in the wage-rate that has to be paid for labour of given quality as more capital enters. Alternatively it may reflect the deteriorating quality of marginal labour. If the product produced were non-traded, or if it were traded but world prices were not given to the country, the curve would also slope downwards because of the negative slope of the demand curve for the product. But we ignore this here.

Given the wage-rate payable when the project actually exists, the total private return is $TMQR$. The cost of the capital required is $TNWR$, so that, in the absence of subsidization, it would not be profitable. But the social product of the project is the larger area $TVQR$. This consists of the private product plus the shaded triangle MVQ which is the intra-marginal producers' surplus or pecuniary externality generated. It is this larger area that has to be set against the cost of the capital $TNWR$ in order to decide the social desirability of the project. If MVQ is greater than $MNWQ$ some subsidization of foreign capital inflow appears to be justified.

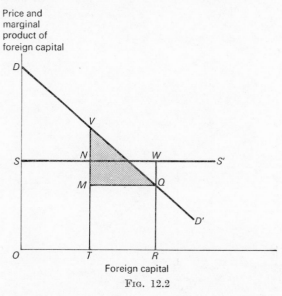

Fig. 12.2

(3) *Upward-sloping Supply Curve of Foreign Capital*

Now remove assumption (3) and allow for the possibility that the supply curve of foreign capital may be rising, as shown in Figure 12.3 by SS'. Assume no external economies or diseconomies and that foreign capital is divisible. The incidence of a tax would now be partly borne by the foreign owners of capital. So clearly there is some case from the national point of view for taxing foreign capital. (This may, of course, be offset by external economies which, in themselves, provide a case for a subsidy.) To maximise its own welfare, the capital-receiving country should therefore behave like a monopsonist and impose a rate of tax which brings the marginal cost of foreign capital into equality with the marginal product of capital.

In Figure 12.3, ss' is the marginal curve to SS'; a monopsonist would buy capital ON, thus restricting inflow from OR to ON. The rate of tax which achieves this result is WK/OW and depends on the elasticity of the supply curve. This is an exact parallel to the orthodox optimum tariff, and has been applied to the foreign investment question by Kemp.[9]

The point is that a tax on foreign capital may be paid

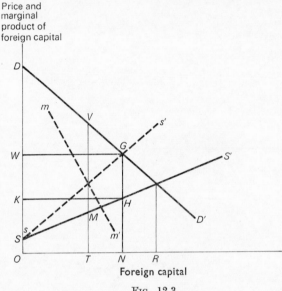

Fɪɢ. 12.3

partly by foreign capitalists, so that it is in the interests of the capital-receiving country to impose the tax—to 'squeeze' foreign capital. But the tax will also discourage capital inflow and so cannot be unlimited. There is in principle an optimal rate of tax which balances the gains from the tax against the loss of capital inflow.

This orthodox optimal tax rate must be distinguished from the tax rate which maximises revenue. The maximum revenue tax rate is VM/VT, being given by the intersection of the marginal revenue curve (mm') with the marginal cost curve

[9] M. C. Kemp, 'The Benefits and Costs of Private Investment from Abroad: Comment', *Economic Record*, 38, March 1962, pp. 108–10.

(ss'). The orthodox optimum welfare tax maximises the sum of revenue and surplus, this being $WGHK + DGW$ at the optimum. The revenue raised is part but not the whole of the gain to the capital-receiving country. It would be the whole gain only if the DD' curve were horizontal, that is, if the marginal product of foreign capital did not decline with increasing quantities.

(4) *Taxation to Raise Revenue*

Now remove assumption (4) and let the raising of revenue to finance government activities be one object of policy. If it were the sole objective then the maximum revenue tax would become the optimum tax. There would then be a case for a tax on foreign capital even when (as in Figure 12.1) there are no external diseconomies and when the supply curve of foreign capital is infinitely elastic, so that no part of the tax is obtained by squeezing the foreign investors. More realistically, the raising of revenue may be one, but not the sole objective. In this case the optimal tax rate would be somewhere between the maximum revenue and the maximum welfare tax rate.

(5) *Taxes on Domestic Capital*

Now remove assumption (5) and allow for taxes on domestic capital. It might be argued that on grounds of *equity* foreign capital should, as far as possible, be taxed at the same rates as similar domestic captial. Such equity considerations do in fact have a considerable impact on popular attitudes to taxation and on the tax system, but will be ignored here.

One approach is to take the tax rate on domestic capital as given. If the aggregate supply of domestic savings and hence domestically-owned capital were unaffected by changes in its profitability after tax, this tax rate on domestic profits would not affect the allocation of resources. The optimum tax on foreign capital would then be independent of the tax rate on domestic capital. There would be no tendency to substitute foreign for domestic capital just because the tax rate on foreign capital is less, or vice versa. If we are thinking of capital in general and not its supply to particular industries, it seems fairly realistic to assume a low or zero elasticity of supply of domestic capital over the relevant range of the curve. So the tax rates on domestic capital can probably be ignored in our discussion.

If there *were* a significant elasticity in the domestic capital supply curve, the rate of tax on domestic capital that is assumed to be given would have to be taken into account in determining the optimal foreign capital tax rate. There would be some case for taxing foreign capital because domestic capital is being taxed. But this optimal rate would not necessarily be equal to the given domestic rate. In fact, we would have a typical second-best situation which could be analysed in the same way as similar problems have been analysed in Chapter 2 for optimum tariff theory.[10]

Another simple approach would be not to take the domestic tax rate as given, but rather to postulate a given revenue target which has to be satisfied by some combination of taxes on domestic and on foreign capital. It can be said in general that, from an efficiency point of view, the lower the elasticity of supply of domestic capital, the more revenue should be raised by taxes on domestic capital, and thus the lower would be the optimum tax on foreign capital. Again, the appropriate analysis is the same as has been used in this book for optimum revenue tariffs.

(6) *Taxes in the Capital-Supplying Country and Double Tax Agreements*

Now remove assumption (6) and allow for taxes in the capital-supplying country on profits of its overseas investments. The position with regard to U.S. capital is that when profits are repatriated they are subject to corporate tax in the U.S. but a credit is given for taxes paid in the capital-receiving country. The latter has prior taxing rights.[11]

For example, consider U.S. capital invested in Australia.

[10] For a full analysis, see Corden, 'Protection and Foreign Investment', pp. 218–20.

[11] See P. B. Musgrave, *United States Taxation of Foreign Investment Income*, The Law School of Harvard University, Cambridge, Mass. 1969; and *United States Income Taxation of Private Investments in Developing Countries*, United Nations, New York (Sales No: E.70.XVI.2), 1970. Changes in U.S. tax laws are always under discussion, including the possibility of abolishing the tax credit system.

If the U.S. tax rate is 48 per cent and the Australian rate is
42·5 per cent, the United States will collect 5·5 per cent of
gross distributed profits in tax. The total tax rate paid by the
enterprise on repatriated profits is determined by the higher
of the two tax rates. It should be noted that the U.S. does not
tax profits of U.S. subsidiaries abroad when these are not repatriated. This is the 'tax deferral' provision in U.S. tax law.
In the case of U.S. capital at least, the complication to be
discussed here applies only to repatriated profits.

Taxes on foreign capital which are imposed in the capital-
supplying country and which are subject to the usual tax
credit arrangements complicate the theory of an optimum
tax policy. In discussing this it will be assumed here realistically that the gross tax rates in the capital-supplying country
are given. It is not then in the interests of Australia (the
typical capital-receiving country) to reduce the Australian tax
rate on foreign capital below the gross rate charged in the
capital-supplying country. For any reduction would simply be
scooped up by the supplying country's tax authorities and
would not increase the flow of capital.

Now suppose that, in the light of the arguments previously
discussed in this paper, the conclusion has been arrived at that
foreign capital should pay a 30 per cent tax. Suppose, further,
that the given gross tax rate in the capital-supplying country
is 48 per cent. A recommendation for a 30 per cent tax would
not take into account the constraint of foreign tax rates.
Australia cannot reduce the total tax payable by foreign
capital below 48 per cent, though it can of course raise the
total tax above this rate. How then can the requirement for a
30 per cent tax be reconciled with this? The answer is clearly
to charge a 48 per cent tax and then give an offsetting subsidy.
But this will work only if the tax authorities in the capital-
supplying country do not add this subsidy, or not the whole of
it, to the tax liability of their firms.

This argument explains why it can conceivably be logical to
subsidize foreign capital, whether directly or indirectly, and
at the same time to tax it. One's first reaction might have been
to say: If foreign capital is to be subsidized, surely the first
step must be to eliminate all taxes on it. But because of the
tax credit system it may not pay to reduce taxes below a

certain level. The point is important because one indirect form of subsidization is through tariffs or quotas.[12]

It has already been mentioned that, owing to the 'tax deferral' provisions in the U.S. tax law, the problem does not arise when U.S.-owned profits are not repatriated. The problem would also disappear if foreign taxes became *deductions* from taxable profits in the U.S. rather than being tax credits. The U.S. Internal Revenue would then charge a 48 per cent tax on that part of profits derived abroad which is left after foreign taxes have been paid. In that case the supply curve of U.S. capital facing the capital-receiving country (*SS'* in Figure 12.3) would simply be shifted to the left by the U.S. tax.

(7) *Sector-specific Capital*

Now remove assumption (7) and allow for heterogeneous, 'sector-specific' capital. This means that there will be many different demand curves, supply curves and prices of foreign capital. For each case separately the arguments developed above for taxing or subsidizing foreign capital apply. Figures 12.1, 12.2 and 12.3 can be reinterpreted to refer not to foreign capital as a whole but to a typical 'sector-specific' category of capital.

Clearly the optimum set of taxes and subsidies on foreign capital will be a differential one, since externalities and supply elasticities will differ as between different types of capital. To apply a tax or subsidy uniformly to all types of capital would be a second-best, though possibly the only feasible, solution. If the only consideration were to 'squeeze' foreign capital monopsonistically, then the appropriate set of taxes could be called the orthodox optimum investment tax structure exactly paralleling the orthodox optimum tariff

---

[12] The discussion here has neglected a provision in the U.S. tax law which gives certain advantages to corporations operating in less-developed countries (the *indirect foreign tax credit*). A reduction in the tax rate in the less-developed country will reduce somewhat the total tax payable by the corporation, U.S. tax not fully making up the difference between the foreign and the standard U.S. rate of 48 per cent. There may thus be some benefit to the government of the less-developed country from charging a rate below 48 per cent. At the most (with a foreign rate of 24 per cent) the corporation's total tax liability can be reduced to 42.24 per cent. See *United States Income Taxation of Private Investments in Developing Countries*, pp. 94–99.

structure which would apply if there were terms of trade effects.

(8) *Monopoly Profits*

Finally, remove assumption (8) to allow for monopoly or excess profits to foreign firms operating in the capital-receiving country.

For each type of direct foreign capital the situation may be as represented in Figure 12.4. There is a minimum expected

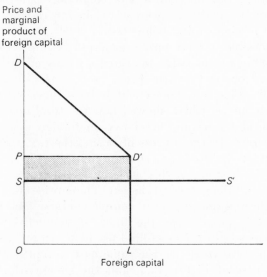

Fɪɢ. 12.4

return which can be represented by a horizontal supply curve (*SS'* in the diagram). This might be thought of as the weighted average of minimum expected returns in different years, taking into account that losses in the initial years will have to be compensated by high returns in later years. There is a demand curve which is vertical at this point of intersection with the supply curve, this being determined by the market limit (*OL* in the diagram). This type of demand curve is an extreme case, but something similar may not be unrealistic. It is assumed here to facilitate exposition, though it is not necessary to the main argument. There is monopolistic price determination and the ruling price is OP which, except in the marginal industry, is above the minimum price OS. Monopoly profits are represented by the shaded area.

Assuming no externalities, the optimum tax in the case represented in the diagram is clearly PS per unit, just sufficient to tax away the monopoly profits. This made-to-measure rate would no doubt be different for every type of capital. External economies would not alter this situation unless OS were greater than LD', in which case it might pay to subsidize until OL of capital has come in. External diseconomies might justify a tax greater than PS, restraining the quantity of foreign capital below OL.

If it is possible to have only a general tax rate on foreign capital, and not to choose the tax rate for each sector, then the analysis must proceed further. From a set of cases such as that represented in Figure 12.4 one could derive a curve, something like a supply curve, which showed how the total quantity of foreign capital fell as a general tax on foreign capital was raised from zero. In the process of raising the tax, monopoly profits would be reduced. At the same time the prices of some types of capital would rise, this being represented in Figure 12.4 by a movement along DD' to the left. These price rises would reflect reductions in prices of complementary factors and (where this is possible in the light of competition from imports) increases in prices of final products. In other words, the broad picture would be the same as was represented by the simple case of Figure 12.3, and the general conclusion that there is an optimum welfare tax on foreign capital remains.

Summary: Principles for Taxing Foreign Capital

Tax rates on foreign capital should, as far as possible, not be less than the gross tax rates payable in the capital-supplying country. In considering whether taxes on foreign capital should be above this level, or whether there should be offsetting subsidies, there are two main considerations.

First, there is the opportunity through taxation to 'squeeze' foreign capital monopsonistically, depending on its elasticity of supply or on the existence of monopoly profits. This is an argument for taxation.

Secondly, externalities may provide an argument for subsidizing or an additional argument for taxing. The net external economies which some people believe to be associated with foreign capital would justify net subsidies (that is, direct and indirect subsidies which are greater than taxes). One could include here the pecuniary externalities associated with large indivisible projects. On the other hand, the various popular arguments hostile to foreign capital inflow, in so far as they have validity, can be embraced in the blanket concept of 'external diseconomies' and might lead to the conclusion that net external economies are negative.

Two other considerations have been mentioned: first, there may be a case for taxing foreign capital as one method of raising revenue for government activities; and secondly, if the elasticity of supply of domestic capital is positive, there is some case for taxing foreign capital because domestic capital is being taxed.

III
TARIFFS AS SECOND-BEST TAXES OR SUBSIDIES ON FOREIGN CAPITAL

It has been shown that direct taxes or subsidies on foreign capital may be required to maximise national welfare. The optimum pattern of taxes and subsidies will take into account all the considerations of the previous section. This would yield a first-best optimum. But it may in fact not be possible to vary taxes and subsidies on foreign capital appropriately.

Thus double tax arrangements may not make it possible to reduce taxes paid by foreign capital below gross tax rates charged in the capital-supplying country (and it is not in the interests of the capital-receiving country for its rates to be below these). Furthermore, it may not be possible to raise tax rates on foreign capital above those charged on domestic capital for reasons of equity or politics. Later it will be noted that it may also be inadvisable to charge tax rates on foreign capital *above* those that would be paid by it in another foreign country because of the possibility of tax evasion.

In any case, conventional attitudes to income and profits taxes make countries reluctant to place varying rates of taxes

on different types of capital even though their supply elasticities or the externalities associated with them differ; the
general principle of discriminating on the basis of industry is
much less acceptable in the framing of an income tax system
than a tariff system (though sometimes there are varying
periods of tax holidays). It follows then that a first-best policy
may conceivably not be possible, so that one should consider a
second-best policy.

The second-best policy to be considered is the use of the tariff
system to do the job that a system of direct taxes and subsidies on foreign capital would really do better. To simplify, it
can be assumed here that to begin with there is a uniform
tariff of x per cent which allows for any external economies
associated with the production of import-competing goods in
general. These externalities are quite independent of whether
production is with domestic or foreign capital. It is assumed
that there are no other reasons for tariffs apart from the foreign
capital complication to be introduced here.

The question is whether particular tariff rates should be
above or below x per cent to allow for the foreign capital
aspects. One must then consider some of the factors discussed
in the previous section. The principles are so familiar that one
can be very brief.

If external economies are associated with particular types of
foreign capital, and a tariff would attract this type of capital.
there may be a second-best case for such a tariff above x per
cent. It is clearly second-best since the tariff may also attract
domestic factors into the industry, even though such factor
movement would not yield any externalities. In addition, the
tariff inflicts the usual consumption cost. A similar argument
for tariffs may apply if the benefits from new investments are
principally the pecuniary externalities they generate, an issue
that only arises if foreign capital enters the country in
significantly indivisible lumps.

There may be an argument for reducing certain tariffs
below x per cent in order to impose the orthodox optimum tax
on foreign capital indirectly. This is a formal way of expressing
the popular view that by protecting foreign-owned industries
a country is transferring income to foreign capitalists. So why
not reduce the tariff and thus transfer some of that income

away from the foreigner? The usual answer is that the tariff reduction may endanger the survival of the industry or discourage its expansion. The orthodox optimum tax, for which the tariff reduction is second-best, balances these two considerations.

Finally, there is the case where a foreign-owned industry earns monopoly or excess profits. This is normally connected with a characteristic of the product market and is not associated in particular with foreign capital. Thus, if this provides an argument for reducing the tariff below the standard x per cent level it would appear on first sight to come under the heading of general arguments for tariffs quite independent of the foreign capital complication. But the community may not be concerned about monopoly profits when these profits go to domestic capitalists. On the other hand, it clearly is in the national interest that foreign profits be 'squeezed' as much as possible without affecting the supply of foreign capital. (If the supply is affected this issue merges into the orthodox optimum tax issue.) Thus the argument for tariff reduction to squeeze excess profits is probably stronger when the profits go to foreigners.

IV

THE EFFECTS OF A GIVEN TARIFF SYSTEM
ON THE GAINS OR LOSSES FROM
FOREIGN INVESTMENT

Let us now ask whether foreign capital inflow should be subsidized, taxed or restricted more or less than otherwise because of a country's tariffs. We hold tariffs constant and regard foreign investment taxes (or controls) as the variables.

It will be assumed that to begin with taxes on foreign investment at a rate of y per cent are imposed to take into account the variety of considerations discussed in Section II. This optimum y per cent may in fact not be a single rate but rather a structure of rates; furthermore, it may be a subsidy or contain elements of subsidy.

It follows immediately that if the existing tariff structure is an optimum one, the foreign capital inflow it will bring forth will also be optimum when it is combined with the y per cent rate on foreign investment. But if the given tariff structure is not optimum and cannot be altered as readily as controls or taxes could be imposed on foreign investment, then there may be a case for second-best taxes, subsidies or controls on foreign capital which would do somewhat inefficiently the job that a change in the tariff structure would do better.

The country may have built into its tariff system protection above optimum for some industries, and yet the relevant tariffs cannot be reduced. The internal price system will then have been distorted and send up wrong signals to investors. A restriction of foreign capital inflow into the over-protected industries (combined with subsidization of inflow into other industries) would then be a second-best optimal policy. It would restrict only foreign and not domestic capital inflow to these industries even though the wrong signals may have influenced both types of capital.

If foreign capital tends to flow into over-protected (above optimum-protected) industries, the gains from the inflow are less than they would otherwise be, and there may indeed be a social loss. The loss is indicated by the fact that the tariff (or the excess tariff) is indirectly subsidizing this capital, even though there is no case on external economies or other grounds for such subsidization.[13] This type of case is probably realistic. In certain cases the gain to a country from foreign capital inflow may well be zero or there may even be a loss because the capital flows into highly protected industries.

V
FOREIGN INVESTMENT AND
THE TERMS OF TRADE

It has already been noted that there is a complete identity between the theory of the orthodox optimum tariff, designed to improve the terms of trade, and the theory of the optimum

[13] Figure 12.1 can be used to illustrate this point. The social marginal product of capital is traced out by a curve which is below the DD′ curve (just as the dd′ curve in the diagram is above it). Tariffs artificially raise the private product curve to DD′, this being the demand curve for capital facing the foreign investors. The effect is the same as if there were net external diseconomies. Given the tariffs, a second-best foreign investment tax policy then requires a compensating tax or control on foreign investment.

tax on foreign investment, designed to squeeze foreign capitalists so as to improve the terms of foreign borrowing. It would thus seem that a country pursuing a first-best policy from a national point of view should impose the orthodox optimum tariff structure so as to obtain desired terms of trade effects and, at the same time, impose the orthodox optimum investment tax structure so as to obtain the desired effects on the terms of foreign borrowing.

Broadly this is indeed correct, though there is more to be said, since tariffs can indirectly affect the terms of foreign borrowing and an investment tax can affect the terms of trade. Furthermore, one must consider second-best situations, where the only instruments of policy are either tariffs or foreign investment taxes.

These relationships have been rigorously explored in articles by Kemp, Jones, and Gehrels,[14] and the following brief discussion, which is wholly intuitive and makes no attempt at rigour or formal proofs, rests on these articles, especially the papers by Jones and Gehrels. These authors employ the standard two-good two-country model, with immobility of labour between countries but mobility of capital. The first two authors use the Heckscher-Ohlin model, with its two factors of production, but Gehrels usefully introduces a third factor so as to avoid some rather unrealistic complications that are shown by Jones to arise in the two-factor model.

(1) *First-Best Policy*

Consider first the simplest case. (1) We have a first-best situation, with both tariffs and investment taxes available as instruments of policy. (2) The home country that imposes the tariffs and taxes faces given demand and supply curves in foreign trade and of capital that are quite independent of each other. The fact that they are given means that retaliation or foreign repercussions of the type discussed in Chapter 7 are

---

[14] M. C. Kemp, 'The Gains from International Trade and Investment: A Neo-Heckscher-Ohlin Approach', *American Economic Review*, 56, September 1966, pp. 788–809; R. W. Jones, 'International Capital Movements and the Theory of Tariffs and Trade', *Quarterly Journal of Economics*, 81, February 1967, pp. 1–38; Franz Gehrels, 'Optimal Restrictions on Foreign Trade and Investment', *American Economic Review*, 59, March 1971, pp. 147–159.

ruled out. Finally, just to focus on the matter in hand, we assume (3) that there are no other complications to be taken into account such as internal income distribution effects and domestic divergences.

The assumption that the foreign demand and supply curves in trade and capital are independent of each other means that any relationship abroad between prices of traded goods and the cost of capital is ruled out. In the two-good two-country model this result will come about if one assumes that the foreign country is completely specialized in the production of one product (a case that Jones has explored). If one is thinking of a multi-country multi-commodity world the assumption is not unreasonable when the home country mainly trades with one set of countries but mainly obtains its capital inflow from another set.

In this simple case the country should indeed, from a national point of view, impose the orthodox optimum tariff and export tax structure as indicated by the standard formulae and at the same time impose the orthodox optimum investment tax structure, as indicated by the same formulae.

The rates of tariff, export tax, and investment tax that emerge will depend on the relevant elasticities, and since these may change along the relevant foreign demand and supply curves, the two optimum structures will indeed be related through the domestic link between the amount of capital in the country and the domestic demand and supply curves of traded goods. The relationship is of the same kind as that which exists between the orthodox optimum tariffs and export taxes on various goods, as explained in Chapter 7.[15] But the main point here is that the standard formulae, quite unmodified, apply. The country is buying and selling goods and buying and selling capital; the argument is simply that with respect to both goods and factor trade it is in its interests to exploit its monopsony and monopoly powers.

Now allow the prices of traded goods in the foreign country and the cost of capital there to be related. Hence we remove assumption (2). For exposition let us confine ourselves to the two-country, two-commodity, two-factor case. Suppose that the foreign country exports capital to the home country, and

[15] Above, pp. 168-9.

that, in addition, it exports its capital-intensive product. If the home country imposes a tariff this will reduce the foreign country's exports and hence shift its pattern of output away from its capital-intensive towards its labour-intensive product. This, in turn, will reduce its domestic demand for capital and will lower the price of capital. Since it exports capital, this will worsen its terms of lending and so improve the terms of borrowing of the home country. It follows that the tariff has not only improved the home country's terms of trade but has also improved its terms of borrowing.

Similarly, if the home country taxes the import of capital, this will force down the price of capital in the foreign country, hence cheapen production of the capital-intensive product, which happens in this case to be the foreign country's export product, and so improve the terms of trade of the home country. It follows that a tax on capital inflow not only improves the home country's terms of foreign borrowing, but also improves its terms of trade.

Taking both effects together, we see then that the trade tax benefits not only the terms of trade but also the terms of borrowing while the investment tax benefits not only the latter but also the terms of trade. The three authors referred to above have shown that when there is this sort of relationship the first-best optimum rate of tariff will be higher than given by the usual formula for the orthodox optimum tariff, and similarly the first-best optimum tax on capital inflow will be higher than that given by the usual formula which ignores these effects. They give the appropriate modifications to the formulae.

The question naturally arises how significant all this is. In practice the relevant elasticities are hardly known anyway, so the formulae are useful mainly as directing attention to some relevant considerations. But the particular case considered here does seem of some interest.

The foreign country is assumed to export a product which is intensive in the capital that it also exports. If we think in sector-specific terms, a foreign country may both export 'chemical-capital' designed to establish a local chemical industry, and export chemical products. The point here is that a tariff on chemical products by the host country may induce more 'chemical' capital exports to it by making production in the foreign capital-supplying country less attractive, and so lower the cost of capital while a tax on capital inflow may keep more capital in the foreign country, and so increase output there and hence lower the price of its exports.

The foreign country may be a capital exporter but export its labour-intensive product, though, since it is presumably well-endowed with capital, this would be contrary to the simple Heckscher-Ohlin model. A tariff will then worsen the terms of foreign borrowing and an investment tax will worsen the terms of trade. The optimum rates of tariff and tax should then be lower than they would be in the absence of these effects, and indeed, if the indirect effects are great enough, the optimum tariff or the optimum investment tax (though never both at the same time) might even be negative.

(2) *Second-best Policies*

Now let us turn to the two second-best cases. We shall ignore the foreign interrelationship discussed above.[16]

One of the second-best cases has already been noted. The only policy instrument available may be the tariff. It should then be manipulated to squeeze foreign capital appropriately. In addition it will affect the terms of trade, and this must be taken into account. The second-best optimum tariff should then be higher or lower than the orthodox optimum tariff rate depending on whether the tariff fosters or discourages capital imports or exports. If the country is a capital importer and if it imports the capital-intensive product a tariff will actually increase the returns to foreign capital domestically. Since it is desired to squeeze the return to foreign capital, in this case the tariff should be reduced below its orthodox optimum level, and its second-best optimum level might even be negative.

The other second-best case, where the only available instrument is the investment tax, is completely symmetrical with the previous case. The point now is that the second-best optimum investment tax must take into account terms of trade effects, the aim being to bias the investment pattern

---

[16] See Jones, op. cit., pp. 20–21, 29–31, where he examines the second-best cases in which the foreign country is completely specialized.

away from the exportable towards the importable, or, more realistically, with many exports, away from those exportables where foreign demand elasticities are especially low and where domestic production is not monopolized.

In the simple two-good model, if the country is a capital-importer and imports the capital-intensive product, a tax on foreign capital inflow will make capital more scarce in the country, so raise the domestic cost of the import-competing product, hence increase imports and so worsen the terms of trade. It follows that in this case the second-best optimum investment tax should be below its orthodox optimum level, and it might even be negative.[17]

VI
MULTINATIONAL CORPORATIONS

One of the striking features of international trade and investment in recent years has been the expansion of the multinational corporations. It has been estimated that in 1965 about half of all United States exports of manufactured goods involved trade by such corporations, whether as buyers or sellers, and about half of this was trade *within* corporations, that is parent-to-subsidiary sales.[18] The multinational corporations completely dominate certain industries, such as oil refining, motor-car production and assembly, computers, and chemicals, but their influence goes well beyond these industries. The explanation of their spread must be similar to the traditional explanations of the growth of large companies within countries: namely in terms of the

---

[17] One could introduce the foreign interrelationships into the second-best cases. Consider the simple case where the home country imports capital and the capital-intensive product, and the tariff is the only policy instrument. The domestic effects of the tariff are then to worsen the terms of foreign borrowing but the foreign effects are to improve them. In Gehrels' formal model (Gehrels, op. cit., pp. 153–155) the foreign effect comes out stronger, so that the second-best optimum tariff should be higher than the orthodox optimum tariff when the latter is defined as ignoring the foreign investment repercussion. But it is not clear that Gehrels' result can be generalised beyond his formal model.

[18] M. T. Bradshaw, 'U.S. Exports to Foreign Affiliates of U.S. Firms', *Survey of Current Business*, 49, May 1969, pp. 37.

motives for horizontal and for vertical integration.[19] In addition, the improvement of international communciations and the decline of nationalism in Europe have also played a part. The question then arises whether the analysis usually used by international trade theorists continues to be appropriate when a significant part of world trade is carried on by multinational corporations. It may be that the theories are still appropriate for that large part of trade which is *not* carried on by the corporations, but for multinational corporation trade it might be necessary to adjust established theories, possibly radically so. This is certainly often asserted. How much of the analysis presented in this chapter, and earlier in this book, needs to be amended to allow for multinational corporations? There are three main aspects to consider.

(1) *Monopoly and Oligopoly*

The multinational corporations usually operate in oligopolistic markets. Certainly one motive for international horizontal integration is to increase the degree of market control. There are only very few cases where the situation comes close to complete international monopoly, but in some cases a particular country faces a dominant supplier, or perhaps a pair of dominant suppliers, both exporting to it and producing domestically. There may still be an element of competition, for sometimes the toughest competitors are oligopolists, but standard theories are unlikely to be appropriate when they assume perfect competition and that foreign demand and supply curves facing a country are infinitely elastic.

For protection theory and policy this has two implications. Firstly, tariffs and import restrictions can sometimes alter the degree of monopoly in a domestic market. This was discussed in Chapter 8.

---

[19] See Richard E. Caves, 'International Corporations: The Industrial Economics of Foreign Investment', *Economica*, 38, February 1971, pp. 1–27 (reprinted in Dunning, *International Investment*); and also Stephen Hymer, 'The Efficiency (Contradictions) of Multinational Corporations', *American Economic Review*, 60, May 1970, pp. 441–8. There is a large literature on multinational corporations, and the most comprehensive and informative discussion is in Raymond Vernon, *Sovereignty at Bay. The Multinational Spread of U.S. Enterprises*, Basic Books, New York, 1971.

Secondly, since one cannot always take import prices as given a country cannot always 'squeeze' a domestic monopolist by reducing a tariff to a made-to-measure level. The domestic monopolist may be a subsidiary of an international corporation, and the principal source of potential imports may be another subsidiary of the same corporation. The domestic price will then not be set by the price of potential imports plus the tariff but rather by whatever would determine the monopolist's price in a closed economy (presumably the principles of price discrimination).

The role of the tariff may be only to force the corporation to stop importing and produce domestically. If it is desired to squeeze the corporation this must be done through an appropriate adjustment of profits tax or of any direct or indirect subsidies that are provided to it.

Yet this argument should not be carried too far. There are few industries where there is a complete international monopoly or where a group of multinational firms behaves like a monopolist. Usually there are alternative suppliers. A tariff may attract multinational corporation X into the country, but multinational corporation Y will remain available to supply the product from abroad at a given price, and corporation X can indeed be kept in line by adjusting the tariff which protects it from corporation Y's product. The applicability of 'made-to-measure' tariffs does not depend on *perfect* competition abroad, but only on some reasonable degree of competition.

How would one analyse the gains or losses to a potential host country which uses a tariff to encourage a multinational corporation to cease exporting a product from overseas and to establish domestic production? Let us consider the special and rather extreme case where this corporation is the only actual and potential foreign supplier and the only potential domestic producer. In the absence of a tariff it produces abroad (in its home country) and exports to the potential host country. This maximizes its profits. A tariff is then imposed which would make production in the host country more profitable than exporting from abroad. The tariff situation may yield the corporation good profits from its subsidiary in the host country but its world-wide profits will fall, for otherwise it would have set up in the country without the tariff. This means that domestic host-country costs of production must be higher than its home-country costs (including costs of transport). If this is true of marginal costs, and not just fixed costs, it will then charge a higher price to domestic host-country consumers than it did before the tariff.

Hence, while some of the burden of the tariff will be borne by the company itself in lower world-wide profits, part of it will be borne by consumers in the host country. This consumer cost might be described as the gross cost of protection to the country. But there may, of course, be gains to set against these, notably profits tax collected from the company and pecuniary externalities owing to an increase in wage-rates induced by the extra demand for labour.

It seems that, if there is a net gain, a tariff may even be first-best in this rather special case. A direct subsidy to the domestic subsidiary of the corporation would certainly avoid the consumption-distortion cost, and would make it possible for the adverse effects to be borne by people other than the particular consumers of the product. But it would increase the corporation's international profits, and so involve some element of subsidization of its profits by the government of the host country. The attraction of a tariff to the country is that it squeezes the overseas activities of the corporation by making them less profitable, this providing the inducement for the corporation to produce domestically. There is an element of the orthodox optimum tariff here, though that concept is strictly applicable only in a perfectly competitive situation.

(2) *Horizontal Integration and the Import of Human Capital*

When a multinational corporation sets up a subsidiary in a country it may well finance itself with imported capital, possibly from its own accumulated resources or from the financial markets of its home country. But it is also possible that it obtains part or all of the capital locally. The import of financial capital is not the essence of a multinational's entry into the country. The essence is that its domestic production will become part of a world-wide organization. If it were just a matter of capital movements one would expect producers of domestic origin to borrow capital from abroad, possibly via the domestic capital market, but to keep control of domestic production facilities.

Apart from financial capital, what the corporation brings in is *human capital*—managerial and technical know-how and experience, including knowledge of marketing methods and contacts with export markets. For various reasons it is often more convenient to import this special type of capital into a country by setting up subsidiaries or branches, or taking over a local company, rather than by selling it through the market. Indeed some of it is hardly marketable. Potential buyers require knowledge itself in order to appreciate its potential value and in order to price it correctly.

It is well-known that one motive for firms' horizontal expansion and integration within a country is to benefit from economies of scale. We have the same consideration here. There are economies of scale in knowledge in the sense that the marginal cost of supplying knowledge to additional beneficiaries is much less than the average cost. Indeed, once knowledge, experience, and so on, have been accumulated the marginal cost of spreading these is practically zero, or at least very low. So this type of capital might be described as a public good in Samuelson's sense.[20] More knowledge for one subsidiary does not mean less knowledge for another.[21]

The question then arises whether our preceding analysis is applicable when the corporation imports human capital or knowledge, rather than financial capital. In practice, as pointed out earlier, it is likely to import a sector-specific package consisting of both. But the analysis appears to be fully applicable, even though some difficulties in measurement of capital and diagrammatic representation might arise. The

---

[20] P. A. Samuelson, 'The Pure Theory of Public Expenditure', *Review of Economics and Statistics*, 36, November 1954, pp. 387–89.

[21] Johnson has focused on this aspect of the multinational corporation, suggesting that it is primarily an instrument for spreading a special kind of *public good*, namely 'knowledge', meaning managerial and technical know-how and so on. See H. G. Johnson, 'The Efficiency and Welfare Implications of the International Corporation', in C. P. Kindleberger (ed.), *The International Corporation*, M.I.T. Press, Cambridge, Mass., 1970 (reprinted in John H. Dunning (ed.), *International Investment*, Penguin Modern Economics Readings, 1972).

fact that the marginal cost of the imported 'capital' to the corporation may be very low or even zero, at least up to a point, is not really of great relevance to the host country, other than on its bearing on the slope of the supply curve of this capital. In principle there are marginal private and marginal social product curves, and there is a supply curve—which may indeed be very steep over a range, so that there is much opportunity for squeezing the corporation's profits. If the knowledge spreads outside the corporation without being charged for there will be externalities that may justify some subsidization, and that have to be set against reasons for taxing. But none of this is new.

(3) *Vertical Integration and the Evasion of Taxes or Tariffs*

The domestic subsidiary of the multinational corporation is likely to import components and other inputs from subsidiaries of the same corporation in other countries. Furthermore, it may export its products to these other subsidiaries. The products of some multinational corporations are made up of components that come from many countries, but all produced within some part of the corporation.

The benefits from such vertical integration across frontiers are obvious. By by-passing the market, the costs of the market are avoided, notably the costs of bilateral bargaining. Of course they are replaced by the costs of coordination within a corporation. Vertical integration reduces the risks of suppliers failing to supply key components at adequate quantity and quality and of markets for components suddenly disappearing. It is also a way of preventing potential competitors obtaining essential supplies or marketing outlets. Yet, while avoiding trade through a market, the corporation can make use of the international division of labour and can reap the gains from international trade.

What are the implications of such international vertical integration for tax and tariff policies? The trade between subsidiaries or branches of an international corporation certainly takes place across frontiers and is recorded in trade statistics. Yet, since it does not pass through the market the prices recorded for tax and customs purposes may not be free-market, or so-called 'arms' length', prices. Rather, they may be chosen deliberately to minimise tax or tariff burdens. This is the problem of transfer prices.[22]

In practice the corporations may be limited in their ability to manipulate recorded or transfer prices to minimise taxes. Market prices for many goods are roughly known, and in countries with reasonably efficient tax and customs authorities it may not be possible for the corporations to record prices in international trade that are too far out of line with free market prices or marginal costs of production. Furthermore, the subsidiaries of many corporations are quite independent-minded, and a subsidiary is likely to oppose non-market pricing if this will reduce its recorded profits. Nevertheless, disregarding such constraints, let us examine the incentives for tax or tariff evasion through manipulation of import or export values.

The issue is well-known in the case of income or profits taxes. If the rate of tax payable in Australia is higher than that payable on profits earned in Britain, then a corporation that operates in both countries and sends components produced by its British subsidiary to its Australian subsidiary will have an incentive to over-price the components, so as to shift profits from Australia to Britain.

The Australian Treasury may find that profits recorded as earned in Australia will decline, perhaps even drastically. This may be so even though there is no change in the operation of the company, and certainly no contraction of its capital stock in Australia. A rise in the Australian tax rate may then lead to a fall in revenue from taxes on these corporations. Hence it may pay Australia to avoid fixing its tax rates much above the rates applying in the parent countries of the corporations. Furthermore, we have seen earlier that it will certainly not be in its interests to charge lower tax rates, so in this case it should fix its rates fairly close to the rates in the corporations' parent countries.

If export taxes are fixed on an *ad valorem* basis, a multinational corporation can reduce their incidence in the same way. If it undervalues its exports, less export tax need be

---

[22] See Vernon, *Sovereignty at Bay*, pp. 137–140; and Thomas Horst, 'The Theory of the Multinational Firm: Optimal Behavior under Different Tariff and Tax Rates', *Journal of Political Economy*, 79, September 1971, pp. 1059–72.

13

paid and recorded profits will be shifted out of the country. This will only be profitable for the corporation if the combination of export tax and profits tax payable domestically (together with any tariff payable in the country of destination) exceeds profits tax payable in the country of destination.

An incentive to over-price components imported from another subsidiary or from the parent company will also arise if there is some possibility that the corporation will not be permitted to remit part or all of its after-tax profits earned in Australia. This has never been a possibility in Australia, but it explains the inducement to over-price in some less-developed countries, such as Colombia.[23] This inducement arises even when the tax-rate in Colombia is the same as in the corporation's parent country. Over-pricing is simply a way of remitting profits and so by-passing exchange control. From the point of view of the corporation, the profits which it is permitted to retain after tax but is not permitted to remit are likely to have a value lower than their nominal money value, so that an exchange control restriction is the equivalent of a tax. Hence, even if the nominal rate of profits tax in Colombia is the same as that in the United States, when exchange control is taken into account the implicit tax rate is higher, and it is this which provides the inducement for overstatement of import values.

It is particularly relevant here to consider the possible evasion of tariffs. A corporation can reduce the incidence of *ad valorem* tariffs by *under*-valuing imported components. But it is not sufficient to look at the amount of customs revenue that the corporation avoids paying in order to assess its gains from the undervaluation. Firstly, some part of the increase in profits that results from the tariff-saving by the corporation will be taxed away. Secondly, recorded profits will be shifted away from the supplying country (Britain) towards the local subsidiary (in Australia). If the rates of

[23] The transfer pricing mechanism has been investigated in detail by Colombian government agencies, especially as it affects pharmaceutical products, and evidence of marked over-pricing has been found (for pharmaceuticals, 87 per cent on average for the period 1966–70). See S. Lall, 'Determinants and Implications of Transfer Pricing by International Firms', *Bulletin of the Oxford Institute of Economics and Statistics*, August 1973; also Vernon, *Sovereignty at Bay*, p. 139.

profits tax payable on earnings in Britain and Australia are the same then this second consideration will have no effect on the profits of the corporation. But if the tax rate is higher in Australia than in Britain there will be a net loss on this account, and this must be set against the gain from saving tariff payments.

Let the true market value of the imports be M, the proportion by which they are undervalued π, the tariff rate t, the rate of tax on profits earned in Australia r_a, and the rate payable on profits earned in Britain r_b. Then the net gain to the company from under-valuing imports is

$$\pi M[t(1-r_a)-(r_a-r_b)],$$

$\pi M \cdot t(1-r_a)$ is the gain from saving tariff payments, allowing for the Australian tax to be paid on this gain, while $\pi M(r_a-r_b)$ is the loss from shifting tax payments from Britain to Australia. The net gain could be negative if r_a were greater than r_b, though on plausible figures (say, $t = 25$ per cent or more, $r_a = 50$ per cent and $r_b = 40$ per cent) it would be positive. With such figures it is in the interests of the company to undervalue the imports as much as possible. If the net gain were negative (i.e. if (r_a-r_b) were greater than $t(1-r_a)$) the firm would have an incentive to overvalue.

If the aim of the tariff is to protect domestic producers of the components a government can counter such evasion of the tariff (through undervaluation) by not accepting the prices quoted by the corporation but rather fixing components prices for duty purposes which are closer to market prices. Alternatively it might impose anti-dumping duties, it might convert the ad valorem tariffs into specific tariffs, or it might replace the tariffs with quantitative import restrictions.

But if the aim is to raise revenue then the tariff-imposing country is actually likely to gain from the understatement of import values induced by an ad valorem tariff. Its loss of customs revenue is likely to be more than offset by a gain in profits taxes. The gain in revenue to the country consists of

$$\pi M[r_a-t(1-r_a)]$$

and this is quite likely to be positive; for example, if $r_a = 50$

per cent it will be positive provided the tariff rate, t, is less than 100 per cent.

It is not the understatement of import values which should concern a country—since this is quite likely to lead to an increase in tax revenue even though customs revenue falls—but rather the overstatement of import values designed to shift profits from a high-tax importing to a low-tax exporting country. But this discussion suggests quite a simple remedy for overstatement, apart from the one already mentioned of equalizing profits taxes in the two countries. The simple remedy is to impose tariffs on those imported components believed to be *over*-priced, and hence to encourage the corporation to *under*-price them instead.

The difficulties in this proposal are two. Firstly, a government will not know in advance whether such tariffs will genuinely cut into the corporation's profits and so possibly make the operation of its local subsidiary unprofitable—as no doubt the corporation will claim—or whether it will simply lead to a change in transfer prices involving a shift in the location of profits, with some net increase in profits tax paid. Secondly, there may be local firms importing these components on the open market, and inevitably these firms would have to pay tariffs on their imports also; and this might provide undesired negative effective protection for these local firms as well as undesired protection for local components producers.

13

THE OPTIMUM TARIFF STRUCTURE AND EFFECTIVE PROTECTION

IT MAY SEEM surprising that so far the concept of effective protection—which was the main theme of *The Theory of Protection*—has played little role in this book. The theorizing in the earlier book was concerned with *positive* analysis—with studying how systems of protection affect a country's resource allocation and consumption patterns. Such positive analysis is certainly an essential foundation for normative analysis. But the great interest in many countries in effective protection, and the efforts that have gone into calculations, can only be explained by the belief that measurements also have *normative* significance. In fact, of course, effective protection does have some role in normative analysis, and this can be seen when we consider the optimal structure of a protective system, to which we now turn.

We have already discussed the optimal structure when raising revenue with minimum-distortion cost is the aim: tariff and export tax rates should be relatively higher on goods with low domestic demand and supply elasticities. We have also discussed the orthodox optimum trade tax structure: in this case trade taxes should be relatively higher when the foreign elasticities are low. It follows that if the aim is both to raise revenue and improve the terms of trade both domestic and foreign elasticities come into play, tax rates being higher when domestic and foreign elasticities are relatively lower. In the remainder of this chapter we ignore these considerations by assuming that revenue-raising is not an aim and foreign prices are given.

We assume now that the protective system is intended to alter domestic output patterns. Furthermore, tariffs, perhaps supplemented by export taxes or subsidies, are the only available instruments of policy. We shall also be concerned with situations where some trade taxes are constrained at non-optimal levels, at least in the short-run, the problem being to determine the second-best optimum for the remainder.

The problem is thus one of second-best optimization, since the reason for intervention is either a domestic divergence or the fact that some trade taxes are constrained. It is in these second-best optimization situations, and when we are concerned not just with partial equilibrium or with a two-sector model, but with proportions between trade taxes on many goods, that effective protection comes into its own. The many limitations of the effective protection concept, as expounded in detail in *The Theory of Protection*, should, of course, be kept in mind here.[1]

I

TARIFF RIGIDITY AND TARIFF DISMANTLING

Let us imagine a situation where free trade for a country would be first-best. Various domestic divergences have been offset by appropriate non-trade interventions. The exchange rate is adjusted so as to maintain external balance while the general level of employment is maintained by fiscal and monetary policies. In spite of the optimality of free trade the country is assumed to have a complex structure of tariffs imposed in the past for a variety of reasons. Possibly there were once valid arguments for protection, but times have changed. First-best policy thus requires the elimination of all tariffs and other trade interventions, making sure at the same time that the exchange rate is adjusted appropriately to maintain external equilibrium and that employment is maintained by fiscal and monetary policies. To make this more realistic we might assume that orthodox optimum export taxes have been imposed and define a 'free trade' policy as one which dismantles all trade interventions by the country concerned other than these export taxes.

[1] There is an extensive literature, much of it somewhat mathematical, in the general area of second-best multi-commodity tax and tariff theory. A consolidation and survey, with full bibliography, is P. J. Lloyd, 'A More General Theory of Price Distortions in Closed and Open Economies', A.N.U., Canberra (to be published). The discussion in the present chapter lays no claim to real rigour but seeks to bring out the main considerations.

If one were just interested in defining the characteristics of the first-best optimum no elaborate analysis would then be necessary. An economic adviser could recommend free trade modified by some export taxes. Furthermore, one would not need to calculate effective protective rates as a guide for policy. Such calculations would be of interest only if one wanted to calculate the cost of protection—that is, the cost of failing to move to the optimum. Only when one introduces some policy constraints does elaborate analysis become necessary and effective protective rate calculations become relevant.

(1) *Some Tariffs not Alterable*

One type of policy constraint is that some tariffs cannot be altered for institutional or political reasons, at least not in the short-run. We have then a typical second-best problem. Given that a part of the protective structure cannot be altered, what is the optimum structure for the remaining part? It is common that effective protection for existing industries, or for products actually being produced in the country, cannot be reduced, but there is freedom in choosing tariffs for new industries or for products that may be produced in the country for the first time. The country's export industries may be receiving very little protection, if any, and, for fiscal or income distribution reasons, this may also be unalterable.

The second-best optimization problem can now be explained with a simple example. Assume first that there are just three products and three industries, all vertically integrated. Established industry E has a given tariff of 50 per cent and export industry X has zero protection, and this is also given. The tariff for the new industry Q is to be determined. The simple answer in that case is that the second-best optimum tariff for Q is likely to be between 50 per cent and zero; the relative substitutability in production and consumption for E and X will determine whether the second-best optimum tariff will be closer to the higher or the lower limit. If Q is a very close consumption substitute for E, and if, on the production side, it competes closely with E for similar factors, the tariff will be close to 50 per cent. But as long as there is some substitutability on the side of either production or consumption relative to X the tariff should be at least somewhat less than 50 per cent, a uniform tariff not being appropriate.

The general nature of the argument is the same as that developed in Chapter 2 in connection with the second-best optimum tariff, based on the method of *Trade and Welfare*. Let us imagine that the tariff on $Q(t_Q)$ is initially at zero and is gradually raised towards equality with the 50 per cent tariff on E. The marginal distortion cost (consumption plus production cost) *relative to X* will then gradually increase, beginning with a zero cost when t_Q is zero. At the same time the distortion *relative to E* will be reduced until it disappears when t_Q has become 50 per cent; hence the marginal gain owing to 'distortion-reversal' relative to E gradually declines. Subtracting the increasing marginal distortion cost (relative to X) from the declining marginal distortion-reversal gain (relative to E) we obtain the net marginal gain which will be zero at some point between $t_Q = 0$ and $t_Q = 50$ per cent. The total gain from imposing a tariff on Q will be maximised when this net marginal gain is zero.[2]

If it were possible to subsidize exports so that the rate of protection for X were not given even though the 50 per cent tariff on E is given, then the first-best solution would be attainable. A 50 per cent tariff on Q combined with a 50 per cent export subsidy for X would attain it. But if E produced various products all with different tariffs, and all unalterable, we would be back in a second-best world. What is clear in this example is that one needs to know the heights of the various protective rates that are *given* to the policy-makers in order to determine the second-best optimum rates for the remaining industries. In addition—and this presents the greatest difficulty in practice—one needs to know the relevant substitution relationships.

The next step is to allow for input-output relationships and so distinguish effective from nominal rates. If consumption-distortion effects can be ignored, the aim being to minimise

---

[2] The model is presented mathematically by R. G. Lipsey and K. Lancaster in 'The General Theory of Second-Best', *Review of Economic Studies*, 24, 1956–57, pp. 18–20. They also allow for complementarity relationships, which have been assumed away in our example. If Q were complementary with E the second-best optimum tariff might actually be negative.

the production-distortion cost of protection, one must calculate effective rates for the various industries where protection is given, and express the second-best optimum tariffs for the industries where protection is under consideration in terms of effective rates. Broadly, if there are existing high effective tariffs for some import-competing industries, and there is little or no protection for exporting industries, the second-best effective rates for the industries or products under consideration will be positive but less than the existing given tariffs. In the absence of precise knowledge of substitution relationships between industries one cannot give a precise answer. But one has more chance of getting close to the right answer if one knows existing and potential effective rates, and not just nominal rates.

When consumption effects are introduced, nominal rates come into their own again. For the new industries there will be a second-best set of effective rates from the point of view of production effects and a second-best set of nominal rates from the point of view of consumption effects, each looked at on its own. So there must be some trading-off in the usual way, and the optimum, taking both into account, will be some compromise between the two.

The general conclusion that emerges is that, while free trade might be best if the whole protective structure were open to variation, given that a part of it is constrained, free trade may no longer be optimal for the remainder. This is a conclusion of some practical importance.[3]

(2) Optimal Tariff Dismantling: Concertina Method

Let us now look at a possible tariff dismantling process. We start with a complex structure of tariffs and want eventually to get to free trade. If possible we want to maximise welfare, or at least Pareto-efficiency, on the way. The optimum path will depend on the particular type of constraint.

One possible constraint is the following. All tariffs may be potentially alterable, but at any given time only some can be altered. One could start with reducing any limited group of tariffs, but not with all of them. To simplify, let us focus on

[3] In W. M. Corden, 'Australian Tariff Policy', *Australian Economic Papers*, 6, December 1967, pp. 131–54, this approach is applied.

the production side only. Effective rates for all products or industries must first be calculated. Which effective rates should be reduced first? If the low tariffs were reduced first, distortions relative to the non-protected (export) sector would be lessened, but distortions relative to the high-protected sector would be increased. The production-distortion cost of protection might then actually increase in the short-term. Hence it is important to reduce the high effective tariffs rather than the low effective tariffs first. A useful decision rule may be to reduce, in the first instance, all tariffs above a certain arbitrary level down to this level.

One could construct an optimal path of tariff reduction. The high tariffs are squeezed down to medium level at the first stage, then these and the existing medium tariffs are squeezed down to a lower level, and so on, until eventually free trade has been attained. This can be called the *concertina* method. In determining this optimum path the scale of effective rates is crucial. It would be quite complicated to convert the time-path of effective rates into a time-path of nominal rates. In principle there is likely (but not certain) to be a solution, but it may yield all sorts of paradoxes. For example, at certain stages some nominal rates may need to increase even though effective rates are always decreasing.[4]

(3) *Optimal Tariff Dismantling: Across-the-Board Reduction*

Next, let us consider another type of constraint. As in the previous case, all tariffs are potentially alterable, but this time the constraint is not that they cannot all be altered at the same time but rather that all alterations must be gradual rather than abrupt. The aim is to allow protected industries time for readjustment.

The main point is now that there is one simple pattern of tariff reduction over time which ensures resource allocation

---

[4] This has been concerned only with the production side. Since nominal tariffs are relevant for consumption effects, the matter is more complicated than indicated here if consumption effects are also important. It should also be noted that the approach assumes that substitution effects dominate complementarity effects. The general argument is developed more formally in Trent J. Bertrand and Jaroslav Vanek, 'The Theory of Tariffs, Taxes and Subsidies: Some Aspects of the Second Best', *American Economic Review*, 61, December 1971, pp. 925–931.

and consumption pattern improvements at every stage. This is *across-the-board tariff reduction*.[5] Every year all tariffs are reduced by an equal percentage, say 10 per cent, and when some low floor, say 5 per cent, has been broken through, the tariffs are reduced to zero. The method has been used in bringing about the gradual elimination of industrial tariffs within the E.E.C., though there were qualifications, and adverse trade diversion effects resulted because the tariff reductions discriminated in favour of fellow E.E.C. members. The outcome of the Kennedy round of international tariff negotiations was also to bring about (broadly) a staged across-the-board tariff reduction for industrial products.

The attractive feature of the across-the-board approach is that it does not require knowledge of substitution relationships to determine the appropriate pattern of tariff reduction. Suppose we start with the unprotected export industry, X, and two import-competing products, with tariffs of 30 per cent and 50 per cent respectively. Resource allocation and consumption distortions result from the three *gaps* or divergences between tariff rates, namely 50 per cent between X and the third product, 30 per cent between X and the second product, and 20 per cent between the two importables (15.4 per cent when measured in relation to the domestic price of the second product). When all the gaps are reduced there must be an improvement, and the more they are reduced, the greater the improvement. Any uniform percentage reduction in the two tariff rates will reduce all three gaps, and the greater the reduction the more the gaps are squeezed.

If the reductions are not across-the-board, an improvement is not certain. Thus, if the 30 per cent tariff is reduced to 15 per cent but the 50 per cent tariff to only 45 per cent, two gaps are indeed reduced, but one is increased. Furthermore, across-the-board reduction is not the only way of ensuring a gain at every stage of the tariff reduction process. Any method that reduces all three gaps, or at least reduces some gaps without raising any others, ensures an improvement.

---

[5] See Michael Bruno, 'Market Distortions and Gradual Reform', *Review of Economic Studies*, 39, July 1972, pp. 373–83. The main propositions below are proven rigorously in this paper. He noted that across-the-board tariff reduction plans were being applied in Israel.

If all nominal rates are reduced by an equal percentage, then all effective rates will fall by the same percentage. This can easily be verified from the effective protection formula. It follows immediately that, assuming no export taxes or subsidies, across-the-board *nominal* tariff reduction must lead to across-the-board *effective* tariff reduction, and the preceding argument can be extended to a model where the tariff rates concerned are effective rates and the products become *activities*. It is not necessary to calculate effective rates to ensure that the tariff reduction is across-the-board in terms of effective rates. But it could be readily shown that this conclusion would need to be modified if export subsidies, or export taxes above orthodox optimum levels, are important and are not included in the across-the-board reductions, or if some tariffs are excluded from the tariff reduction process.

(4) *Tariff Dismantling: The Exchange Rate*

It would be naive to suggest that these simple approaches provide sufficient guidelines for a country wishing to make a transition to free trade or to an optimal or a preferred system of protection. The adverse effects of trade tax reductions on government revenue, and the need to find alternative sources, must be borne in mind. Furthermore, income distribution considerations, and especially the requirements of the conservative social welfare function, are likely to be crucial. Changes need to be gradual, and, if radical, to take place in an environment of high employment, high growth rates or improving terms of trade so as to avoid substantial falls in real incomes to any significant section of the population.

One might apply an unusual cost-benefit approach to any proposed change. The benefit is the resource allocation (Pareto-efficiency) gain—which consists of the gains of the gainers minus the losses of the losers measured in the usual way—and the cost is the absolute fall in real incomes of the damaged parties, measured in some way, possibly as a proportion of the fall in their monetary incomes, allowing for price changes facing them. The lower the relevant supply and demand elasticities, and hence the greater the income distribution relative to the resource allocation effect, the higher the cost-benefit ratio. With zero elasticities there will be only income redistribution without any resource allocation effect, and hence all cost and no benefit.

Tariff reductions may lead to temporary unemployment, and this should also enter the calculus. Any given tariff reduction may then yield a long-term Pareto-efficiency gain and a short-term loss. Gains and losses at various time-periods might then be adjusted to allow for absolute falls in real incomes of losers, and finally a discounted present value of a tariff reduction would be obtained. Making this calculation for various levels of tariff change, a tariff would be reduced until the discounted present value of a change is zero.

It is also necessary for the exchange rate to adjust appropriately to maintain external balance. If protective tariffs are to be gradually removed it would seem then that there should be a steady depreciation during the transition process. But such a continuous depreciation might provoke speculative capital movements. In *Industry and Trade in Some Developing Countries*[6] a practical solution to this problem is suggested which calls only for a once-for-all devaluation and hence would avoid undesired effects on the capital account. It makes use of the fact that a given protective effect can be attained by various combinations of tariffs and export taxes and subsidies, provided only that each set of protective rates is associated with the appropriate exchange rate. This is the 'symmetry of various protective structures'.[7] Let us explain it here, using Figures 13.1a and 13.1b. We assume just one import and one export and given world prices, though the argument can be generalized to a multi-commodity model and is unaffected by terms of trade effects.

Figure 13.1a shows the price and quantity of the import, *DD'* being the import demand curve, and Figure 13.1b similarly refers to the export, *HH'* being the export supply curve. Units are so chosen that at the initial exchange rate the domestic price of both is 100 rupees. There is a 100 per cent

---

[6] Little et al., *Industry and Trade in Some Developing Countries*, pp. 364–367. The whole of Chapter 10 of this book is an excellent discussion of the problem of transition from a non-optimal protection system to a more rational system of industrial promotion.

[7] *The Theory of Protection*, pp. 119–122.

tariff, so that the tariff-inclusive domestic price of the import
is 200 rupees. If the tariff were removed the rupee would have
to be devalued sufficiently to maintain balance of payments
equilibrium. This is brought about by a devaluation that
raises the duty-free price of imports and the price of exports to
133 rupees, imports and exports rising by the same value
(the two shaded areas are equal).

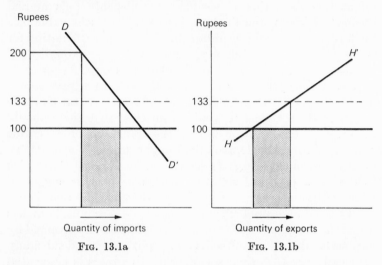

FIG. 13.1a   FIG. 13.1b

The recommendation in *Industry and Trade in Some
Developing Countries* is then that in the first instance this
devaluation, which establishes the free trade equilibrium
exchange rate, should take place, and it should be associated
with (a) a reduction of the previous 100 per cent tariff to a
50 per cent tariff, and (b) the imposition of a 25 per cent export
tax. The 50 per cent tariff will ensure that the domestic price
of imports stays at 200 rupees (since 50 per cent added on to
the new duty-free price of 133 rupees yields the original
domestic price facing producers and consumers of 200 rupees),
and the 25 per cent export tax will similarly ensure that the
domestic price of exports stays at 100 rupees (which is 75 per
cent of the new tax-free price of exports of 133 rupees). Thus a
tariff of 50 per cent combined with an export tax of 25 per cent
is symmetrical with the initial set of a tariff of 100 per cent
combined with a zero export tax.

Subsequently the tariff can be gradually lowered from 50 per cent to zero and at the same time the export tax should be reduced from 25 per cent to zero so that, at every point in time, the increase in exports brought about by the export tax reduction is equal to the increase in imports resulting from the tariff reduction. Gradually imports and exports will increase, as indicated by the two arrows, and the exchange rate can stay constant at its new level.

The practical problem is to judge the exchange rate appropriate to free trade equilibrium. This is particularly difficult when the changes are taking place during a period of general inflation; it may then be desirable to have some devaluation to allow for any divergence between rates of inflation abroad and at home, quite apart from the devaluation required by the transition to free trade.

II
UNIFORM EFFECTIVE PROTECTION

Let us now turn to a different problem. This time the government is free to vary all tariffs and is concerned not with dismantling a tariff system but rather with the construction of an optimal system in a situation where generalized tariff protection would be at least a second-best optimum.

We assume now that it is desired to foster value-added in manufacturing *in general* relative to agriculture and minerals. The reason for such generalized protection may be some kind of infant industry argument applying to the whole of manufacturing industry, or perhaps a non-economic argument, though various other economic arguments, all second-best or worse, can clearly be envisaged. In any case, there may then be some case for uniform effective protection for manufactures. This should ideally be provided for manufactured exports as well as for manufactures sold on the home market.

The general idea has had a good deal of appeal to economists who have accepted some case for protection of manufacturing, or have felt compelled to accept the reality of a government's desire for protection, and have at the same time sought both administrative simplicity and an industrial selection role for the price mechanism. The basic case for some degree of uniformity of protection in such a situation is rather obvious, though there are important qualifications which have been analysed rigorously in *The Theory of Protection*.[8] Here we can be brief; the analysis will be developed further below to take into account made-to-measure tariff-making.

If the argument for protection were truly general, so that there was little point in discriminating between different manufacturing activities, and if subsidy disbursement costs were low and uniform, uniform effective protection for manufacturing provided by subsidies would indeed be first-best. The assumptions are, of course, rather severe. If, in addition, one ignored consumption distortions the desired uniform effective protection could also be brought about by tariffs combined with export subsidies for manufactures.

If one allows for consumption effects and wishes to minimise the by-product consumption distortion effects of tariffs and export subsidies, some departure from effective rate uniformity may be desirable. Further, such departure may be desirable if protected products are inputs into exportables, and especially if only tariffs, and not export subsidies, can be used. It is then particularly important to modify the adverse effects that tariffs on inputs have on exports of manufactures by keeping nominal tariffs on inputs into exportables low. But here let us focus not on the qualifications (which have been explained fully in *The Theory of Protection*) but on the main implications of the basic uniform-effective-protection idea.

The main implication is that it is desirable to calculate the effective rates yielded by the existing protective structure. It can then be seen in which direction adjustments of tariffs

[8] *The Theory of Protection*, Ch. 8. This chapter analyses not only the case for uniform protection when the reason for intervention is a uniform domestic divergence in production, but also when there are terms of trade effects on the export side, or the aim is to improve the balance of payments and the exchange rate is fixed.

It can be argued that the qualifications to the uniform effective protection argument are so important (notably those concerned with consumption-distortion) that there is no case at all for using the uniform tariff approach even as a starting point for policy analysis. The present discussion and Chapter 8 of *The Theory of Protection* expound many of the qualifications as well as arguments in favour; a policy advisor will have to make up his mind in the light of the particular circumstances of the country concerned. The present discussion should be interpreted as analysis, not advocacy.

must be made. The aim might be gradually to reduce the high effective rates and at the same time to raise low rates, until uniformity is attained.

The actual rate of uniform effective protection that should be aimed at can be determined in either of two ways. The first approach is to start with some target size of manufacturing value-added (at free trade prices), and adjust the rate gradually until this target is attained. The second approach requires an estimate to be made of the average external economies to which manufacturing is thought to give rise—that is, the marginal divergence between private and social cost or private and social benefit. This should be expressed as a percentage of value-added at free trade prices. The target uniform rate of effective protection should then be equal to this rate.

In a country dedicated to protection of manufacturing for non-economic reasons, or for unsound economic reasons, a distortion cost of protection is inevitable. A policy of uniform effective protection may then reduce these costs. At the same time it must be realized that, because of the complications created by consumption-distortion effects, and by the effects of tariff protection in raising the domestic prices of inputs into exportable goods, the protective structure that would actually minimize the cost of achieving a given amount of manufacturing value-added would not be precisely uniform. Nevertheless, it may be politically and administratively useful to maintain a simple principle of effective rate uniformity, at least as a reference point. It may not be ideal, and may call for some qualifications in particular cases, but may well be preferable to the alternative of a haphazard tariff structure which, in the absence of simple guide-lines, is readily amenable to influence from pressure groups.

It must also be stressed that effective rate uniformity is not really a prescription for simplicity in tariff-making. Only nominal rate uniformity is really simple. A very complex structure of nominal rates may be required to attain effective rate uniformity for the range of industries to be protected. If there are some imports, say of basic materials, where it is not desired to foster import-replacement by domestic production, then those protected activities which have a high content of these unprotected materials inputs will need relatively low nominal rates if they are to have the same effective protection as other protected activities which use less of the unprotected inputs. Similarly, those protected activities which have exportable inputs will need lower nominal rates than other activities which do not have such inputs.

An additional complication is that changes in input-output coefficients over time call for changes in nominal tariffs so as to maintain the uniform effective tariff. Furthermore, if uniform effective protection is provided only for those activities where there is some actual domestic production, an extension of the range of domestic production will reduce the effective rates of products which use the newly protected product as input; so *their* nominal rates will have to be increased appropriately, leading to yet further nominal rate changes for products that use the latter products as inputs.

For all these reasons any fanaticism about effective rate uniformity should be avoided. There is a lot to be said for some tendency towards nominal rate uniformity. It may sometimes be best to have a small stock of three or four basic nominal rates out of which all effective rates are constructed, the aim being only to avoid excessive non-uniformity in effective rates without any pretence that complete uniformity in effective rates can or should be achieved.

III
THE SELECTION OF INDUSTRIES TO PROTECT AND THE MADE-TO-MEASURE PRINCIPLE

One can have some doubts about providing a uniform rate of protection, whether effective or nominal, on the following grounds. What is the point of giving each industry or activity the same rate of protection when some need much more than others? Will we not just be distributing unnecessary largesse to some of them, impose undue costs on users and consumers of the excessively protected industries and give the latter opportunities to exploit local consumers?

These ideas underlie the *made-to-measure principle* which was introduced in Chapter 8.[9] It was pointed out that this principle underlies thinking about tariff-making and the provision of

---

[9] See above, pp. 219–23.

subsidies and tax concessions in many countries, and helps to explain why tariff structures are often so complicated. It explains why the desire to protect particular industries or products has not generally led to across-the-board tariffs or subsidies but rather to tariffs or subsidies limited rather precisely to the purpose in hand so as to minimise adverse effects on consumers, taxpayers, or other industries.

The question then arises how the preceding analysis is affected when the made-to-measure principle is applied. A uniform effective tariff is a way of selecting products to be produced and hence industries to be established or maintained in the country. Those products that are economical at that tariff will be produced and those that are not will not be produced. If, on the other hand, the made-to-measure principle is applied some other criterion must be used to select industries that are to be protected; once an industry is selected it can then be given its made-to-measure tariff or subsidy.

What criterion is to be used? Let us assume, as before, that there is some general argument for protecting manufacturing industry, there being a uniform rate of marginal divergence between private and social costs applying to all of them (say 20 per cent), or alternatively some non-economic case for protection measured by a 20 per cent margin. We shall ignore consumption effects of tariffs as well as the possibility of exporting manufactures, and hence are setting up the same assumptions that would lead to a recommendation of a uniform effective tariff on the basis of the earlier discussion.

Two approaches are possible. Both make use of effective protection.

First there is the *planner's approach*. Ideally he would look at every potential industry, and then calculate the made-to-measure effective tariff required. He would then establish all those industries where the calculation yields a figure of 20 per cent or less, giving each its relevant tariff.

Secondly, there is the *private enterprise approach*. The tariff-making authority sets 20 per cent as an upper limit to the effective tariff that will be provided. It then invites entrepreneurs to apply for protection, indicating that they can never expect to get more than 20 per cent. If the system is consistently applied and entrepreneurs have full knowledge of potentialities, this should lead to the correct result. It is not sufficient that the highest effective tariff will be 20 per cent; it is also necessary that all potential industries that could manage with 20 per cent or less are actually established and protected up to their made-to-measure level. Unless this is done the country will have failed to establish some socially-desirable industries.

If the aim is to attain a particular target of protected manufacturing production, or of imports replaced, the two approaches can be restated as follows.

The *planner's approach* would be to consider a number of potential industries which are believed to require relatively moderate protection, place them in order according to the rate of effective protection required, and then, beginning with the one with the lowest protection requirement and working upwards, see how many are needed to yield the protection or import-replacement target. These would then be protected, giving each its made-to-measure tariff or subsidy.

The *private enterprise approach* would be for the tariff authorities to set an upper limit tariff, offering to provide made-to-measure tariffs up to that limit, and see how many industries come forth as a result. If this upper limit leads to insufficient import-replacement the upper limit would be raised. The target would thus be attained by trial and error.

The protection or import-replacement target might be described as a 'foreign exchange saving' target. This is the way the planner's approach is sometimes put by writers concerned with industrial planning in an open economy.[10] It allows for the possibility of exporting, as well as for import-replacement. The problem is to minimise the 'domestic

[10] On this general subject, see B. Balassa and D. M. Schydlowsky, 'Effective Tariffs, Domestic Cost of Foreign Exchange, and the Equilibrium Exchange Rate', *Journal of Political Economy*, 76, May/June 1968, pp. 348–60; Michael Bruno, 'Domestic Resource Costs and Effective Protection: Clarification and Synthesis', *Journal of Political Economy*, 80, Jan./Feb. 1972, pp. 16–33; Anne O. Krueger, 'Evaluating Restrictionist Trade Regimes: Theory and Measurement', *Journal of Political Economy*, 80, Jan/Feb. 1972, pp. 48–62; B. Balassa and D. M. Schydlowsky, 'Domestic Resource Costs and Effective Protection Once Again', *Journal of Political Economy*, 80, Jan./Feb. 1972, pp. 63–69. The approach described in the text as the 'planner's approach', using effective protection, can be found particularly in the first of the articles listed here, but is also similar to the approach of Bruno and Krueger (though they allow for various other complications).

resource cost' of a given foreign exchange saving. This could be restated as minimising the production-distortion cost of protection when the object of protection is non-economic or, at least, is not allowed for in the 'cost-of-protection' calculation. The recommendation is then to use the effective protection criterion, or something very similar, preferring industries that require low effective rates over those requiring high rates.

All this depends on a *general* argument for protection. One must obviously qualify these recommendations when special arguments apply to particular industries, so that the upper limit available to an industry may be above or below 20 per cent.

Various disadvantages of the made-to-measure method were noted earlier.[11] Here must be added the disadvantages that arise when it is attempted to combine the made-to-measure principle with a comprehensive policy of tariff protection based on some *general* argument for protection. Made-to-measure tariff-making leads to a highly complex tariff system. Every time one industry's nominal tariff is adjusted to give it the effective tariff it 'needs' and no more, nominal tariffs that protect the products of its customers should ideally be altered, for otherwise their effective rates would get out of line. The system can thus make heavy demands not only on the honesty but also on the efficiency of the tariff-making machinery. Furthermore, a made-to-measure effective tariff system is also likely to lead to greater variations among nominal rates than a uniform effective tariff system. Hence it is likely to lead to more distortions in consumption and factor-use patterns. Perhaps in practice it is thus often best to aim at a uniform tariff system which is adjusted only in a few particular cases for made-to-measure considerations.

IV
UNIFORM EFFECTIVE PROTECTION FOR LABOUR

Let us now assume that the reason for tariff protection is that the wage rate facing import-competing industry exceeds the social opportunity cost of labour. Hence the actual wage exceeds the shadow wage. Possible reasons for this have been

[11] See above, pp. 222–3.

discussed in Chapter 6 (disguised unemployment or wage-differential) and Chapter 10 (inadequate savings). Another reason might be that external economies, perhaps through labour training, are associated with the employment of labour (Chapter 9).

First-best policy is then to subsidize the use of labour and combine this with free trade. But we are now in a second-best world. If the labour-intensity of the value-added content of the various protected industries were identical, we would have a case for uniform effective protection, at least if we ignore consumption-distortion effects and do not apply the made-to-measure principle. But once labour-intensities differ, such effective rate uniformity will no longer be appropriate. One should provide higher effective rates for the more labour-intensive industries.

It is of course just common sense that when the wage-rate facing industry is too high from a social point of view, the system of protection should make up for this by favouring relatively labour-intensive industries. The question is whether more precision can be given to this general prescription, and in particular whether the effective rate concept remains useful. It must be stressed that the discussion to follow will ignore consumption-distortion effects. These require the simple results to be qualified, perhaps severely so, at least when tariffs, rather than subsidies, are the protective devices.

(1) Effective Protective Rate for Labour

If the divergence between private and social wage-cost is x *per man*, irrespective of the particular wage-rate, the principle is to provide for each activity a set of tariffs which is equivalent to a subsidy of x per man employed. If the divergence is x per cent of the actual wage, the tariffs to be provided will yield a uniform *rate* of subsidy-equivalent, the rate relating the subsidy-equivalent of the tariffs to the actual wage-bill. We now consider the implications of the latter case.

The relevant concept now becomes, what has been called, the *effective protective rate for labour* or EPR_L. The ordinary effective rate expresses the nominal tariff on the final good minus the weighted average of the tariffs on its inputs as a proportion of the value-added per unit at free trade prices, where value-added includes the non-traded content. By contrast, the EPR_L expresses the same nominal tariff on the final good minus input tariffs as a proportion not of the whole value-added per unit at free trade prices but only of part of it. In the familiar effective rate formula it groups the non-labour part of value-added, notably the cost of capital, with the traded inputs.[12]

The simple answer to our problem is now the following. If the actual wage facing all import-competing industries exceeds the shadow-wage (indicating the social opportunity cost of labour) by a uniform percentage, say 20 per cent, then a uniform EPR_L of 20 per cent is required. If labour-intensities differ between industries this prescription will automatically yield non-uniform ordinary effective rates. The ordinary effective rates will be higher for relatively labour-intensive industries than for industries which use relatively less labour.

The basic idea is explained in Figure 13.2. Fixed proportions between the various inputs are assumed and, purely for simplicity of exposition, there are no tariffs on the inputs. We make the small country assumption. The cost of the various traded inputs required per unit of the final product is OG. The cost of capital per unit of the final product and of any non-labour content in non-traded inputs is shown by the vertical distance between RR' and GG'. This cost is assumed to rise as output increases if the industry faces a rising supply curve of capital and other non-labour inputs. The subsequent argument would remain unaltered if the supply curve were horizontal. The supply curve of labour facing the industry (including the labour content in non-traded inputs) is also assumed to be upward-sloping and is added on to RR' to yield HH', which is the ordinary private supply curve of the

---

[12] A formula is given in the Appendix to this chapter. For further discussion in the context of positive economics, see *The Theory of Protection*, pp. 171–2. The concept comes originally from Giorgio Basevi, 'The United States Tariff Structure: Estimates of Effective Rates of Protection of United States Industries and Industrial Labor', *Review of Economics and Statistics*, 48, May 1966, pp. 147–60. In the positive analysis of resource allocation effects of a tariff structure this concept only has significance when the elasticity of supply to an industry of the non-labour factors is infinite, and even then (as shown in *The Theory of Protection*, pp. 170–172) the term 'effective protective rate of labour' does not describe it correctly. But for the normative analysis here it is not necessary to assume an infinitely elastic supply of non-labour factors.

product. The given world price is OS and in the absence of any tariff, output would be OA.

Suppose that a tariff of ST is provided. Output will increase to OA'. The ordinary effective rate is ST/GS. But the EPR_L is ST/VS. It expresses the rise in the domestic price brought about by the tariff (namely ST) as a proportion of what the labour cost would have to be if the non-labour cost were given (at OV) and if the free trade price of OS prevailed. A formula

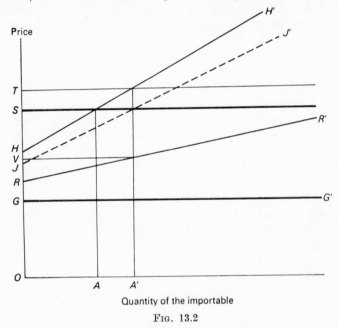

Quantity of the importable

Fɪɢ. 13.2

for it, which also takes into account tariffs on the traded inputs, is given in the Appendix to this chapter.

Next, let us revalue the cost of labour at the shadow wage. Adding the resulting social supply curve of labour on to RR' we obtain the supply curve JJ' that shows the social cost of production, labour being valued at the shadow-wage and not the actual wage. Because the shadow-wage is assumed to be below the actual wage, the curve JJ', showing social cost of production, is below the curve HH', which shows private cost of production. With that particular shadow-wage the tariff ST happens to be optimal since the marginal social cost of the product at output OA' equals its value at world prices. The actual private cost of labour at the optimal output exceeds the marginal social cost (as indicated by the cost calculated at the shadow-wage) by a proportion ST/VS and, when the tariff is optimal (as drawn) this is equal to the EPR_L. Hence, when the EPR_L is equal to the proportion by which the actual wage exceeds the shadow-wage we know that the tariff is at the correct rate.

(2) *Practical Implications and Complications*

It follows that if the divergence between the actual wage and the shadow-wage facing all import-competing industries is uniformly 20 per cent their EPR_Ls ought also to be 20 per cent. This is shown in the Appendix. The policy implication is that, in any given actual situation, one should calculate the various EPR_Ls, and where they turn out to be above 20 per cent they should be lowered appropriately and where they turn out to be below 20 per cent they should be raised.

While this prescription sounds simple, it is complicated in practice once one allows for tariffs on inputs and for general equilibrium relationships. A change in one nominal tariff may raise the EPR_L for one product and lower it for others that use it as input. Hence one cannot just reduce the nominal tariffs of all products that have EPR_Ls that are too high while raising nominal tariffs of products with EPR_Ls that are too low. This type of complication is familiar from the theory of effective protection. It seems likely that industries that have EPR_Ls above the required uniform rate should contract in size while those that have rates below the uniform rate should expand. But one cannot be absolutely sure about this. In Figure 13.2 the supply curve facing the industry concerned is likely to shift when tariffs on other goods alter. Because of general equilibrium interrelationships (discussed in *The Theory of Protection*) paradoxical effects on the output pattern of establishing a uniform EPR_L are certainly conceivable.

(3) *Substitution of Labour for Capital*

So far we have ruled out the possibility of substitution of labour for capital in any given industry by assuming fixed proportions between labour and non-labour inputs. Let us now remove this assumption. It might be thought (on the basis of the argument of Chapter 2) that the use of the device of the tariff rather than the wage subsidy would mean that industries would have no incentive to increase the labour-intensity of their production methods.

This is true if one takes the tariff for each industry as given. But the rule that the tariff for each industry should be adjusted so that its EPR_L is equal to the proportional excess of the private over the social wage cost (20 per cent in our example) means that, for any given set of input prices and input tariffs, an industry should get a higher nominal tariff for its own output the greater the labour-content in its value-added. Hence increasing labour-intensity should be rewarded with a higher tariff, so that, in principle at least, an inducement for using labour-intensive techniques is provided. In practice, an estimate of an industry's labour-intensity might be made when an industry is set up, or when the appropriate set of tariffs for all industries is being worked out, but subsequently there is no inducement for the industry either to ensure that its estimated labour-intensity is maintained or to increase it, unless tariffs are continually reviewed in the light of industries' changing labour-use.

APPENDIX
EFFECTIVE PROTECTION FOR LABOUR AND THE DIVERGENCE BETWEEN PRIVATE AND SOCIAL WAGE COST

For one unit of a final product, define as follows:
P_j = free trade price of final product, P_i = free trade cost of the bundle of traded inputs required by one unit of final product, T_i = weighted average of the tariffs on these inputs.

The following must be defined as per unit of final product *in the optimal situation*.

P_k = cost of capital and other non-labour factors, W = private or actual wage cost, S = marginal social wage cost (shadow wage), C = private cost of production, C_s = marginal social cost of production (which must be equal to the free trade price in the optimal situation), T_j = nominal tariff on final product required to make possible optimal output, given the private cost C, EPR_L = effective rate of protection for labour in optimal situation, s = proportional excess of private wage cost over social wage cost.

First we wish to show that $EPR_L = s$. Hence, if s is the same for all protected industries, EPR_L must be also.

$$EPR_L = \frac{T_j - T_i}{P_j - P_i - P_k} \tag{1}$$

$$s = \frac{W - S}{S} \tag{2}$$

$$C = P_i + T_i + P_k + W \tag{3}$$

$$T_j = C - P_j \tag{4}$$

$$C_s = P_i + P_k + S \tag{5}$$

$$C_s = P_j \tag{6}$$

From (3) and (4)

$$T_j - T_i = W - (P_j - P_i - P_k) \tag{7}$$

From (5) and (6)

$$S = P_j - P_i - P_k \tag{8}$$

From (7) and (8)

$$T_j - T_i = W - S \tag{9}$$

From (1), (2), (8), (9)

$$EPR_L = s \tag{10}$$

Next, we wish to obtain a formula for EPR_L which incorporates the tariff *rates* t_j and t_i and the input shares at protection prices in the optimal situation, namely a'_{ij} for the traded input and a'_{kj} for the cost of capital and other non-labour factors.

$$t_j = T_j / P_j \tag{11}$$

$$t_i = T_i / P_i \tag{12}$$

$$a'_{ij} = (P_i + T_i)/(P_j + T_j) \tag{13}$$

$$a'_{kj} = P_k/(P_j + T_j) \tag{14}$$

From (1), (11), (12), (13), (14)

$$EPR_L = \frac{\dfrac{t_j}{1+t_j} - \dfrac{a'_{ij} t_i}{1+t_i}}{\dfrac{1}{1+t_j} - \dfrac{a'_{ij}}{1+t_i} - a'_{kj}} \tag{15}$$

This can be compared with the formula for the ordinary effective rate, g_j, in which capital and other non-labour costs are grouped with labour cost and hence not treated the same as the traded input. Thus we have

$$g_j = \frac{T_j - T_i}{P_j - P_i} \tag{1.1}$$

and we then obtain from (1.1), (11), (12) and (13)

$$g_j = \frac{\dfrac{t_j}{1+t_j} - \dfrac{a'_{ij} t_i}{1+t_i}}{\dfrac{1}{1+t_j} - \dfrac{a'_{ij}}{1+t_i}} \tag{16}$$

which is formula (6.1) on p. 37 of *The Theory of Protection*. From (15) and (16)

$$\frac{EPR_L}{g_j} = \frac{1}{1 - \dfrac{a'_{kj}}{\dfrac{1}{1+t_j} - \dfrac{a'_{ij}}{1+t_i}}} \tag{17}$$

Various conclusions about the relationship between the effective rate of protection for labour and the ordinary effective rate can be derived from (17). For example, if the industry's labour intensity rises—so that a'_{kj} falls—and if EPR_L is kept constant, g_j must increase numerically.

14

PROTECTION THEORY AND COST-BENEFIT ANALYSIS

IT SHOULD be possible to relate the normative theory of protection to the principles of cost-benefit analysis. Both are forms of applied welfare economics, and when cost-benefit analysis is concerned with investment decisions in the traded-goods sector of an open economy they are obviously concerned with the same sort of thing. Many of the issues discussed in this book arise in a practical way in cost-benefit analysis. The relationship is of interest both because cost-benefit analysis is the technique appropriate for many practical decisions and because it has become a growing and sophisticated study in recent years. Of particular interest here are the 'Little-Mirrlees' *Manual of Industrial Project Analysis in Developing Countries* and the U.N.I.D.O. *Guidelines for Project Evaluation*.[1]

A particular project will have a stream of expected benefits and costs. A private enterprise, or a public enterprise which operates like a private enterprise, would evaluate these at expected market prices. For each year there will then be an expected net cash flow, positive or negative, this being the excess of expected gross benefits over costs, both evaluated at market prices. The net flows for the different years must

---

[1] Ian Little and James Mirrlees, *Manual of Industrial Project Analysis in Developing Countries*, Development Centre of the O.E.C.D., Paris, 1969; revised new edition, *Project Analysis and Planning*, Heinemann, London, 1973. Also, Partha Dasgupta, Stephen Marglin and Amartya Sen, (for U.N.I.D.O.) *Guidelines for Project Evaluation*, United Nations, New York (Sales No. E. 72. II. B. 11), 1972. A special issue of the *Bulletin* of the Oxford Institute of Economics and Statistics (Vol. 34, Feb. 1972) was devoted to discussion of the Little-Mirrlees *Manual*, and all the contributions are relevant. The paper by Vijay Joshi, 'The Rationale and Relevance of the Little-Mirrlees Criterion', is the best analysis (at time of writing) of the Little-Mirrlees method, and the present chapter has been much influenced by it.

then be added, discounting each at the market rate of interest. In this way a present value of the project is obtained. This is the method of *discounted cash flow*. Private decision-makers would evaluate each year's flow at the market prices expected to rule in that year, and would discount at the market rate of interest.

In the presence of domestic and trade divergences, market prices will not indicate social benefits and costs correctly. Similarly the market rate of interest may diverge from the social discount rate. Social cost-benefit analysis consists of evaluating the various expected streams of benefits and costs at prices representing social benefits and opportunity costs— that is, at *shadow-prices*—and then discounting at the *social* discount rate, rather than the market rate of interest. The present *social* values of proposed projects can then be obtained, and these would form the basis of investment decisions.

If all costs, including capital costs, have been included in the calculation, as well as all benefits, then a project should be undertaken if the present social value is positive. If there is a constraint on capital funds, a shadow-price of capital must be established that will just ensure the use of all the available capital. We are not concerned here with the possible divergence between the social discount rate and the market rate of interest, and shall assume that they are identical. The problem is to determine the shadow-prices.

I

THE PROBLEM OF IMPLEMENTATION

A problem which sometimes tends to get neglected in the shadow-pricing approach is the problem of implementation. It is closely related to the method of protection.

A private sector project may not be privately profitable at expected market prices but would be socially profitable on the basis of shadow-prices. If it is to be undertaken—as is socially desirable—a private loss must be avoided. It is not sufficient to make the shadow-pricing calculation. The market price calculation must also be made. This will tell us how much subsidization or protection will be required. Tariff protection can be seen as one way of implementing a project which is found to have a positive social present value but which would have a negative private present value in the absence of protection.

Yet even this is too simple a way of putting the matter since it suggests that the social cost-benefit calculation could be independent of the private or market cost-benefit calculation. The method of protection or subsidization will have by-product costs; these costs may increase with the ex-ante private loss that has to be covered or eliminated, and they must enter the social cost-benefit calculation. For example, if protection is by tariff there will be a by-product consumption-distortion cost, as well as some collection costs, while if protection is by subsidy, the raising of revenue will give rise to distortion and collection costs.

If the project is in the public sector—and the cost-benefit literature is mainly concerned with public sector projects—the issue is rather similar. If the project would make a loss at market prices, provision has to be made either for protecting the product, perhaps through a tariff, or for covering the loss out of public funds. This protection or financing will create by-product distortions that must be taken into account. If such provision is not made, or cannot be made for institutional or political reasons, then the project is not feasible irrespective of its profitability in shadow-pricing terms.

These problems do not arise, of course, if a project is profitable at market prices but unprofitable (negative present value) at shadow-prices. If it is a public sector project it would just not be started. If it is a private sector project then the question is whether there is a system of investment licensing—as in India, for example. With such a system the project would not be allowed to go ahead, the purpose of social cost-benefit analysis being to screen projects that the private sector wishes to undertake. Alternatively, a tax of some kind might be imposed which would cease to make the project privately profitable.

Similarly, in many countries there are development banks which provide finance for private projects at rates of interest below market rates and hence involving some implicit subsidy. The role of social cost-benefit analysis can then be to screen private applicants for funds. The social discount rate which should be used may well differ both from the free market rate and from the actual rate charged to private borrowers by the bank.

II
SHADOW-PRICING AND DOMESTIC DIVERGENCES

(1) *The Shadow Wage-rate and the Effective Protective Rate for Labour*

In the literature of cost-benefit analysis for developing countries much emphasis has been put on the divergence between the market and the shadow wage-rate, the latter measuring the social opportunity cost of labour to a project. Let us assume here that there are no trade divergences or distortions (world prices given and no trade taxes or subsidies) and only one domestic divergence: namely the market wage rate exceeds the shadow wage-rate by 20 per cent.

Consider an example. The present value of the project's gross private benefits is $100m. This is assumed to coincide with social benefits. The social or *shadow* wage cost is $60m. and the private wage cost $72m. Other costs, including normal profits, interest charges and so on, are $35m. Thus, while the project is socially profitable, with a net present value of $5m., it is privately unprofitable (net present value − $7m.). The usual cost-benefit analysis tells us it should be started.

How does this relate to the theory of protection? We could have provided an *effective rate of protection for labour* of 20 per cent. If the ratio of wage cost to gross benefit were the same every year this would have involved a constant nominal tariff rate of 12 per cent.[2] If this ratio varied the nominal rate would also vary, but in any case the net result would be to raise the present value of the benefits evaluated at domestic market prices by $12m. and to convert a private loss (present value − $5m.) to a private gain (present value of $7m.). Alternatively, a made-to-measure tariff or subsidy might be provided every year, so that the present value of the net benefit would become zero or just marginally positive.

[2] See pp. 382–5 above for explanation of the concept of the *effective protective rate to labour*.

One must first make a calculation to see whether the project should be protected at all. This can be done in two arithmetic ways that come to the same thing: either one makes the social cost-benefit calculation described, using a shadow wage, or one works out whether the project would survive if it obtained an assured effective protective rate for labour of 20 per cent, this rate being equal to the percentage by which the market wage rate exceeds the shadow wage. If the project's survival will depend on protection the private investment decision may hinge on the degree of certainty that attaches to the promise of protection. This promise, whether in the form of a particular nominal rate or an assurance of 'adequate' or made-to-measure protection, will have a private present value that has been discounted for risk.

The calculation of the shadow wage-rate raises all the issues discussed in Chapter 6, and in addition the questions of income distribution and of optimal savings. It should also be remembered that a sufficiently large indivisible project may significantly raise the wage that the project has to pay. In that case, even if there is no domestic marginal divergence, so that the market wage correctly indicates the *marginal* social opportunity cost of labour, the private wage cost for the whole project would exceed the social wage cost by the producers' surplus that is generated. The market wage would equal the social opportunity cost of the *last* man employed, but the shadow wage has to be equal to the *average* social opportunity cost of all the men employed on the project.

(2) *Externalities and Infant Industries*

Anything discussed in this book can be introduced into a social cost-benefit calculation. For example, a project may yield external benefits or costs. Cost-benefit analysis requires these to be valued, so that they can be given a shadow-price and included in an obvious way in the calculation. This presents, of course, many problems, discussed in the standard literature on the subject.

If external economies attach to production in other industries from which the project draws labour then the social cost of this labour to the project exceeds the private cost by the extent of the externalities foregone. Conceivably a particular type of

labour may create an equal value of externalities per man in
the industries from which it is drawn as in the project. In
that case both the social cost and the social benefit of the
labour will exceed respectively the private cost and the private
benefit, and the two divergences will just cancel out in their
effect on the project's net benefit.

The project may have to go through a learning or infancy
period. This will simply be reflected in private costs exceeding
private benefits in the early period and benefits exceeding
costs later. The cost-benefit calculation requires net costs and
benefits in each period to be discounted appropriately. If there
are dynamic external economies then, in the periods when
external learning effects, knowledge-diffusion or favourable
social atmosphere-creation are bearing fruit, social benefits
will exceed private benefits. The time-paths of net social
costs and benefits may then diverge significantly from the
paths of benefits and costs calculated from market prices
facing the project.

III
INVESTMENT DECISIONS IN THE PRESENCE
OF TRADE DISTORTIONS

We now come to an issue on which there is much confused
thinking and which is closely related to the discussion in the
previous chapter.[3] The country has a set of given non-optimal
tariffs and other trade interventions and the problem is to
make a second-best optimum investment decision in a
particular case subject to this constraint. What shadow prices
should be used? The issue is of practical importance since in
many less-developed countries trade distortions of this kind are
probably the main source of large divergences between private
and social costs and benefits.

It is desirable to set up a very precise and limited model
identical to the one used in the previous chapter; the conclusions can be readily generalized. We have two existing
industries, E and X, and one potential industry, Q, the
project to be considered being the establishment of Q. The
country faces given world prices (small country assumption)

[3] See above, pp. 367–9.

and all three products are traded products, X being the exportable and E and Q importables. There are no domestic divergences. There is a given 50 per cent tariff on E, this being the one distortion in the economy.

The orthodox approach, based on the theory of the second-best and expounded in the previous chapter, is that, while the first-best optimum would require zero tariffs for E and Q, if the tariff on E is given at 50 per cent, the second-best optimum tariff for Q is likely to be positive but less than 50 per cent, say 30 per cent. Hence, once free trade has been departed from in one sector of the economy, free trade is no longer likely to be optimal in the remaining sectors.

It can be shown that two other approaches to the problem lead essentially to the same result.

(1) *The Shadow Exchange Rate*

One way of making essentially the same point is the use of the concept of the shadow exchange rate. Suppose the actual exchange rate is $1R = \$1$ where R is the domestic currency and \$ the foreign currency. This is the rate that determines domestic production of X at the margin. But in the case of E the effective 'exchange rate' allowing for the tariff is $1 \cdot 5R = \$1$. This is really a misuse of the term 'exchange rate' since we are simply saying that the domestic price which governs production is raised by the 50 per cent tariff, but it is a way the term is often used. It follows that on the export side foreign exchange effectively exchanges at the rate of $1R = \$1$ and on the import side, at the rate $1 \cdot 5R = \$1$. When some resources are withdrawn from X and some from E to produce Q, the foreign exchange cost of this withdrawal in the form of reduced exports and more imports will be obtained by translating the domestic resource cost at a rate which is an appropriately weighted average of $1R = \$1$ and $1 \cdot 5R = \$1$. The weights depend on the shares of X and E in supplying the required resources.

Hence we may obtain a shadow exchange rate of $1 \cdot 3R = \$1$. It is the true cost of foreign exchange required for expanding the Q project. We then apply this rate to the project. While we value the resources it uses at the domestic market prices resulting from the actual exchange rate, we value the output of the project at the shadow exchange rate, not the actual rate. This is equivalent to evaluating the project as if it had a 30 per cent tariff. The shadow exchange rate is thus in this example the actual exchange rate modified by the second-best optimum tariff.

This shadow exchange rate is often confused with the free trade equilibrium exchange rate, but they are really quite different.[4] The shadow exchange rate is concerned with a second-best situation while the free trade equilibrium rate is the rate that would rule in the first-best situation. Both are likely to be between $1R = \$1$ and $1 \cdot 5R = \$1$, but this is the extent of their similarity. They could, of course, by chance be close together or even identical.

The shadow rate is concerned with a situation where a project is established that will draw resources both out of the existing protected industry E and out of the unprotected industry X, so that output of X is likely to decline. By contrast, the equilibrium exchange rate is concerned with a situation where the protected industry, E, will contract and the two unprotected industries, Q and X, will expand. Thus, in one case output of X contracts, and in the other it expands.

The usual approach of cost-benefit analysis is to consider policies towards one project or industry, or perhaps a limited number of them, with protection elsewhere given. The shadow exchange rate, as defined above, would then be appropriate. This can be contrasted with a situation where the sizes of all industries are subject to project appraisal and where one could thus focus on a first-best analysis. One could estimate, or guess at, the equilibrium rate appropriate to free trade and (in the absence of terms of trade effects) this would indeed then be the shadow rate. One would use it to reconstruct the free trade situation. One practical variant of this approach that is sometimes suggested is to reach the first-best solution by a series of project decisions based on the free-trade equilibrium rate, not worrying too much about the short-term

---

[4] See Michael Bruno, 'Domestic Resource Costs and Effective Protection: Classification and Synthesis', *Journal of Political Economy*, 80, Jan,/Feb. 1972, pp. 16–23; Bela Balassa and Daniel M. Schydlowsky, "Domestic Resource Costs and Effective Protection Once Again', *Journal of Political Economy*, 80, Jan./Feb. 1972, pp. 63–69. I am also indebted to unpublished papers by R. M. Parish.

non-optimality of decisions reached during the approach to the first-best.[5]

(2) *Little-Mirrlees Method*

An alternative approach has been popularized by the 'Little-Mirrlees' *Manual of Industrial Project Analysis in Developing Countries*. Its method is to value all the inputs and output of a project at the border prices, excluding tariffs, taxes, subsidies and so on, facing the country. The output is valued in terms of the imports it replaces or the exports it generates, and similarly all the inputs are valued in terms of the extra imports they generate, whether directly or indirectly, or the reduction in exports they bring about. The method is striking in the simplicity of its basic principle, though it has often been found puzzling and been misunderstood, and some of its simplicity may be deceptive.

Let us consider here a project (establishing industry Q) that has a constant stream of expected benefits and costs, so that we can take any one year as typical. The project draws resources from both E and X. In each case the resources can be valued at the given foreign prices. One works out by how much the value of exports will decrease and the value of imports will increase as a result of the movement of domestic resources out of X and E. One can express these values in terms of the foreign currency ($s) or translate them into domestic 'accounting prices,' using any accounting rate of exchange, say 1R = $1. It is most convenient to use the existing exchange rate, and we shall do so now. Similarly, the gross benefits are valued at the given foreign price of Q, also translated into the domestic currency (Rs) at the chosen accounting rate of exchange.

The gross benefits consist of the value of imports replaced. The net benefit is the net gain in foreign exchange, which the country is then free to use for consumption or investment, or keep as foreign reserves, as it pleases.

When there are no taxes or subsidies on exports, as we are

assuming here, the cost at domestic market prices of marginal resources drawn out of X will be identical with their value at Little-Mirrlees shadow prices. Thus they might cost $40 ($= 40R$). This is the value of the fall in exports resulting from the resource movement. But the marginal resources drawn from E, with its 50 per cent tariff, have a cost at domestic market prices higher than the cost evaluated at shadow or accounting prices. While at accounting prices the cost might be $60 ($= 60R$), this being the value of the extra imports required to replace exactly the reduced domestic output of E, at domestic market prices the value is 90R, namely 50 per cent higher than at accounting prices because of the tariff. Furthermore, let us assume that the value of the imports replaced by the project is $100 or 100R, or marginally above this. Here there is initially no divergence between the accounting price and the market price.

Thus a Little-Mirrlees calculation would show that the project would be just on the margin of social profitability (costs being $100 or 100R and benefits just marginally above this). But it would make a private loss of 30R, essentially because the resources drawn from E have had their domestic market prices raised by the 50 per cent tariff. It follows that the project should be started, and to be viable should have a 30 per cent tariff or subsidy.

The result is thus identical with our earlier approaches. If resources for Q come from both X and E, and there is a 50 per cent tariff on E and no protection for X, a tariff or subsidy for the project which is positive but less than 50 per cent will be appropriate.

It is sometimes thought that the Little-Mirrlees method of evaluating everything at border prices means that free trade policy is considered optimal for the project under consideration even when there are given tariffs elsewhere. But this is clearly not so. The model used is one where free trade is first-best (subject to considerations discussed in the previous and the next sections), but when some trade distortions are given, particular projects may justify tariffs or subsidies. The point is the same as that made in the previous chapter. Furthermore, it is sometimes misunderstood as meaning that some sort of virtue attaches to border prices, even though these may reflect the possibly distorting or nefarious activities of foreign governments, corporations, and so on. But in fact it means only that these prices are taken as given. They represent a set of constraints subject to which a country must pursue its own objectives.[6]

(3) *Are Given Tariffs Distorting, and Why are they Given?*

The methods just discussed assume that the given tariffs and other taxes and subsidies are distorting rather than correcting with regard to resource allocation. Let us focus on the Little-Mirrlees method in examining this, though the issue is the same for all three methods. Because of this assumption the market prices generated by tariffs, taxes etc., are being deliberately ignored in making the cost-benefit analysis. If the latter were correcting—for example, if tariffs were second-best devices for offsetting externalities—then the method would have to be modified. One could still use world prices in the first instance, but would then have to attach some value to external benefits or costs forgone when resources move out of protected industries.

If the tariffs and other trade interventions are distorting, there are still four possibilities. One is that the purpose of tariffs is to raise revenue. The tariffs may then indeed be first-best ways of raising given revenue. The changes in domestic market prices to which they give rise will lead to the familiar by-product distortion costs. It is then quite rational to ignore market prices in the case of investment appraisal, operating as if there were no tariffs. This method will reduce the by-product distortion costs. For the projects concerned

---

[6] When a project uses non-traded inputs, or uses labour that comes out of activities producing non-traded goods, it is no longer possible to evaluate everything at border (world) prices. To some extent exports elsewhere may decline or imports increase as a result of the resources absorbed by the project and these effects can indeed be evaluated at border prices, but there may also be increases in domestic prices of goods, causing reduction of consumption and requiring evaluation in terms of something like consumers' surplus (willingness to pay).

This also applies if there is an import quota on the product which the project will produce or on imports which it uses, and if the quota is not so adjusted as to keep domestic consumption of the controlled good constant.

The simple version of the Little-Mirrlees method requires all goods to be 'fully-traded'. See Joshi, 'The Rationale and Relevance of the Little-Mirrlees Criterion'.

the false signals sent out by the distorted prices are simply ignored.

Secondly, the purpose of tariffs may be to restrict final consumption of some goods. Consumption taxes applying uniformly to imports and domestic production of these goods would be first-best in the absence of collection costs, but conceivably, allowing for high collection costs on domestic excises, tariffs may be first-best. Alternatively, there may be political or window-dressing reasons for preferring a second-best system of tariffs.

In any case, even if the tariff-induced price changes facing consumers are desirable, the price changes facing producers will lead to by-product production-distortion costs. As in the case of revenue tariffs, it is then quite rational to ignore market prices in the case of investment appraisal, so ignoring the false signals sent out and reducing the by-product distortion costs.

The third possibility is that the tariffs have been imposed to achieve certain purposes concerned with the production pattern with which the appraisers of investment projects are not in sympathy, such as to shift income distribution towards the protected industries or to foster certain types of production for the sake of externalities believed to be generated. The project appraisers cannot get the tariffs reduced explicitly, but they can modify or even eliminate their effects by providing what is, essentially, offsetting protection for the project industries.

We may have a situation where one part of the government —one department or minister, or one set of advisers—has different objectives from another part, and each has control of part of the decision-making machinery. Instead of a single government policy being pursued, one department of government then maximizes its 'objective function'—conceived of here as depending on efficiency of resource allocation (Pareto criterion), qualified perhaps for income distribution considerations—subject to the constraint of policies imposed by other departments and their backers.[7]

---

[7] See A. K. Sen, 'Control Areas and Accounting Prices: An Approach to Economic Evaluation', *Economic Journal*, 82, March 1972. (Supplement). pp, 486–501; and also various articles in the *Bulletin* of the Oxford Institute, op. cit.

Finally, the fourth possibility is that the distorting tariffs are in some sense inevitable even though they are not really desired by the government. One could say that they are irrational, but one must then ask why the distortions are not removed. The answer may be that pressure groups will oppose their removal, or that many people, including economically uneducated politicians and government officials, believe them to be beneficial, but are simply ignorant of the true economic effects. It may be that all the main policy-makers are convinced it would be better if they were removed, but they are unable to persuade Parliament or the general public of this. But this then raises the question of whom one means by 'main policy-makers' and why Parliament or the general public are opposed.

One cannot get away from the fact that trade taxes and controls are imposed by governments, and social cost-benefit analysis is also assumed to be done by governments. One must then be clear in one's mind why one expects a government to carry out and act upon a social cost-benefit analysis, while not being able to persuade it to impose first-best trade taxes. Perhaps the answer is that one must operate on both fronts, and the recommendation one makes on the investment appraisal front needs to take into account the degree of success in persuasion expected on the commercial policy front.

IV
INVESTMENT DECISIONS WITH TERMS OF TRADE EFFECTS

Let us now remove the small country assumption and so allow for the effects of a project on the terms of trade. This yields two distinct complications.

(1) Social Adjustment for Terms of Trade

The first problem arises because of the familiar divergence between domestic market prices and marginal social (or *national*) values created by terms of trade effects. One can formally get around this problem by assuming that a set of orthodox optimal trade taxes is being imposed, so that this particular divergence between market and shadow prices has been corrected. But this is not a very operational approach.

One could also say in very general terms that evaluation must be at *marginal* border prices not actual prices. But this is rather vague. A helpful approach can be expounded more clearly in terms of Figures 14.1 and 14.2.

Figure 14.1 refers to the product produced by the proposed project, assumed to be an export product. DD' is the foreign demand curve for this export and OH are exports of it initially. The project will increase output and exports by HG

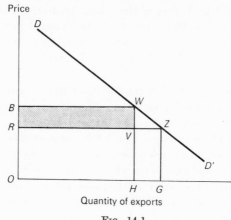

Fig. 14.1

and hence bring down the price from OB to OR. It is assumed here that the fall in price leaves the output from capacity existing before the project, as well as domestic consumption, unaffected, so that the project's output represents the net increase in exports.

In calculating the gross benefit from the project one cannot use the initial border price of OB. So, firstly, one must make the terms of trade adjustment that even a private producer would make, namely to allow for the fall in price that will result from the project, in order to assess the private or market value of the project. The relevant border price is obviously the price that rules once the project is in production. The gross private benefit is thus $HVZG$.

In addition one must allow for the *social adjustment for terms of trade* or *SATT*. This is the adverse effect that the project has on prices received on initial pre-project output of

OH. This SATT is the shaded area *RBWV* and is a social cost. If domestic production of the product concerned is not monopolized, so that pre-project output is not produced by the same firm that is responsible for the project, this SATT is *not* a private cost. It is rather like an external diseconomy, and must enter the project's cost-benefit calculation.

Figure 14.2 refers to an exportable input used by a project. Before the project, exports of this input are *OH*. But the project requires *GH* of the input and so reduces exports to this

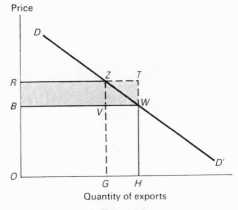

Fig. 14.2

extent. As a result the price of the input will rise from *OB* to *OR*. It is assumed that the price rise does not affect domestic consumption or production of the input, apart from the extra consumption by the project itself, so that the project's extra demand is equal to the net decrease in exports.

The private or market cost of the input will depend on the new price *OR* and is thus *GZTH*. The social adjustment for improved terms of trade, or SATT, is this time a benefit, and is the whole of the shaded area, *BRTW*. It has two components: one part is the rise in the price of remaining exports *BRZV*, and the other part is the excess of the private cost of the exports forgone over the initial value of these exports, namely *VZTW*.

A similar analysis could be applied when the project produces importables, or uses imported inputs, and when border prices change as a result of changes in the country's import demand.

In general, the SATT is the change in the border price applied to the initial quantity of exports or imports.

(2) *Changes in Domestic Production and Consumption*

The second complication that results from border prices changing in response to the project is the following. Domestic consumption, as well as production outside the output of the project itself, will be affected by demand and supply

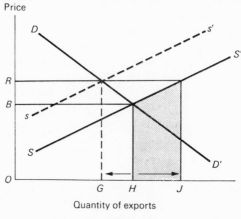

Fig. 14.3

changes caused by the project. This was explicitly assumed away in the preceding discussion.

Let us consider the case where a project uses an exportable input, and by increasing the demand for the input raises its price. This was represented in Figure 14.2. Now we allow for the fact that, while (i) exports of this input will indeed fall, in addition (ii) domestic consumption of the product will decline, and (iii) domestic production will increase. These domestic consumption and production changes must then be valued correctly.

Figure 14.3, illustrates the main point. The pre-project export supply curve for the input is SS'. The project increases domestic demand for the input at any given price by adding its own demand to the initial demand, and so it causes the export supply curve to shift to the left, to ss', the horizontal distance between ss' and SS' indicating the project demand. Hence the price rises from OB to OR. Exports fall by GH; the consequences of this we have already analysed. In addition, the price rise causes domestic production to increase and consumption from other sources to fall, the sum of these two being HJ.

Hence the total project demand of GJ is met from two sources (as shown by the two arrows), a decline in exports and a movement along the original supply curve. If the curve SS' correctly indicated the marginal social cost of the extra domestic supply, or consumption forgone, then the value of the latter element is the shaded area in Figure 14.3. But if there are taxes or subsidies, a wage differential, or other divergences, this element may need to be analysed in more detail. The fall in consumption might be valued in terms of willingness to pay, while the rise in production might be broken down into various elements of cost, including traded input, wage cost revalued at the shadow-wage, and so on.

V

THE WEIGHTING OF COSTS AND BENEFITS

It is possible in a social cost-benefit calculation to attach weights to different elements of cost and benefit which take into account such matters as the fiscal problem, the possible need to foster savings, and income distribution effects. This is a simple operational way of incorporating in an investment appraisal many of the considerations discussed earlier in this book, including the implementation problem.

(1) *Weighting Technique: The Numeraire and Shadow Prices*

Suppose one evaluates the costs and benefits of a project taking into account domestic divergences such as the wage differential, and trade distortions, such as existing tariffs. For the latter one might use either the Little-Mirrlees method or the method of the shadow exchange rate. There are, let us suppose, only two main types of income affected, namely income of wage earners and government revenue. The present value of the expected rise in wage income is $25m. and the present value of the fall in government revenue, owing to subsidization required, is $24m. On this unweighted basis, then, the project shows positive present value of $1m.

Now we introduce weights. A change in government revenue is given a heavier weight than an equal change in wage income because of the fiscal problem: hence society, or the project evaluators, consider an extra $4 of government revenue to be equal to an extra $5 of wage income.

We must now arbitrarily choose a numeraire. We could make government revenue the numeraire. In that case we revalue the rise in wage income in our example downwards to $20m., and so get a net benefit of − $4m. Alternatively we use wage income as the numeraire, and so revalue the fall in government revenue to − $30m., getting a net benefit of − $5m. The sign of the net benefit must be the same whichever is the numeraire, so it does not really matter which we choose. No value judgement attaches to the choice of numeraire. The Little-Mirrlees *Manual* chooses government revenue as the numeraire and the U.N.I.D.O. *Guidelines* choose consumption.

In practice one could make a distinction not just between government revenue and wage income, but could allow for many types of income. One could distinguish the wages of skilled workers from those of the unskilled. One could obviously distinguish private profits from wages, and income of the rich from income of the poor. One could distinguish wages earned in one region from those in another, and so could introduce any regional discrimination one desired. If one wished to favour savings over consumption one could attach one weight to income expected to be consumed and another to income likely to be saved. The higher the weight given to any particular type of income the lower the shadow price of the factor which earns this income. For example, if it is desired to favour unskilled workers, their shadow-wages would have to be reduced. If it is desired to *dis*favour wage-earners because of their low savings propensities, the shadow-wage would have to be raised.

(2) *The Cost of Raising Government Revenue*

If the raising of revenue were costless, with no-by-product distortion and collection costs, and if one could assume that the government pursued first-best policies, one should not introduce differential weights. The government would be collecting sufficient taxes to finance socially desirable public services and to ensure that the desired disposable income distribution between contemporaries, including regional distribution, is attained. Furthermore, it would raise enough taxes to ensure that enough savings are generated as required by the social discount rate and the social productivity of capital. If private investment is inadequate the government could, for example, use tax revenue to provide finance for private firms through development banks. Given such government policies, Pareto-efficiency criteria alone should then govern investment project appraisal. The point is familiar.

But of course the raising of revenue is not costless. It has a marginal cost. This, as we saw in Chapter 3, is crucial for first-best analysis. One should then attach an extra weight to government revenue, the size of the weight depending on the marginal cost of raising revenue.

Suppose that for every $100 of revenue raised at the margin, private consumption and investment have to be reduced by $110. Hence, in the process of transferring funds to the government, $10 'gets lost' or is used up in 'transport'. In a first-best situation the marginal value of $110 of private income will then be equal to $100 of income at the disposal of the government. This then gives us the appropriate weight for investment project appraisal.

Once we allow for the fact that the raising of government revenue is not costless there is some case for taking income distribution considerations into account in resource allocation decisions. This point was stressed in Chapter 3. Similarly there may be a case for allowing for the problem of fostering savings relative to consumption. There is a case then for introducing differential weights. One is justified in departing somewhat from narrow Pareto-efficiency criteria in project appraisal because one can no longer assume that the government—even though it pursues first-best policies—deals with these matters through the budget to the point of offsetting or eliminating the domestic income distribution and savings divergences.

(3) *Second-best Situation*: *Government Revenue Inadequate*

This argument for using differential weights may be greatly strengthened if the country is not pursuing first-best policies.

The government may not be raising as much revenue as it should, given the marginal cost of raising revenue, and as a result may be failing to carry out adequate income distribution policies or to generate adequate savings.

In a first-best situation revenue would be raised up to the point where the social utility of $100 of government revenue is equal to the social utility of $110 in the hands of the public. But political and institutional considerations which the project evaluators may not wish to build into their own social welfare functions but which they treat as constraints may prevent revenue being raised to this extent. The social utility of $100 of government revenue may thus be much greater than $110 of private spending, so that the weight to be given to money in the hands of the government needs to be much higher.

If the raising of government revenue to finance income redistribution and other socially desirable policies were held back only by the marginal cost of raising revenue it might be held back very little, since in many countries, this cost is not likely to be very high, perhaps, at most, ten per cent of revenue raised. One could then perhaps ignore the need for differential weights. But it is another matter entirely when political and institutional considerations hold down the level of taxation and so make government revenue inadequate.

The question then arises how one would decide that government revenue is inadequate and that this should be taken into account in project appraisal. Making an estimate of the marginal cost of raising revenue is one thing; estimating the extent to which the government is departing from first-best policy owing to political and institutional constraints is another. The issues raised earlier are again relevant: if the department of government engaged in cost-benefit analysis is expected to be rational, why are other departments of government assumed to be irrational, or, in some sense, pursuing undesirable policies?

(4) *Three Approaches to Differential Weighting*

It is a matter of some controversy whether differential weights should be attached to the incomes involved in a cost-benefit exercise. Even if the principle is accepted the choice of weights is bound to be controversial. Essentially there are three points of view on this subject.[8]

First, there is the approach of attaching particular differential weights in each exercise. This philosophy of differential weights is central to the Little-Mirrlees *Manual* and the UNIDO *Guidelines*, both of which focus on the need for government policy to foster savings by the government, the latter also stressing income distribution effects. Earlier, in *Trade and Welfare*, Meade made distributional weighting a centre-point of his analysis.

The case for this approach must rest primarily on the second-best nature of governments and on the ability of cost-benefit analysts to choose roughly appropriate weights. A particular danger is that value judgements will lie hidden amidst the arithmetic or the impressive mathematical formulae—though some people may regard this as an advantage if they happen to agree with the value judgements of the particular analysts concerned.

A second approach is to take a neutral attitude on income distribution effects, not by ignoring these effects but rather by presenting the political decision-makers and the general public with figures which enumerate the gains and losses to various classes of citizens. In this view it seems much better *not* to add up gains and losses in a single figure.[9] A slight variant of this approach—and one that the present author would favour—is to present *both* the likely gains and losses to various classes, *and* a number of alternative 'added-up' figures, each figure using a different weighting system. The latter approach is also suggested in the UNIDO *Guidelines*. If one believes that explicitness and rational decision-making should be fostered it is important to make the choice of weights quite explicit in each case.

There is a third approach. Both Mishan and Harberger have explicitly argued that cost-benefit calculations should not include allowances for income distribution effects, though they are not suggesting that these effects should necessarily

---

[8] We return here to the issues discussed on pp. 104–7.

[9] This position is argued by Meade in his review of Mishan's *Cost-Benefit Analysis*, *Economic Journal*, 82, March 1972, pp. 344–6.

be ignored in making decisions.[10] Mishan has written that 'the rationale of cost-benefit analysis is based on that of a potential Pareto improvement', and later that 'the tradition of separating the cost-benefit calculation from the distributional effects continues in practice and is unlikely to change in the foreseeable future'.[11] As a factual description of the usual practice of cost-benefit analysts in the United States and Britain this latter remark does seem at present broadly correct.

Harberger has pointed out that the general use of differential distributional weights would mean that each cost-benefit analyst might come up with a different figure for any particular project. Economists could not check each others' opinions, since they are not professionally qualified to make income distribution judgements. It would therefore be difficult to establish a professional consensus, something which is presumably desirable if applied welfare economics is to be well-regarded.

The approach of ignoring income distribution and focusing on Pareto-efficiency may have some conceivable plausibility in the case of developed countries. The marginal cost of raising revenue may be very small and perhaps one can assume that governments will pursue (what they consider) first-best policies with regard to such matters as income distribution and optimum savings. Furthermore, the effects of many investment decisions may be broadly diffused, so that there may be no clear or significant income redistribution effects.

But it is clearly much less reasonable in the case of less-developed countries. Even in the case of developed countries it is not really acceptable. Mishan himself has stressed the importance of income distribution considerations, so there seems little point in producing figures that simply ignore them.

[10] E. J. Mishan, *Cost-Benefit Analysis*, Allen and Unwin, London 1971; Arnold C. Harberger, 'Three Basic Postulates for Applied Welfare Economics: An Interpretive Essay', *Journal of Economic Literature*, 9, September 1971, pp. 785–97. For a general discussion of the issues, see also P. D. Henderson, 'Some Unsettled Issues in Cost-Benefit Analysis', in Paul Streeten (ed.), *Unfashionable Economics*, Weidenfeld and Nicolson, London, 1970.

[11] Mishan, op. cit. pp. 3–4.

15

CONCLUSION

LET us briefly draw together the strands of the normative analysis in this book.

A principal theme has been the argument that generally the first-best policy for correcting a *domestic* divergence is likely to be some kind of non-trade intervention, such as a subsidy or tax close to the point of the original divergence. The basic analysis was set out in Chapter 2. In the remainder of the book numerous widely-used arguments for protection have been analysed with the methods of Chapter 2, distinguishing first-best policy, second-best policy, and so on, and noting especially where tariffs fit into the hierarchy. A key role has been played by the concept of the *by-product distortion* (pp. 12–14).

It has thus been shown that desired income distribution changes are best brought about directly, not through trade policies (Chapter 3), that problems created by real wage rigidity should be dealt with by direct subsidies (pp. 121–2), and that if the wage-rate payable by manufacturing industries in developing countries exceeds the social opportunity cost of labour, wage subsidies are required, rather than tariffs (Chapter 6). If an increase in savings is desired, it should be fostered by direct taxation (Chapter 10), and, similarly, problems created by fluctuations in export earnings are best overcome by fiscal policy (p. 315).

In some circumstances one might say that the source of the divergence is not strictly domestic in nature. For example, the exchange rate may be rigid for institutional reasons (though this must be combined with money wage rigidity to be relevant), or the divergence may concern foreign investment (Chapter 12). But it continues to be true in these cases that a tariff or similar device is unlikely to be first-best.

In the course of analysing oft-used arguments for protection it has not only been shown that trade policies may not be appropriate first-best or even second- or third-best devices, but that there are weaknesses in the fundamental arguments. Thus, much of the book has been devoted to considering the need for intervention more generally, quite apart from the need for trade policies specifically. The main issue under discussion in Chapter 6 was the case for subsidising employment in the manufacturing industries of less-developed countries, since the case for trade protection as a third-best or fourth-best device rests on this first-best case for wage subsidisation. Similarly, in Chapter 9 the case for subsidising infant industries and the fashionable argument for assisting advanced-technology industries (pp. 275–9) were examined rather sceptically. Generally, the underlying arguments, even for non-trade intervention, do not rest very securely.

There are clearly numerous circumstances when trade policies could be second-, third- or fourth-best and, given various constraints, these may sometimes justify actual policies. Once one departs from first-best and is willing to go well down the hierarchy of policies a great deal can be justified. In any particular case one has to justify the need to go down the hierarchy—and this depends on the particular constraints and non-economic considerations one is prepared to accept.

But sometimes trade policies may indeed be first-best. Let us then bring together the various cases or circumstances that emerged in this book where this is so. Subsequently they will be classified in a number of ways.

(1) If a country can improve its terms of trade by using tariffs or export taxes it may be first-best to impose the appropriate set of taxes (Chapter 7). It is vital to stress that this orthodox optimum trade tax argument takes a purely national point of view, though, given certain assumptions about a desirable world income distribution and the unavailability of other instruments, one can work out a case for an optimum set of restrictions from a cosmopolitan welfare point of view (pp. 191–2). The orthodox optimum tax rates will be lower the higher the foreign trade elasticities the country faces, and these elasticities may rise as time for adjustment is allowed.

But as long as the elasticities are less than infinite there is some argument for trade taxes on these grounds.

The simple version of the argument also assumes (a) no foreign retaliation and (b) that private industry is perfectly competitive. When these assumptions are removed (pp. 171–6) a first-best argument may still remain, but it is no longer so assured or simple. Finally, for most countries it is likely to be a first-best argument mainly for export taxes rather than tariffs (p. 184–5).

(2) The market for a product may be dominated by one or a few foreign firms producing abroad, while actual or potential competitors are mainly domestic. If some increase in competition is desired, there may then be a case for a tariff designed to reduce or avoid monopoly. This is one version of the *dominant firm effect* (pp. 217–18). It may also be first-best to counter short-term predatory dumping with tariffs, as this is an instrument of foreign monopoly. The latter argument assumes that it is practicable to distinguish the motive for low-priced imports (p. 247).

(3) A tariff on imports produced by a multinational corporation abroad may induce it to set up domestic production facilities, and it is possible (but by no means certain) that a net gain to the country would result (pp. 357–8).

(4) High collection costs for non-trade taxes compared with trade taxes may make trade taxes first-best for raising revenue (pp. 64–7). This could be quite a significant argument for the use of trade taxes in less-developed countries, expecially the least-developed among them.

(5) Tariffs on imports by multinational corporations may prevent tax evasion by these firms through transfer pricing (p. 364).

(6) Collection costs for protective tariffs may be very low compared with collection costs for revenue tariffs and for other taxes that would finance protective subsidies. A protective tariff may then be a first-best device when there is a domestic divergence that requires expansion of an import-competing industry (p. 46).

(7) High subsidy disbursement costs may make trade intervention first-best when there are domestic divergences (pp. 48–50).

(8) If the conservative social welfare function applies and import prices have fallen, a tariff may be first-best, at least temporarily. Similarly, there may be a first-best case for reducing tariffs or other trade restrictions gradually rather than suddenly. This case hinges essentially on an information problem (pp. 109–10).

(9) It may be desired to make the country more self-sufficient for nationalistic reasons (p. 31–2).

(10) The domestic income distribution effects of trade intervention may not always be as obvious as the effects of more direct taxes and subsidies. Thus trade intervention has *cosmetic* attractions and may be first-best from some governments' points of view—that is, given a social welfare function that places some value on obfuscation.

The last two—rather unattractive—arguments are non-economic, and perhaps should not be included in the list. They argue that there will be economic costs from achieving certain objectives, and trade policies may minimise these costs.

Some arguments are likely to be of much less practical relevance than others. In general, empirical judgements have been avoided here, but arguments (3), (5) and (6) are probably relatively unimportant, so that they should not be put on a par with the other arguments. Arguments (1), (2), (3) and (9) are essentially concerned with *trade* divergences, though this is not quite clear in the case of argument (2). Arguments (4) and (5) concern the fiscal role of trade taxes. Only arguments (6) and (7) apply to the case of domestic divergences as this term is usually understood, and only these two arguments seriously qualify the theme of Chapter 2 that trade policy is never first-best for correcting domestic divergences. Argument (8) concerns an *income distribution divergence* which originates from a trade change; it might be described as either a domestic or a trade divergence.

Quite distinct from the question of whether trade policies should be used at all is another question. What is the optimum tariff structure or, more generally, the optimum trade tax structure given that, for one reason or another, trade taxes are used? This problem has turned up several times and has been presented in various forms.

(1) What set of tariffs and export taxes will raise a given

amount of revenue with minimum distortion costs? The broad answer (pp. 70–74) is that it will be a set where tax rates are relatively high on goods with relatively low *domestic* demand and supply elasticities.

(2) What set of trade taxes will maximise the nation's benefit from improved terms of trade, taking into account the losses from reducing trade as well? The answer (pp. 168–9) is that the orthodox optimum trade tax structure is one where rates are relatively high on goods with relatively low *foreign* demand and supply elasticities.

(3) What is the optimum tariff structure when the purpose of tariffs is to foster a given amount of domestic value-added in the import-competing sector, or to deal with the consequences of a wage-rate facing the import-competing sector that exceeds the social opportunity cost of labour? The answers to these questions are not simple, but make use (in Chapter 13) of the concepts of uniform effective protection and of the uniform effective rate to labour.

(4) The concept of the *made-to-measure* tariff (or subsidy) is also relevant both for understanding actual protective structures and for constructing second-best optimal structures (pp. 219–23, pp. 378–81). It does not provide an argument for the use of a tariff or subsidy, but suggests a way in which a tariff or subsidy provided for various reasons may conceivably avoid some undesired by-product effects, notably creation of monopoly profits, excessive fragmentation of domestic production, and undue distortion or income distribution effects resulting from financing subsidies.

(5) Finally, one may need to take some tariffs or quotas as given. The problem is then to determine the optimum tariff for a particular product when some of the constrained tariffs are fixed at non-optimal levels. This involves standard exercises in the theory of the second-best (pp. 367–9, pp. 394–9).

This is a book about the theory of economic policy. Nevertheless, it does not yield direct policy conclusions automatically applicable to all countries at all times. It is undoubtedly true that much of the discussion implies that governments could frequently benefit their peoples by removing trade restrictions, perhaps gradually, especially if they replace them with other, more suitable, forms of intervention. But this can hardly be a firm, dogmatic conclusion since there are too many qualifications. To underline the purpose of this type of book it seems appropriate to conclude with Keynes' words: "The theory of economics does not furnish a body of settled conclusions immediately applicable to policy. It is a method rather than a doctrine, an apparatus of the mind, a technique of thinking which helps its possessors to draw correct conclusions."[1]

[1] John Maynard Keynes, in his Introduction to the *Cambridge Economic Handbooks*.

INDEX OF PERSONAL NAMES

INDEX OF SUBJECTS